Digital Signal Processing System-Level Design Using LabVIEW™

Digital Signal Processing System-Level Design Using LabVIEW™

by Nasser Kehtarnavaz and Namjin Kim
University of Texas at Dallas

AMSTERDAM • BOSTON • HEIDELBERG • LONDON
NEW YORK • OXFORD • PARIS • SAN DIEGO
SAN FRANCISCO • SINGAPORE • SYDNEY • TOKYO

Newnes is an imprint of Elsevier

Newnes is an imprint of Elsevier
30 Corporate Drive, Suite 400, Burlington, MA 01803, USA
Linacre House, Jordan Hill, Oxford OX2 8DP, UK

Copyright © 2005, Elsevier Inc. All rights reserved.

No part of this publication may be reproduced, stored in a retrieval system, or transmitted in any form or by any means, electronic, mechanical, photocopying, recording, or otherwise, without the prior written permission of the publisher.

Permissions may be sought directly from Elsevier's Science & Technology Rights Department in Oxford, UK: phone: (+44) 1865 843830, fax: (+44) 1865 853333, e-mail: permissions@elsevier.com.uk. You may also complete your request on-line via the Elsevier homepage (http://elsevier.com), by selecting "Customer Support" and then "Obtaining Permissions."

 Recognizing the importance of preserving what has been written, Elsevier prints its books on acid-free paper whenever possible.

Library of Congress Cataloging-in-Publication Data

(Application submitted)

British Library Cataloguing-in-Publication Data
A catalogue record for this book is available from the British Library.

ISBN: 0-7506-7914-X

For information on all Newnes publications
visit our Web site at www.books.elsevier.com

05 06 07 08 09 10   10 9 8 7 6 5 4 3 2 1

Printed in the United States of America

Contents

Preface ix
What's on the CD-ROM? xi
Chapter 1: Introduction 1
   1.1 Digital Signal Processing Hands-On Lab Courses 2
   1.2 Organization 3
   1.3 Software Installation 3
   1.4 Updates 4
   1.5 Bibliography 4
Chapter 2: LabVIEW Programming Environment 5
   2.1 Virtual Instruments (VIs) 5
   2.2 Graphical Environment 7
   2.3 Building a Front Panel 8
   2.4 Building a Block Diagram 10
   2.5 Grouping Data: Array and Cluster 12
   2.6 Debugging and Profiling VIs 13
   2.7 Bibliography 14
Lab 1: Getting Familiar with LabVIEW: Part I 15
   L1.1 Building a Simple VI 15
   L1.2 Using Structures and SubVIs 23
   L1.3 Create an Array with Indexing 27
   L1.4 Debugging VIs: Probe Tool 28
   L1.5 Bibliography 30
Lab 2: Getting Familiar with LabVIEW: Part II 31
   L2.1 Building a System VI with Express VIs 31
   L2.2 Building a System with Regular VIs 37
   L2.3 Profile VI 41
   L2.4 Bibliography 42

Chapter 3: Analog-to-Digital Signal Conversion 43
    3.1 Sampling 43
    3.2 Quantization 49
    3.3 Signal Reconstruction 51

Lab 3: Sampling, Quantization and Reconstruction 55
    L3.1 Aliasing 55
    L3.2 Fast Fourier Transform 59
    L3.3 Quantization 64
    L3.4 Signal Reconstruction 68
    L3.5 Bibliography 72

Chapter 4: Digital Filtering 73
    4.1 Digital Filtering 73
    4.2 LabVIEW Digital Filter Design Toolkit 77
    4.3 Bibliography 78

Lab 4: FIR/IIR Filtering System Design 79
    L4.1 FIR Filtering System 79
    L4.2 IIR Filtering System 85
    L4.3 Building a Filtering System Using Filter Coefficients 90
    L4.4 Filter Design Without Using DFD Toolkit 91
    L4.5 Bibliography 94

Chapter 5: Fixed-Point versus Floating-Point 95
    5.1 Q-format Number Representation 95
    5.2 Finite Word Length Effects 99
    5.3 Floating-Point Number Representation 100
    5.4 Overflow and Scaling 102
    5.5 Data Types in LabVIEW 102
    5.6 Bibliography 104

Lab 5: Data Type and Scaling 105
    L5.1 Handling Data types in LabVIEW 105
    L5.2 Overflow Handling 107
    L5.3 Scaling Approach 111
    L5.4 Digital Filtering in Fixed-Point Format 113
    L5.5 Bibliography 122

Chapter 6: Adaptive Filtering 123
    6.1 System Identification 123
    6.2 Noise Cancellation 124
    6.3 Bibliography 126

Lab 6: Adaptive Filtering Systems 127
    L6.1 System Identification 127
    L6.2 Noise Cancellation 134

Contents

    L6.3 Bibliography 138
Chapter 7: Frequency Domain Processing 139
    7.1 Discrete Fourier Transform (DFT) and Fast Fourier Transform (FFT) 139
    7.2 Short-Time Fourier Transform (STFT) 140
    7.3 Discrete Wavelet Transform (DWT) 142
    7.4 Signal Processing Toolset 144
    7.5 Bibliography 145
Lab 7: FFT, STFT and DWT 147
    L7.1 FFT versus STFT 147
    L7.2 DWT 152
    L7.3 Bibliography 156
Chapter 8: DSP Implementation Platform: TMS320C6x Architecture and Software Tools 157
    8.1 TMS320C6X DSP 157
    8.2 C6x DSK Target Boards 161
    8.3 DSP Programming 163
    8.4 Bibliography 166
Lab 8: Getting Familiar with Code Composer Studio 167
    L8.1 Code Composer Studio 167
    L8.2 Creating Projects 167
    L8.3 Debugging Tools 173
    L8.4 Bibliography 182
Chapter 9: LabVIEW DSP Integration 183
    9.1 Communication with LabVIEW: Real-Time Data Exchange (RTDX) 183
    9.2 LabVIEW DSP Test Integration Toolkit for TI DSP 183
    9.3 Combined Implementation: Gain Example 184
    9.4 Bibliography 190
Lab 9: DSP Integration Examples 191
    L9.1 CCS Automation 191
    L9.2 Digital Filtering 193
    L9.3 Fixed-Point Implementation 202
    L9.4 Adaptive Filtering Systems 206
    L9.5 Frequency Processing: FFT 211
    L9.6 Bibliography 220
Chapter 10: DSP System Design: Dual-Tone Multi-Frequency (DTMF) Signaling 221
    10.1 Bibliography 224
Lab 10: Dual-Tone Multi-Frequency 225
    L10.1 DTMF Tone Generator System 225

Contents

    L10.2 DTMF Decoder System 228
    L10.3 Bibliography 230

Chapter 11: DSP System Design: Software-Defined Radio 231
    11.1 QAM Transmitter 231
    11.2 QAM Receiver 234
    11.3 Bibliography 238

Lab 11: Building a 4-QAM Modem 239
    L11.1 QAM Transmitter 239
    L11.2 QAM Receiver 242
    L11.3 Bibliography 252

Chapter 12: DSP System Design: MP3 Player 253
    12.1 Synchronization Block 254
    12.2 Scale Factor Decoding Block 256
    12.3 Huffman Decoder 257
    12.4 Requantizer 259
    12.5 Reordering 261
    12.6 Alias Reduction 261
    12.7 IMDCT and Windowing 262
    12.8 Polyphase Filter Bank 264
    12.9 Bibliography 266

Lab 12: Implementation of MP3 Player in LabVIEW 267
    L12.1 System-Level VI 267
    L12.2 LabVIEW Implementation 268
    L12.3 Modifications to Achieve Real-Time Decoding 281
    L12.4 Bibliography 286

Index 287

Preface

For many years, I have been teaching DSP (Digital Signal Processing) lab courses using various TI (Texas Instruments) DSP platforms. One question I have been getting from students in a consistent way is, "Do we have to know C to take DSP lab courses?" Until last year, my response was, "Yes, C is a prerequisite for taking DSP lab courses." However, last year for the first time, I provided a different response by saying, "Though preferred, it is not required to know C to take DSP lab courses." This change in my response came about because I started using LabVIEW to teach students how to design and analyze DSP systems in our DSP courses.

The widely available graphical programming environments such as LabVIEW have now reached the level of maturity that allow students and engineers to design and analyze DSP systems with ease and in a relatively shorter time as compared to C and MATLAB. I have observed that many students taking DSP lab courses, in particular at the undergraduate level, often struggle and spend a fair amount of their time debugging C and MATLAB code instead of placing their efforts into understanding signal processing system design issues. The motivation behind writing this book has thus been to avoid this problem by adopting a graphical programming approach instead of the traditional and commonly used text-based programming approach in DSP lab courses. As a result, this book allows students to put most of their efforts into building DSP systems rather than debugging C code when taking DSP lab courses.

One important point that needs to be mentioned here is that in order to optimize signal processing algorithms on a DSP processor, it is still required to know and use C and/or assembly programming. The existing graphical programming environments are not meant to serve as optimizers when implementing signal processing algorithms on DSP processors or other hardware platforms. This point has been addressed

Preface

in this book by providing two chapters which are dedicated solely to algorithm implementation on the TI family of TMS320C6000 DSP processors.

It is envisioned that this alternative graphical programming approach to designing digital signal processing systems will allow more students to get exposed to the field of DSP. In addition, the book is written in such a way that it can be used as a self-study guide by DSP engineers who wish to become familiar with LabVIEW and use it to design and analyze DSP systems.

I would like to express my gratitude to NI (National Instruments) for their support of this book. In particular, I wish to thank Jim Cahow, Academic Resources Manager at NI, and Ravi Marawar, Academic Program Manager at NI, for their valuable feedback. I am pleased to acknowledge Chuck Glaser, Senior Acquisition Editor at Elsevier, and Cathy Wicks, University Program Manager at TI, for their promotion of the book. Finally, I am grateful to my family who put up with my preoccupation on this book-writing project.

Nasser Kehtarnavaz
December 2004

What's on the CD-ROM?

- The accompanying CD-ROM includes all the lab files discussed throughout the book. These files are placed in corresponding folders as follows:
    - Lab01: Getting familiar with LabVIEW: Part I
    - Lab02: Getting familiar with LabVIEW: Part II
    - Lab03: Sampling, Quantization, and Reconstruction
    - Lab04: FIR/IIR Filtering System Design
    - Lab05: Data Type and Scaling
    - Lab06: Adaptive Filtering Systems
    - Lab07: FFT, STFT, and DWT
    - Lab08: Getting Familiar with Code Composer Studio
    - Lab09: DSP Integration Examples
    - Lab10: Building Dual Tone Multi Frequency System in LabVIEW
    - Lab11: Building 4-QAM Modem System in LabVIEW
    - Lab12: Building MP3 Player System in LabVIEW
- To run the lab files, the National Instruments LabVIEW 7.1 is required and assumed installed. The lab files need to be copied into the folder "C:\LabVIEW Labs\".

What's on the CD-ROM?

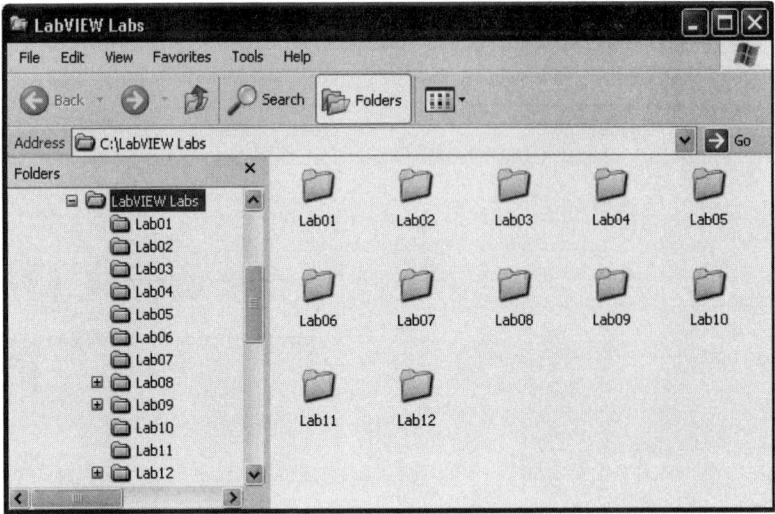

- For Lab 8 and Lab 9, the Texas Instruments Code Composer Studio 2.2 (CCStudio) is required and assumed installed in the folder "C:\ti\". The subfolders correspond to the following DSP platforms:
  - DSK 6416
  - DSK 6713
  - Simulator (configured as DSK6713 as shown below)

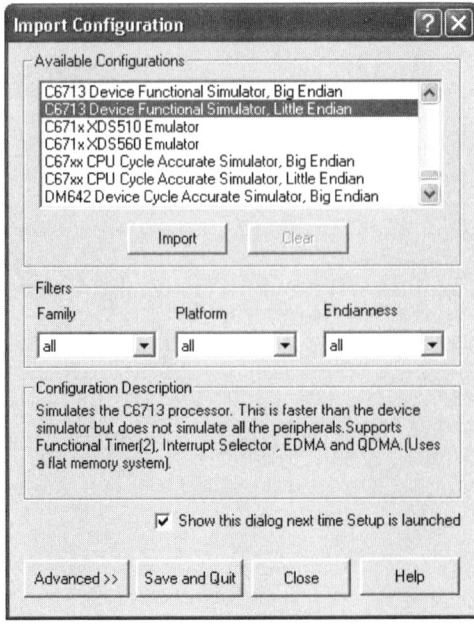

CHAPTER 1

Introduction

The field of digital signal processing (DSP) has experienced a considerable growth in the last two decades, primarily due to the availability and advancements in digital signal processors (also called DSPs). Nowadays, DSP systems such as cell phones and high-speed modems have become an integral part of our lives.

In general, sensors generate analog signals in response to various physical phenomena that occur in an analog manner (that is, in continuous time and amplitude). Processing of signals can be done either in the analog or digital domain. To perform the processing of an analog signal in the digital domain, it is required that a digital signal is formed by sampling and quantizing (digitizing) the analog signal. Hence, in contrast to an analog signal, a digital signal is discrete in both time and amplitude. The digitization process is achieved via an analog-to-digital (A/D) converter. The field of DSP involves the manipulation of digital signals in order to extract useful information from them.

There are many reasons why one might wish to process an analog signal in a digital fashion by converting it into a digital signal. The main reason is that digital processing allows programmability. The same processor hardware can be used for many different applications by simply changing the code residing in memory. Another reason is that digital circuits provide a more stable and tolerant output than analog circuits—for instance, when subjected to temperature changes. In addition, the advantage of operating in the digital domain may be intrinsic. For example, a linear phase filter or a steep-cutoff notch filter can easily be realized by using digital signal processing techniques, and many adaptive systems are achievable in a practical product only via digital manipulation of signals. In essence, digital representation (zeroes and ones) allows voice, audio, image, and video data to be treated the same for error-tolerant digital transmission and storage purposes.

Chapter 1

1.1 Digital Signal Processing Hands-On Lab Courses

Nearly all electrical engineering curricula include DSP courses. DSP lab or design courses are also being offered at many universities concurrently or as follow-ups to DSP theory courses. These hands-on lab courses have played a major role in student understanding of DSP concepts. A number of textbooks, such as [1-3], have been written to provide the teaching materials for DSP lab courses. The programming language used in these textbooks consists of either C, MATLAB®, or Assembly, that is text-based programming. In addition to these programming skills, it is becoming important for students to gain experience in a block-based or graphical (G) programming language or environment for the purpose of designing DSP systems in a relatively short amount of time. Thus, the main objective of this book is to provide a block-based or system-level programming approach in DSP lab courses. The block-based programming environment chosen is LabVIEW™.

LabVIEW (Laboratory Virtual Instrumentation Engineering Workbench) is a graphical programming environment developed by National Instruments (NI), which allows high-level or system-level designs. It uses a graphical programming language to create so-called Virtual Instruments (VI) blocks in an intuitive flowchart-like manner. A design is achieved by integrating different components or subsystems within a graphical framework. LabVIEW provides data acquisition, analysis, and visualization features well suited for DSP system-level design. It is also an open environment accommodating C and MATLAB code as well as various applications such as ActiveX and DLLs (Dynamic Link Libraries).

This book is written primarily for those who are already familiar with signal processing concepts and are interested in designing signal processing systems without needing to be proficient C or MATLAB programmers. After familiarizing the reader with LabVIEW, the book covers a LabVIEW-based approach to generic experiments encountered in a typical DSP lab course. It brings together in one place the information scattered in several NI LabVIEW manuals to provide the necessary tools and know-how for designing signal processing systems within a one-semester structured course. This book can also be used as a self-study guide to design signal processing systems using LabVIEW.

In addition, for those interested in DSP hardware implementation, two chapters in the book are dedicated to executing selected portions of a LabVIEW designed system on an actual DSP processor. The DSP processor chosen is TMS320C6000. This processor is manufactured by Texas Instruments (TI) for computationally intensive signal processing applications. The DSP hardware utilized to interface with

LabVIEW is the TI's C6416 or C6713 DSK (DSP Starter Kit) board. It should be mentioned that since the DSP implementation aspect of the labs (which includes C programs) is independent of the LabVIEW implementation, those who are not interested in the DSP implementation may skip these two chapters. It is also worth pointing out that once the LabVIEW code generation utility becomes available, any portion of a LabVIEW designed system can be executed on this DSP processor without requiring any C programming.

1.2 Organization

The book includes twelve chapters and twelve labs. After this introduction, the LabVIEW programming environment is presented in Chapter 2. Lab 1 and Lab 2 in Chapter 2 provide a tutorial on getting familiar with the LabVIEW programming environment. The topic of analog to digital signal conversion is presented in Chapter 3, followed by Lab 3 covering signal sampling examples. Chapter 4 involves digital filtering. Lab 4 in Chapter 4 shows how to use LabVIEW to design FIR and IIR digital filters. In Chapter 5, fixed-point versus floating-point implementation issues are discussed followed by Lab 5 covering data type and fixed-point effect examples. In Chapter 6, the topic of adaptive filtering is discussed. Lab 6 in Chapter 6 covers two adaptive filtering systems consisting of system identification and noise cancellation. Chapter 7 presents frequency domain processing followed by Lab 7 covering the three widely used transforms in signal processing: fast Fourier transform (FFT), short time Fourier transform (STFT), and discrete wavelet transform (DWT). Chapter 8 discusses the implementation of a LabVIEW-designed system on the TMS320C6000 DSP processor. First, an overview of the TMS320C6000 architecture is provided. Then, in Lab 8, a tutorial is presented to show how to use the Code Composer Studio™ (CCStudio) software development tool to achieve the DSP implementation. As a continuation of Chapter 8, Chapter 9 and Lab 9 discuss the issues related to the interfacing of LabVIEW and the DSP processor. Chapters 10 through 12, and Labs 10 through 12, respectively, discuss the following three DSP systems or project examples that are fully designed via LabVIEW: (i) dual-tone multi-frequency (DTMF) signaling, (ii) software-defined radio, and (iii) MP3 player.

1.3 Software Installation

LabVIEW 7.1, which is the latest version at the time of this writing, is installed by running *setup.exe* on the LabVIEW 7.1 Installation CD. Some lab portions use the LabVIEW toolkits 'Digital Filter Design,' 'Advanced Signal Processing,' and 'DSP

Chapter 1

Test Integration for TI DSP.' Each of these toolkits can be installed by running *setup.exe* located on the corresponding toolset CD.

If one desires to run parts of a LabVIEW designed system on a DSP processor, then it is necessary to install the Code Composer Studio software tool. This is done by running *setup.exe* on the CCStudio CD. The most updated version of CCStudio at the time of this writing, CCStudio 2.2, is used in the DSK-related labs.

The accompanying CD includes all the files necessary for running the labs covered throughout the book.

1.4 Updates

Considering that any programming environment goes through enhancements and updates, it is expected that there will be updates of LabVIEW and its toolkits. To accommodate for such updates and to make sure that the labs provided in the book can still be used in DSP lab courses, any new version of the labs will be posted at the website http://www.utdallas.edu/~kehtar/LabVIEW for easy access. It is recommended that this website be periodically checked to download any necessary updates.

1.5 Bibliography

[1] N. Kehtarnavaz, *Real-Time Digital Signal Processing Based on the TMS320C6000*, Elsevier, 2005.

[2] S. Kuo and W-S. Gan, *Digital Signal Processors: Architectures, Implementations, and Applications*, Prentice-Hall, 2005.

[3] R. Chassaing, *DSP Applications Using C and the TMS320C6x DSK*, Wiley Inter-Science, 2002.

CHAPTER 2

LabVIEW Programming Environment

LabVIEW constitutes a graphical programming environment that allows one to design and analyze a DSP system in a shorter time as compared to text-based programming environments. LabVIEW graphical programs are called virtual instruments (VIs). VIs run based on the concept of data flow programming. This means that execution of a block or a graphical component is dependent on the flow of data, or more specifically a block executes when data is made available at all of its inputs. Output data of the block are then sent to all other connected blocks. Data flow programming allows multiple operations to be performed in parallel since its execution is determined by the flow of data and not by sequential lines of code.

2.1 Virtual Instruments (VIs)

A VI consists of two major components; a front panel (FP) and a block diagram (BD). An FP provides the user-interface of a program, while a BD incorporates its graphical code. When a VI is located within the block diagram of another VI, it is called a subVI. LabVIEW VIs are modular, meaning that any VI or subVI can be run by itself.

2.1.1 Front Panel and Block Diagram

An FP contains the user interfaces of a VI shown in a BD. Inputs to a VI are represented by so-called controls. Knobs, pushbuttons and dials are a few examples of controls. Outputs from a VI are represented by so-called indicators. Graphs, LEDs (light indicators) and meters are a few examples of indicators. As a VI runs, its FP provides a display or user interface of controls (inputs) and indicators (outputs).

A BD contains terminal icons, nodes, wires, and structures. Terminal icons are interfaces through which data are exchanged between an FP and a BD. Terminal icons

Chapter 2

correspond to controls or indicators that appear on an FP. Whenever a control or indicator is placed on an FP, a terminal icon gets added to the corresponding BD. A node represents an object which has input and/or output connectors and performs a certain function. SubVIs and functions are examples of nodes. Wires establish the flow of data in a BD. Structures such as repetitions or conditional executions are used to control the flow of a program. Figure 2-1 shows what an FP and a BD window look like.

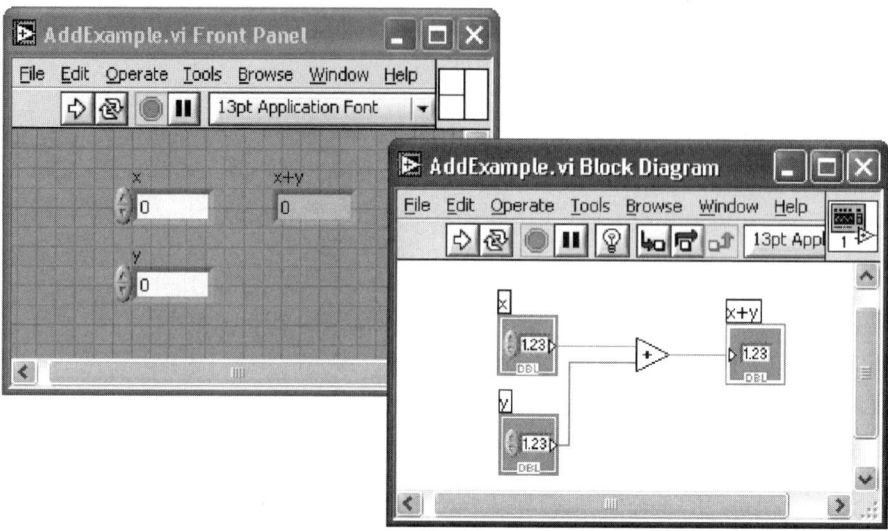

Figure 2-1: LabVIEW windows: front panel and block diagram.

2.1.2 Icon and Connector Pane

A VI icon is a graphical representation of a VI. It appears in the top right corner of a BD or an FP window. When a VI is inserted in a BD as a subVI, its icon gets displayed.

A connector pane defines inputs (controls) and outputs (indicators) of a VI. The number of inputs and outputs can be changed by using different connector pane patterns. In Figure 2-1, a VI icon is shown at the top right corner of the BD and its corresponding connector pane having two inputs and one output is shown at the top right corner of the FP.

LabVIEW PROGRAMMING ENVIRONMENT

2.2 Graphical Environment

2.2.1 Functions Palette

The Functions palette, see Figure 2-2, provides various function VIs or blocks for building a system. This palette can be displayed by right-clicking on an open area of a BD. Note that this palette can only be displayed in a BD.

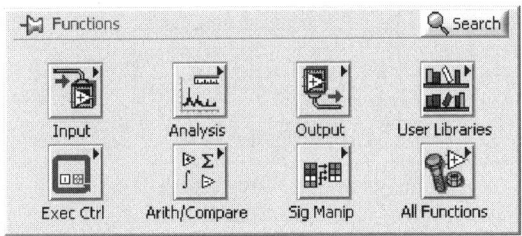

Figure 2-2: Functions palette.

2.2.2 Controls Palette

The Controls palette, see Figure 2-3, provides controls and indicators of an FP. This palette can be displayed by right-clicking on an open area of an FP. Note that this palette can only be displayed in an FP.

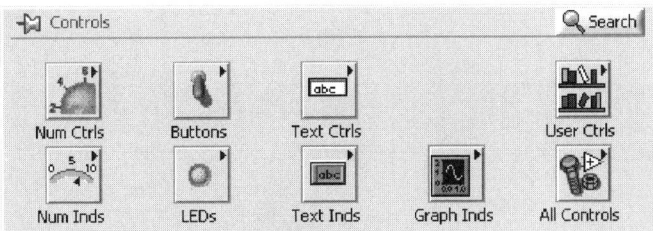

Figure 2-3: Controls palette.

2.2.3 Tools Palette

The Tools palette provides various operation modes of the mouse cursor for building or debugging a VI. The Tools palette and the frequently-used tools are shown in Figure 2-4.

Each tool is utilized for a specific task. For example, the Wiring tool is used to wire objects in a BD. If the automatic tool selection mode is enabled by clicking the **Automatic Tool Selection** button, LabVIEW selects the best matching tool based on a current cursor position.

Chapter 2

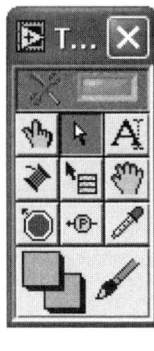
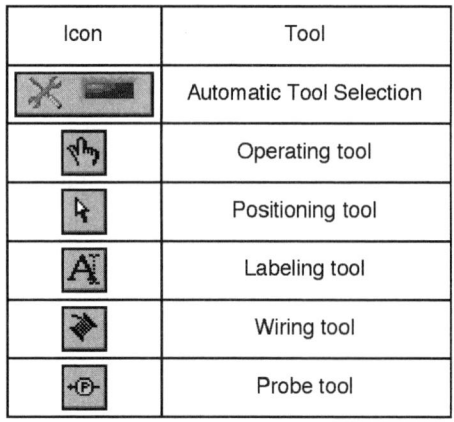

Figure 2-4: Tools palette.

2.3 Building a Front Panel

In general, a VI is put together by going back and forth between an FP and a BD, placing inputs and outputs on the FP and building blocks on the BD.

2.3.1 Controls

Controls make up the inputs to a VI. Controls grouped in the Numeric Controls palette are used for numerical inputs, grouped in the Buttons & Switches palette for Boolean inputs, and grouped in the Text Controls palette for text and enumeration inputs. These control options are displayed in Figure 2-5.

2.3.2 Indicators

Indicators make up the outputs of a VI. Indicators grouped in the Numeric Indicators palette are used for numerical outputs, grouped in the LEDs palette for Boolean outputs, grouped in the Text Indicators

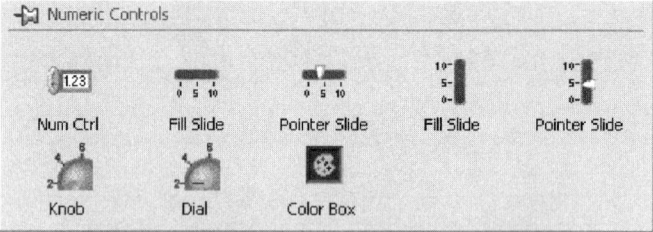
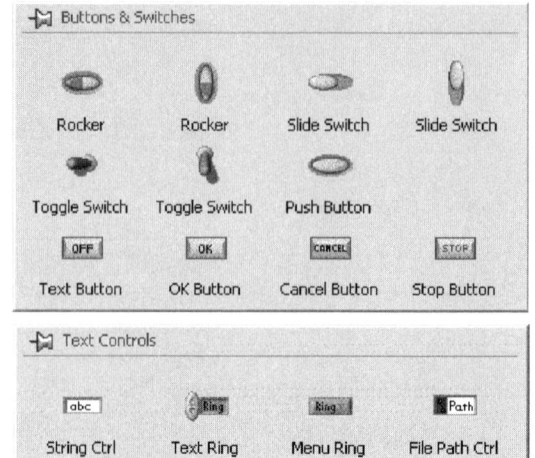

Figure 2-5: Control palettes.

8

LabVIEW Programming Environment

palette for text outputs, and grouped in the Graph Indicators palette for graphical outputs. These indicator options are displayed in Figure 2-6.

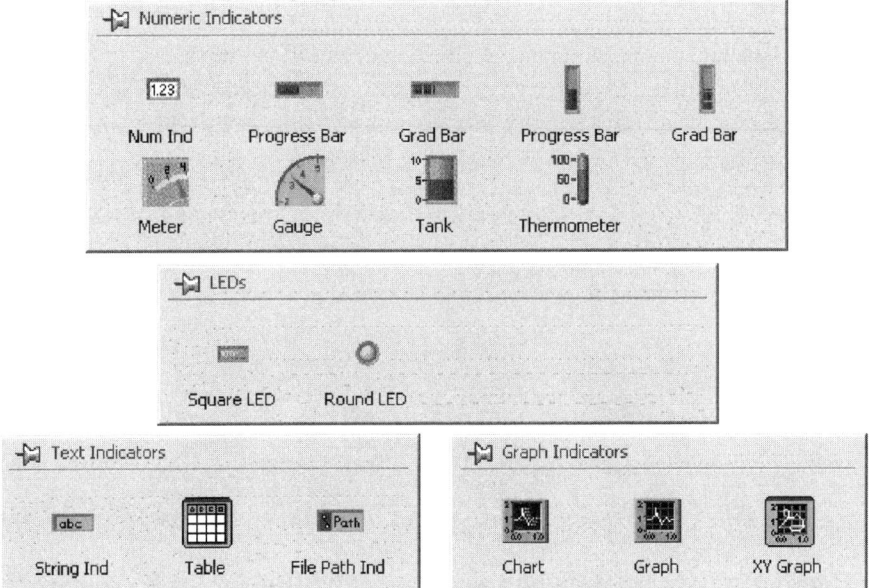

Figure 2-6: Indicator palettes.

2.3.3 Align, Distribute and Resize Objects

The menu items on the toolbar of an FP, see Figure 2-7, provide options to align and distribute objects on the FP in an orderly manner. Normally, after controls and indicators are placed on an FP, one uses these options to tidy up their appearance.

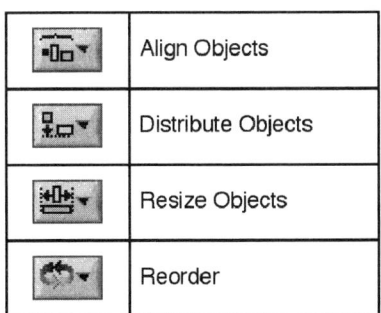

Figure 2-7: Menu for align, distribute, resize, and reorder objects.

Chapter 2

2.4 Building a Block Diagram

2.4.1 Express VI and Function

Express VIs denote higher-level VIs that have been configured to incorporate lower-level VIs or functions. These VIs are displayed as expandable nodes with a blue background. Placing an Express VI in a BD brings up a configuration dialog window allowing adjustment of its parameters. As a result, Express VIs demand less wiring. A configuration window can be brought up by double-clicking on its Express VI.

Basic operations such as addition or subtraction are represented by functions. Figure 2-8 shows three examples corresponding to three types of a BD object (VI, Express VI, and function).

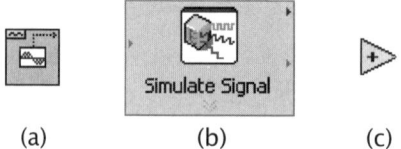

Figure 2-8: Block Diagram objects: (a) VI, (b) Express VI, and (c) function.

Both subVI and Express VI can be displayed as icons or expandable nodes. If a subVI is displayed as an expandable node, the background appears yellow. Icons are used to save space in a BD, while expandable nodes are used to provide easier wiring or better readability. Expandable nodes can be resized to show their connection nodes more clearly. Three appearances of a VI/Express VI are shown in Figure 2-9.

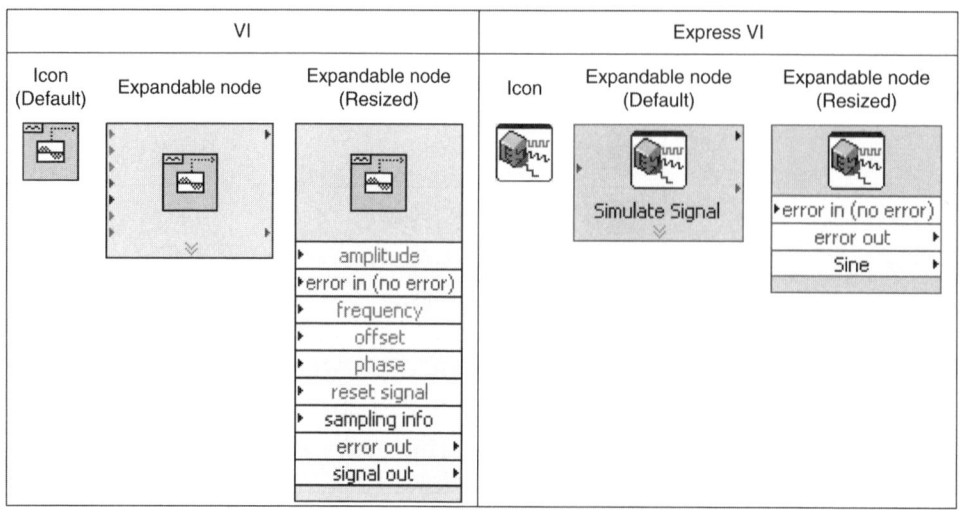

Figure 2-9: Icon versus expandable node.

LabVIEW Programming Environment

2.4.2 Terminal Icons

FP objects are displayed as terminal icons in a BD. A terminal icon exhibits an input or output as well as its data type. Figure 2-10 shows two terminal icon examples consisting of a double precision numerical control and indicator. As shown in this figure, terminal icons can be displayed as data type terminal icons to conserve space in a BD.

	Control	Indicator
Terminal Icons	 1.23 DBL	 1.23 DBL
Data Type Terminal Icons	 DBL	 DBL

Figure 2-10: Terminal icon examples displayed in a BD.

2.4.3 Wires

Wires transfer data from one node to another in a BD. Based on the data type of a data source, the color and thickness of its connecting wires change.

Wires for the basic data types used in LabVIEW are shown in Figure 2-11. Other than the data types shown in this figure, there are some other specific data types. For example, the dynamic data type is always used for Express VIs, and the waveform data type, which corresponds to the output from a waveform generation VI, is a special cluster of waveform components incorporating trigger time, time interval, and data value.

Wire Type	Scalar	1D Array	2D Array	Color
Numeric				Orange (Floating point) Blue (Integer)
Boolean				Green
String				Pink

Figure 2-11: Basic types of wires [2].

2.4.4 Structures

A structure is represented by a graphical enclosure. The graphical code enclosed by a structure is repeated or executed conditionally. A loop structure is equivalent to a for loop or a while loop statement encountered in text-based programming languages, while a case structure is equivalent to an if-else statement.

2.4.4.1 For Loop

A For Loop structure is used to perform repetitions. As illustrated in Figure 2-12, the displayed border indicates a For Loop structure, where the count terminal N represents the number of times the loop is to be repeated. It is set by wiring a value from outside of the loop to it. The iteration terminal i denotes the number of completed iterations, which always starts at zero.

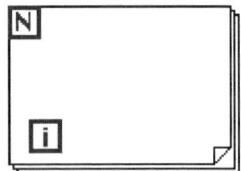

Figure 2-12: For Loop.

2.4.4.2 While Loop

A While Loop structure allows repetitions depending on a condition, see Figure 2-13. The conditional terminal initiates a stop if the condition is true. Similar to a For Loop, the iteration terminal i provides the number of completed iterations, always starting at zero.

Figure 2-13: While Loop.

2.4.4.3 Case Structure

A Case structure, see Figure 2-14, allows running different sets of operations depending on the value it receives through its selector terminal, which is indicated by ?. In addition to Boolean type, the input to a selector terminal can be of integer, string, or enumerated type. This input determines which case to execute. The case selector shows the status being executed. Cases can be added or deleted as needed.

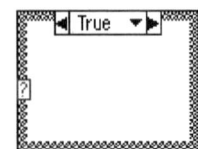

Figure 2-14: Case structure.

2.5 Grouping Data: Array and Cluster

An array represents a group of elements having the same data type. An array consists of data elements having a dimension up to $2^{31} - 1$. For example, if a random number is generated in a loop, it makes sense to build the output as an array since the length of the data element is fixed at 1 and the data type is not changed during iterations.

A cluster consists of a collection of different data type elements, similar to the structure data type in text-based programming languages. Clusters allow one to reduce the number of wires on a BD by bundling different data type elements together and passing them to only one terminal. An individual element can be added to or extracted from a cluster by using the cluster functions such as `Bundle by Name` and `Unbundle by Name`.

2.6 Debugging and Profiling VIs

2.6.1 Probe Tool

The Probe tool is used for debugging VIs as they run. The value on a wire can be checked while a VI is running. Note that the Probe tool can only be accessed in a BD window.

The Probe tool can be used together with breakpoints and execution highlighting to identify the source of an incorrect or an unexpected outcome. A breakpoint is used to pause the execution of a VI at a specific location, while execution highlighting helps one to visualize the flow of data during program execution.

2.6.2 Profile Tool

The Profile tool can be used to gather timing and memory usage information, in other words, how long a VI takes to run and how much memory it consumes. It is necessary to make sure a VI is stopped before setting up a Profile window.

An effective way to become familiar with LabVIEW programming is by going through examples. Thus, in the two labs that follow in this chapter, most of the key programming features of LabVIEW are presented by building some simple VIs. More detailed information on LabVIEW programming can be found in [1-5].

2.7 Bibliography

[1] National Instruments, *LabVIEW Getting Started with LabVIEW*, Part Number 323427A-01, 2003.

[2] National Instruments, *LabVIEW User Manual*, Part Number 320999E-01, 2003.

[3] National Instruments, *LabVIEW Performance and Memory Management*, Part Number 342078A-01, 2003.

[4] National Instruments, *Introduction to LabVIEW Six-Hour Course*, Part Number 323669B-01, 2003.

[5] Robert H. Bishop, *Learning With Labview 7 Express*, Prentice Hall, 2003.

Lab 1: Getting Familiar with LabVIEW: Part I

The objective of this first lab is to provide an initial hands-on experience in building a VI. For detailed explanations of the LabVIEW features mentioned here, the reader is referred to [1]. LabVIEW7.1 can get launched by double-clicking on the LabVIEW 7.1 icon. The dialog window shown in Figure 2-15 should appear.

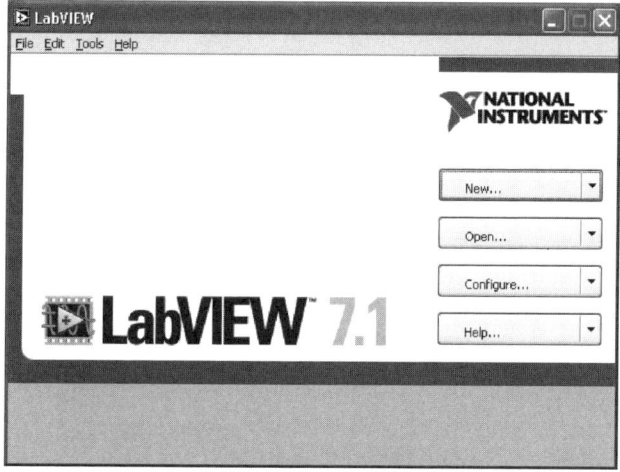

Figure 2-15: Starting LabVIEW.

L1.1 Building a Simple VI

To become familiar with the LabVIEW programming environment, it is more effective if one goes through a simple example. The example presented here consists of calculating the sum and average of two input values. This example is described in a step-by-step fashion below.

L1.1.1 VI Creation

To create a new VI, click on the arrow next to **New…** and choose **Blank VI** from the pull down menu. This step can also be done by choosing **File** → **New VI** from the menu. As a result, a blank FP and a blank BD window appear, as shown in Figure 2-16. It should be remembered that an FP and a BD coexist when building a VI.

Lab 1

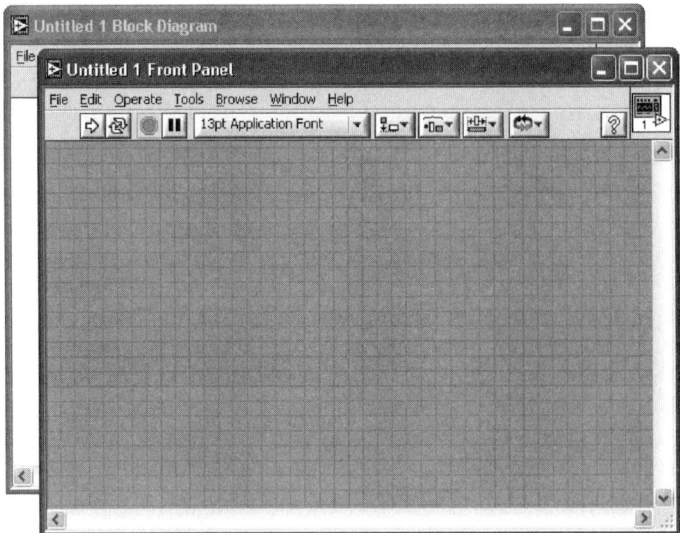

Figure 2-16: Blank VI.

Clearly, the number of inputs and outputs to a VI is dependent on its function. In this example, two inputs and two outputs are needed, one output generating the sum and the other the average of two input values. The inputs are created by locating two Numeric Controls on the FP. This is done by right-clicking on an open area of the FP to bring up the Controls palette, followed by choosing **Controls → Numeric Controls → Numeric Control**. Each numeric control automatically places a corresponding terminal icon on the BD. Double-clicking on a numeric control highlights its counterpart on the BD, and vice versa.

Next, let us label the two inputs as x and y. This is achieved by using the Labeling tool from the **Tools** palette, which can be displayed by choosing **Window → Show Tools Palette** from the menu bar. Choose the Labeling tool and click on the default labels, Numeric and Numeric 2, in order to edit them. Alternatively, if the automatic tool selection mode is enabled by clicking **Automatic Tool Selection** in the **Tools** palette, the labels can be edited by simply double-clicking on the default labels. Editing a label on the FP changes its corresponding terminal icon label on the BD, and vice versa.

Similarly, the outputs are created by locating two Numeric Indicators (**Controls → Numeric Indicators → Numeric Indicator**) on the FP. Each numeric indicator automatically places a corresponding terminal icon on the BD. Edit the labels of the indicators to read Sum and Average.

For a better visual appearance, objects on an FP window can be aligned, distributed, and resized using the appropriate buttons appearing on the FP toolbar. To do this,

Getting Familiar with LabVIEW: Part I

select the objects to be aligned or distributed and apply the appropriate option from the toolbar menu. Figure 2-17 shows the configuration of the FP just created.

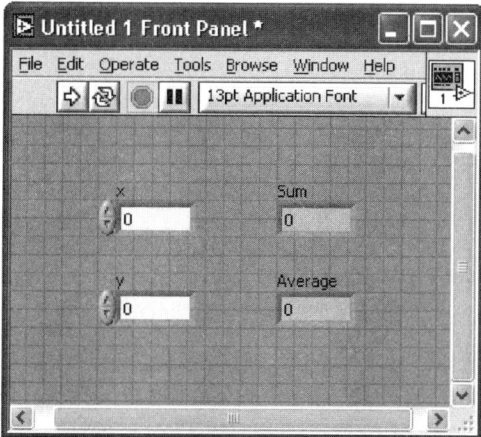

Figure 2-17: FP configuration.

Now, let us build a graphical program on the BD to perform the summation and averaging operations. Note that <Ctrl + E> toggles between an FP and a BD window. If one finds the objects on a BD are too close to insert other functions or VIs in between, a horizontal or vertical space can be inserted by holding down the <Ctrl> key to create space horizontally and/or vertically. As an example, Figure 2-18 (b) illustrates a horizontal space inserted between the objects shown in Figure 2-18 (a).

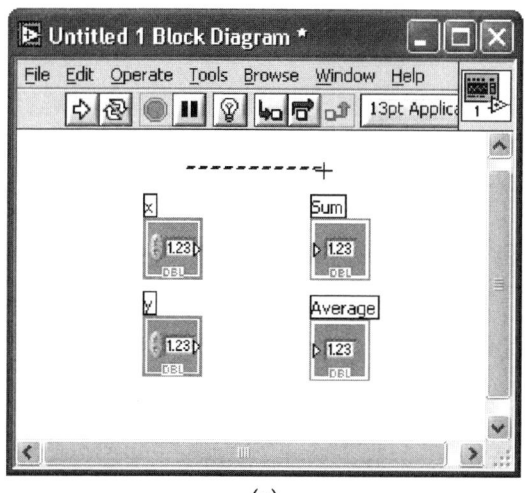
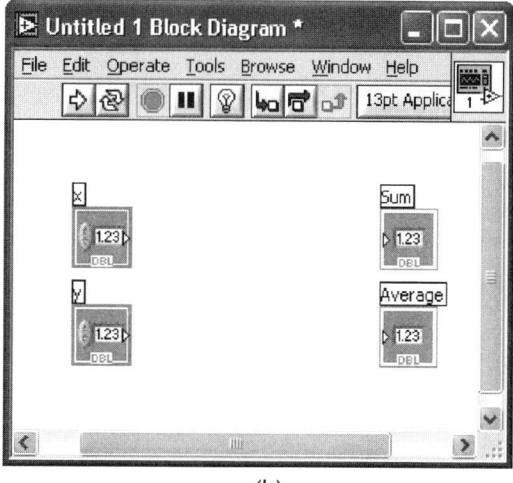

(a)   (b)

Figure 2-18: Inserting horizontal/vertical space: (a) creating space while holding down the <Ctrl> key, and (b) inserted horizontal space.

Lab 1

Next, place an Add function (**Functions** → **Arithmetic & Comparison** → **Express Numeric** → **Add**) and a Divide function (**Functions** → **Arithmetic & Comparison** → **Express Numeric** → **Divide**) on the BD. The divisor, in our case 2, needs to be entered in a Numeric Constant (**Functions** → **Arithmetic & Comparison** → **Express Numeric** → **Numeric Constant**) and connected to the y terminal of the Divide function using the Wiring tool.

To have a proper data flow, functions, structures and terminal icons on a BD need to be wired. The Wiring tool is used for this purpose. To wire these objects, point the Wiring tool at a terminal of a function or a subVI to be wired, left click on the terminal, drag the mouse to a destination terminal and left click once again. Figure 2-19 illustrates the wires placed between the terminals of the numeric controls and the input terminals of the add function. Notice that the label of a terminal is displayed whenever the cursor is moved over it if the automatic tool selection mode is enabled. Also, note that the Run button ⇨ on the toolbar remains broken until the wiring process is completed.

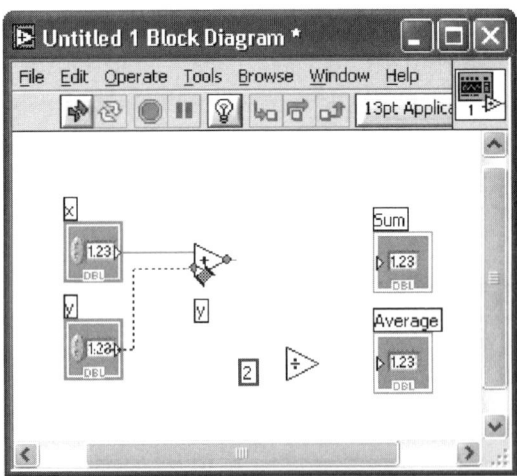

Figure 2-19: Wiring BD objects.

For better readability of a BD, wires which are hidden behind objects or crossed over other wires can be cleaned up by right-clicking on them and choosing **Clean Up Wire** from the shortcut menu. Any broken wires can be cleared by pressing <Ctrl + B> or **Edit** → **Remove Broken Wires**.

The label of a BD object, such as a function, can be shown (or hidden) by right-clicking on the object and checking (or unchecking) **Visible Items** → **Label** from the shortcut menu. Also, a terminal icon corresponding to a numeric control or indicator

Getting Familiar with LabVIEW: Part I

can be shown as a data type terminal icon. This is done by right-clicking on the terminal icon and unchecking **View As Icon** from the shortcut menu. Figure 2-20 shows an example where the numeric controls and indicators are shown as data type terminal icons. The notation DBL represents double precision data type.

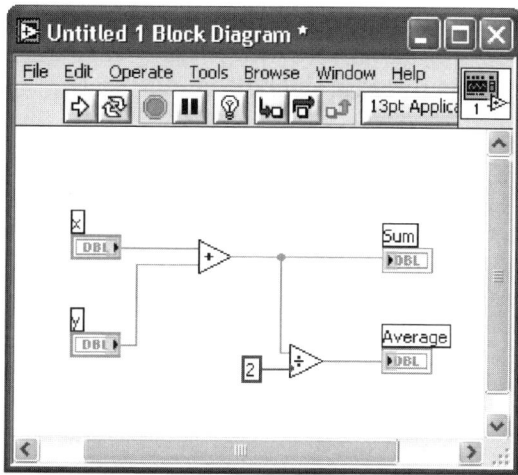

Figure 2-20: Completed BD.

It is worth pointing out that there exists a shortcut to build the above VI. Instead of choosing the numeric controls, indicators or constants from the Controls or Functions palette, the shortcut menu **Create**, activated by right-clicking on a terminal of a BD object such as a function or a subVI, can be used. As an example of this approach, create a blank VI and locate an Add function. Right-click on its x terminal and choose **Create** → **Control** from the shortcut menu to create and wire a numeric control or input. This locates a numeric control on the FP as well as a corresponding terminal icon on the BD. The label is automatically set to x. Create a second numeric control by right-clicking on the y terminal of the Add function. Next, right-click on the output terminal of the Add function and choose **Create** → **Indicator** from the shortcut menu. A data type terminal icon, labeled as x+y, is created on the BD as well as a corresponding numeric indicator on the FP.

Next, right-click on the y terminal of the Divide function to choose **Create** → **Constant** from the shortcut menu. This creates a Numeric Constant as the divisor and wires its y terminal. Type the value 2 in the numeric constant. Right-click on the output terminal of the Divide function, labeled as x/y, and choose **Create** → **Indicator** from the shortcut menu. If a wrong option is chosen, the terminal does not

Lab 1

get wired. A wrong terminal option can easily be changed by right-clicking on the terminal and choosing **Change to Control** or **Change to Constant** from the shortcut menu.

To save the created VI for later use, choose **File → Save** from the menu or press <Ctrl + S> to bring up a dialog window to enter a name. Type Sum and Average as the VI name and click **Save**.

To test the functionality of the VI, enter some sample values in the numeric controls on the FP and run the VI by choosing **Operate → Run**, by pressing <Ctrl + R>, or by clicking the Run button on the toolbar. From the displayed output values in the numeric indicators, the functionality of the VI can be verified. Figure 2-21 illustrates the outcome after running the VI with two inputs 10 and 30.

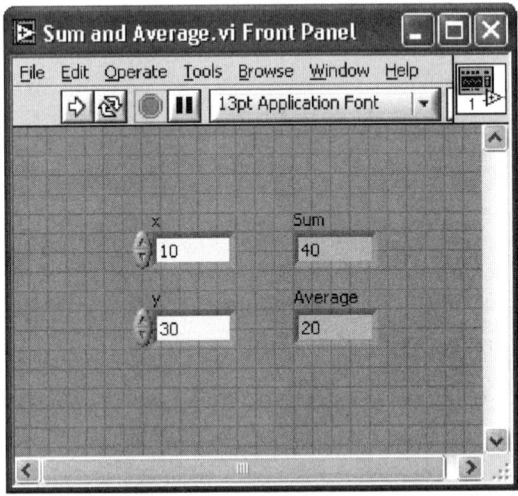

Figure 2-21: VI verification.

L1.1.2 SubVI Creation

If a VI is to be used as part of a higher level VI, its connector pane needs to be configured. A connector pane assigns inputs and outputs of a subVI to its terminals through which data are exchanged. A connector pane can be displayed by right-clicking on the top right corner icon of an FP and selecting **Show Connector** from the shortcut menu.

The default pattern of a connector pane is determined based on the number of controls and indicators. In general, the terminals on the left side of a connector pane pattern are used for inputs, and the ones on the right side for outputs. Terminals can be added to or removed from a connector pane by right-clicking and choosing **Add**

Getting Familiar with LabVIEW: Part I

Terminal or **Remove Terminal** from the shortcut menu. If a change is to be made to the number of inputs/outputs or to the distribution of terminals, a connector pane pattern can be replaced with a new one by right-clicking and choosing **Patterns** from the shortcut menu. Once a pattern is selected, each terminal needs to be reassigned to a control or an indicator by using the Wiring tool, or by enabling the automatic tool selection mode.

Figure 2-22(a) illustrates assigning a terminal of the Sum and Average VI to a numeric control. The completed connector pane is shown in Figure 2-22(b). Notice that the output terminals have thicker borders. The color of a terminal reflects its data type.

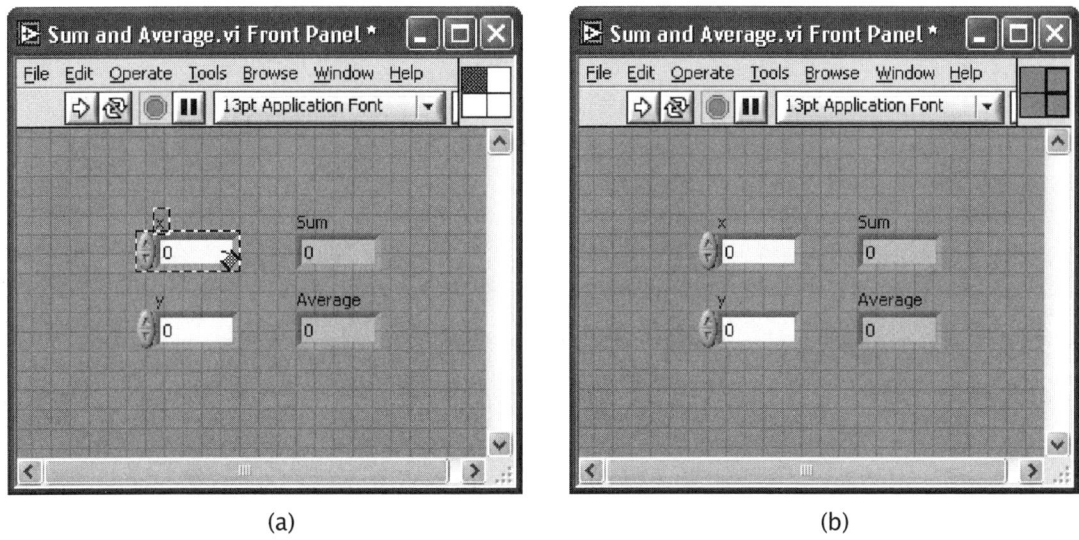

Figure 2-22: Connector pane: (a) assigning a terminal to a control, and (b) terminal assignment completed.

Considering that a subVI icon is displayed on the BD of a higher level VI, it is important to edit the subVI icon for it to be explicitly identified. Double-clicking on the top right corner icon of a BD brings up the Icon Editor. The tools provided in the Icon Editor are very similar to those encountered in other graphical editors, such as Microsoft Paint. An edited icon for the Sum and Average VI is illustrated in Figure 2-23.

A subVI can also be created from a section of a VI. To do so, select the nodes on the BD to be included in the subVI, as shown in Figure 2-24(a). Then, choose **Edit** → **Create SubVI**. This inserts a new subVI icon. Figure 2-24(b) illustrates the BD with an

Lab 1

Figure 2-23: Editing subVI icon.

inserted subVI. This subVI can be opened and edited by double-clicking on its icon on the BD. Save this subVI as *Sum and Average.vi*. This subVI performs the same function as the original Sum and Average VI.

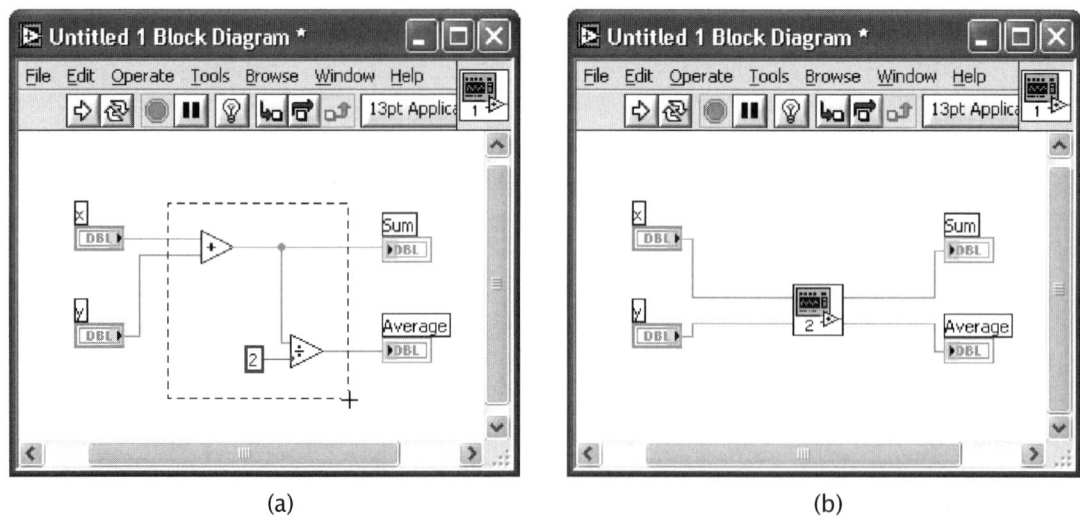

(a)   (b)

Figure 2-24: Creating a subVI: (a) selecting nodes to make a subVI, and (b) inserted subVI icon.

In Figure 2-25, the completed FP and BD of the Sum and Average VI are shown.

Getting Familiar with LabVIEW: Part I

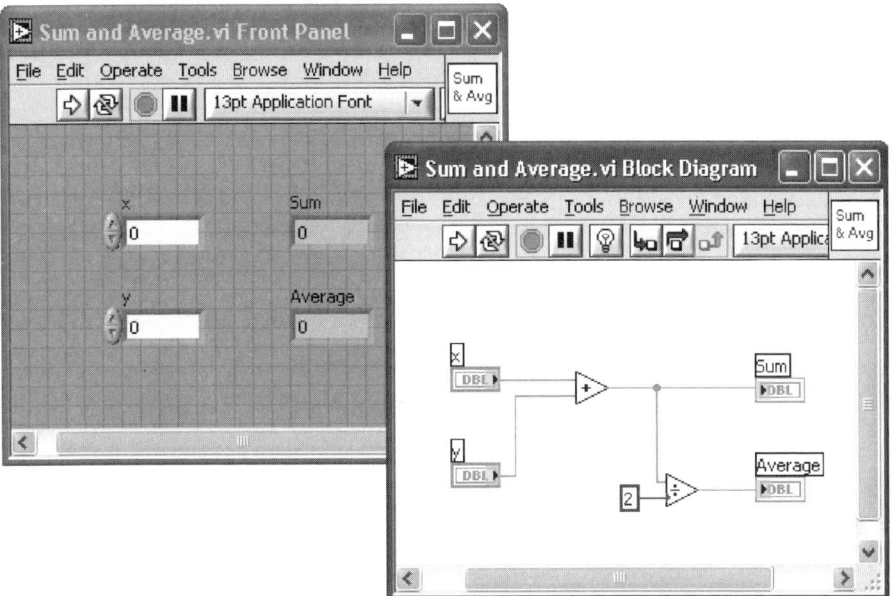

Figure 2-25: Sum and Average VI.

L1.2 Using Structures and SubVIs

Let us now consider another example to demonstrate the use of structures and subVIs. In this example, a VI is used to show the sum and average of two input values in a continuous fashion. The two inputs can be altered by the user. If the average of the two inputs becomes greater than a preset threshold value, a LED warning light is lit.

As the first step towards building such a VI, build an FP as shown in Figure 2-26(a). For the inputs, use two Knobs (**Controls** → **Numeric Controls** → **Knob**). Adjust the size of the knobs by using the Positioning tool. Properties of knobs such as precision and data type can be modified by right-clicking and choosing **Properties** from the shortcut menu. A Knob Properties dialog box is brought up and an **Appearance** tab is shown by default. Edit the label of one of the knobs to read Input 1. Select the **Data Range** tab, and click **Representation** to change the data type from double precision to byte by selecting **Byte** among the displayed data types. This can also be achieved by right-clicking on the knob and choosing **Representation** → **Byte** from the shortcut menu. In the **Data Range** tab, a default value needs to be specified. In this example, the default value is considered to be 0. The default value can be set by right-clicking on the control and choosing **Data Operations** → **Make Current Value Default** from the shortcut menu. Also, this control can be set to a default value by right-clicking and choosing **Data Operations** → **Reinitialize to Default Value** from the shortcut menu.

Lab 1

Label the second knob as Input 2 and repeat all the adjustments as done for the first knob except for the data representation part. The data type of the second knob is specified to be double precision in order to demonstrate the difference in the outcome. As the final step of configuring the FP, align and distribute the objects using the appropriate buttons on the FP toolbar.

To set the outputs, locate and place a Numeric Indicator, a Rounded LED (**Controls → LEDs → Rounded LED**), and a Gauge (**Controls → Numeric Indicators → Gauge**). Edit the labels of the indicators as shown in Figure 2-26(a).

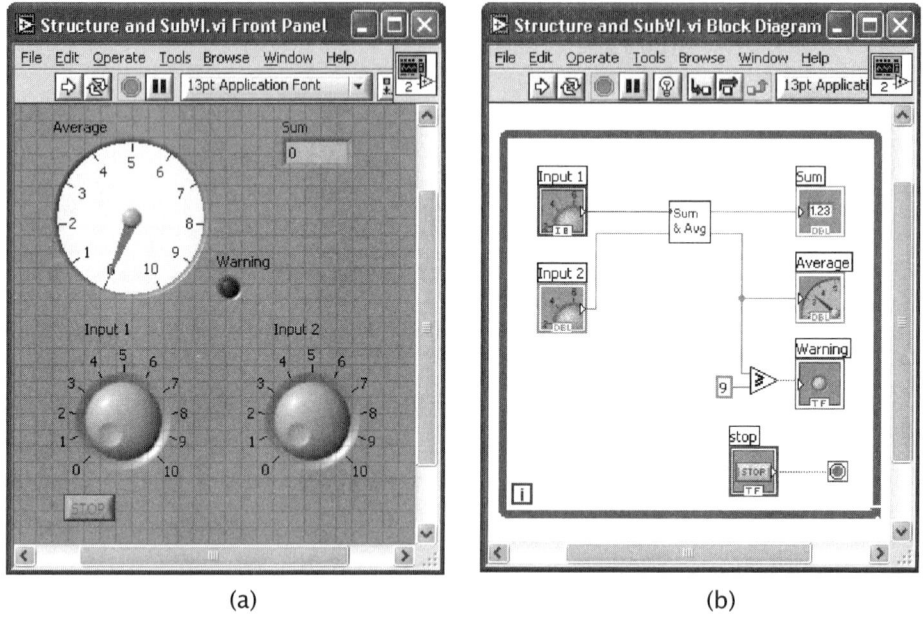

Figure 2-26: Example of structure and subVI: (a) FP, and (b) BD.

Now, let us build the BD. There are five control and indicator icons already appearing on the BD. Right-click on an open area of the BD to bring up the **Functions** palette and then choose **All functions → Select a VI…**. This brings up a file dialog box. Navigate to the Sum and Average VI in order to place it on the BD. This subVI is displayed as an icon on the BD. Wire the numeric controls, Input 1 and Input 2, to the x and y terminals, respectively. Also, wire the Sum terminal of the subVI to the numeric indicator labeled Sum, and the Average terminal to the gauge indicator labeled Average.

A Greater or Equal? function is located from **Functions → Arithmetic & Comparisons → Comparison → Greater or Equal?** in order to compare the average

Getting Familiar with LabVIEW: Part I

output of the subVI with a threshold value. Create a wire branch on the wire between the `Average` terminal of the subVI and its indicator via the Wiring tool. Then, extend this wire to the x terminal of the `Greater or Equal?` function. Right-click on the y terminal of the `Greater or Equal?` function and choose **Create → Constant** in order to place a `Numeric Constant`. Enter 9 in the numeric constant. Then, wire the `Rounded LED`, labeled as `Warning`, to the `x>=y?` terminal of this function to provide a Boolean value.

In order to run the VI continuously, a While Loop structure is used. Choose **Functions → Execution Control → While Loop** to create a `While Loop`. Change the size by dragging the mouse to enclose the objects in the `While Loop` as illustrated in Figure 2-27.

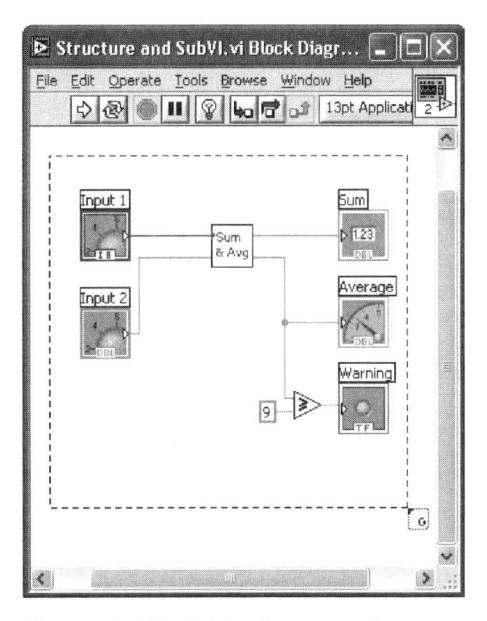

Figure 2-27: While Loop enclosure.

Once this structure is created, its boundary together with the loop iteration terminal [i], and conditional terminal [◉] are shown on the BD. If the `While Loop` is created by using **Functions → All Functions → Structures → While Loop**, then the `Stop Button` is not included as part of the structure. This button can be created by right-clicking on the conditional terminal and choosing **Create → Control** from the shortcut menu. A Boolean condition can be wired to a conditional terminal, instead of a stop button, in order to stop the loop programmatically.

25

Lab 1

As the final step, tidy up the wires, nodes and terminals on the BD using the **Align object** and **Distribute object** options on the BD toolbar. Then, save the VI in a file named *Strucure and SubVI.vi*.

Now run the VI to verify its functionality. After clicking the Run button on the toolbar, adjust the knobs to alter the inputs. Verify whether the average and sum are displayed correctly in the gauge and numeric indicators. Note that only integer values can be entered via the Input 1 knob while real values can be entered via the Input 2 knob. This is due to the data types associated with these knobs. The Input 1 knob is set to byte type, that is, I8 or 8 bit signed integer. As a result, only integer values within the range −128 and 127 can be entered. Considering that the minimum and maximum value of this knob are set to 0 and 10 respectively, only integer values from 0 to 10 can thus be entered for this input.

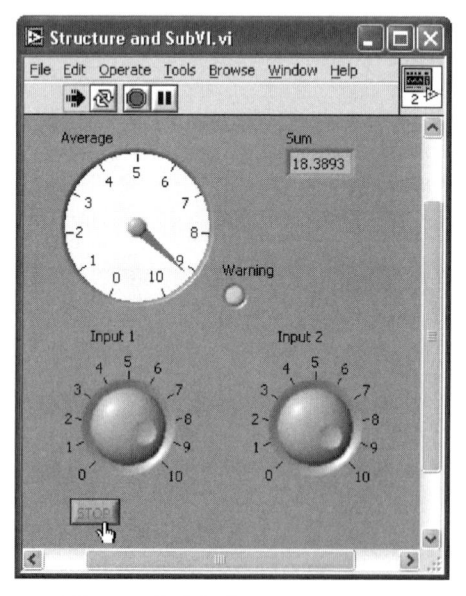

Figure 2-28: FP as VI runs.

When the average value of the two inputs becomes greater than the preset threshold value of 9, the warning LED will light up, see Figure 2-28. Click the stop button on the FP to stop the VI. Otherwise, the VI keeps running until the conditional terminal of the While Loop becomes true.

L1.3 Create an Array with Indexing

Auto-indexing enables one to read/write each element from/to a data array in a loop structure. That feature is covered in this section.

Let us first locate a For Loop (**Functions → All Functions → Structures → For Loop**). Right-click on its count terminal and choose **Create → Constant** from the shortcut menu to set the number of iterations. Enter 10 so that the code inside it gets repeated ten times. Note that the current loop iteration count, which is read from the iteration terminal, starts at index 0 and ends at index 9.

Place a Random Number (0-1) function (**Functions → Arithmetic & Comparison → Express Numeric → Random Number (0-1)**) inside the For Loop and wire the output terminal of this function, number (0 to 1), to the border of the For Loop to create an output tunnel. The tunnel appears as a box with the array symbol [ ] inside it. For a For Loop, auto-indexing is enabled by default whereas for a While Loop, it is disabled by default. Create an indicator on the tunnel by right-clicking and choosing **Create → Indicator** from the shortcut menu. This creates an array indicator icon outside the loop structure on the BD. Its wire appears thicker due to its array data type. Also, another indicator representing the array index gets displayed on the FP. This indicator is of array data type and can be resized as desired. In this example, the size of the array is specified as 10 to display all the values, considering that the number of iterations of the For Loop is set to be ten.

Create a second output tunnel by wiring the output of the Random Number (0-1) function to the border of the loop structure, then right-click on the tunnel and choose **Disable indexing** from the shortcut menu to disable auto-indexing. By doing this, the tunnel becomes a filled box representing a scalar value. Create an indicator on the tunnel by right-clicking and choosing **Create → Indicator** from the shortcut menu. This sets up an indicator of scalar data type outside the loop structure on the BD.

Next, create a third indicator on the Number (0 to 1) terminal of the Random Number (0-1) function located in the For Loop to observe the values coming out. To do this, right-click on the output terminal or on the wire connected to this terminal and choose **Create → Indicator** from the shortcut menu.

Place a Time Delay Express VI (**Functions → Execution Contol → Time Delay**) to delay the execution in order to have enough time to observe a current value. A configuration window is brought up to specify the delay time in seconds. Enter the value 0.1 to wait 0.1 seconds at each iteration. Note that the Time Delay Express VI is shown as an icon in Figure 2-29 in order to have a more compact display.

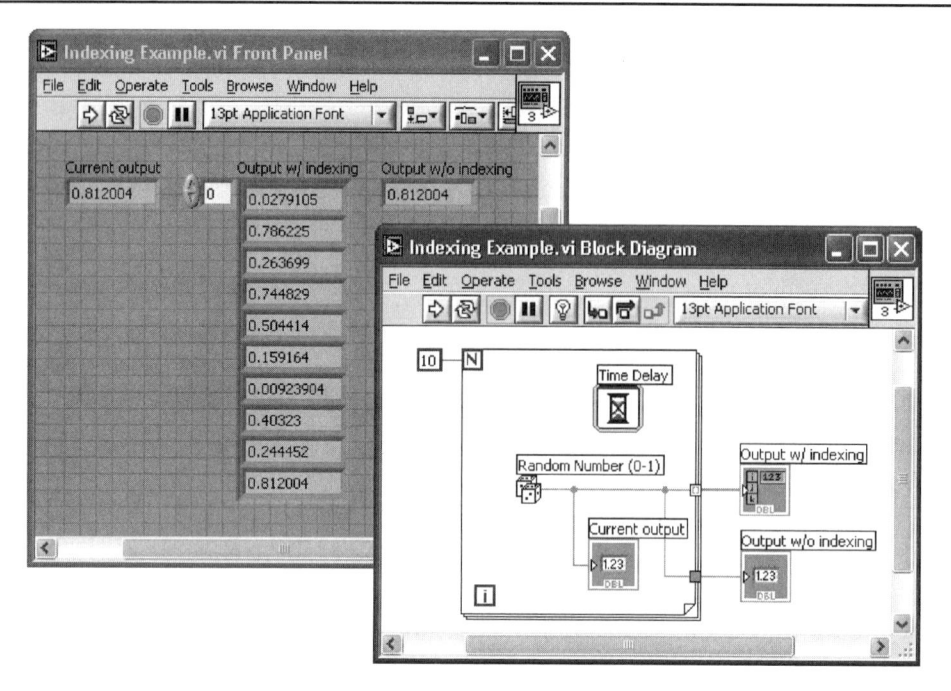

Figure 2-29: Creating array with indexing.

Save the VI as *Indexing Example.vi* and run it to observe its functionality. From the output displayed on the FP, a new random number should be displayed every 0.1 seconds on the indicator residing inside the loop structure. However, no data should be available from the indicators outside the loop structure until the loop iterations end. An array of 10 elements should be generated from the indexing-enabled tunnel while only one output, the last element of the array, should be passed from the indexing-disabled tunnel.

L1.4 Debugging VIs: Probe Tool

The Probe tool is used to observe data that are being passed while a VI is running. A probe can be placed on a wire by using the Probe tool or by right-clicking on a wire and choosing **Probe** from the shortcut menu. Probes can also be placed while a VI is running.

Placing probes on wires creates probe windows through which intermediate values can be observed. A probe window can be customized. For example, showing data of array data type via a graph makes debugging easier. To do this, right-click on the wire where an array is being passed and choose **Custom Probe → Controls → Graph Indicators → Waveform Graphs** from the shortcut menu.

Getting Familiar with LabVIEW: Part I

As an example of using custom probes, a `Waveform Chart` is used here to track the scalar values at probe location 1, a `Waveform Graph` to monitor the array at probe location 2, and a regular probe window at probe location 3 to see the values of the `Indexing Example` VI. These probes and their locations are illustrated in Figure 2-30.

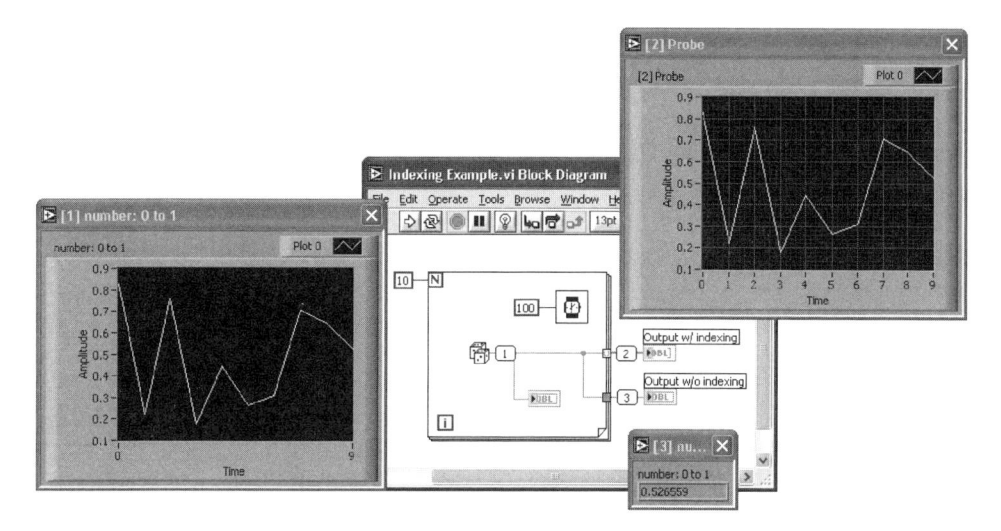

Figure 2-30: Probe tool.

L1.5 Bibliography

[1] National Instruments, *LabVIEW User Manual*, Part Number 320999E-01, 2003.

Lab 2: Getting Familiar with LabVIEW: Part II

Now that an initial familiarity with the LabVIEW programming environment has been acquired in Lab 1, this second lab covers an example where a simple DSP system is built, thus enhancing the familiarity of the reader with LabVIEW. This example involves a signal generation and amplification system. The shape of the input signal (sine, square, triangle, or sawtooth), as well as its frequency and gain, are altered by using appropriate FP controls.

The system is built with Express VIs first, then the same system is built with regular VIs. This is done in order to illustrate the advantages and disadvantages of Express VIs versus regular VIs for building a system.

L2.1 Building a System VI with Express VIs

The use of Express VIs allows less wiring on a BD. Also, it provides an interactive user-interface by which parameter values can be adjusted on the fly. The BD of the signal generation system using Express VIs is shown in Figure 2-31.

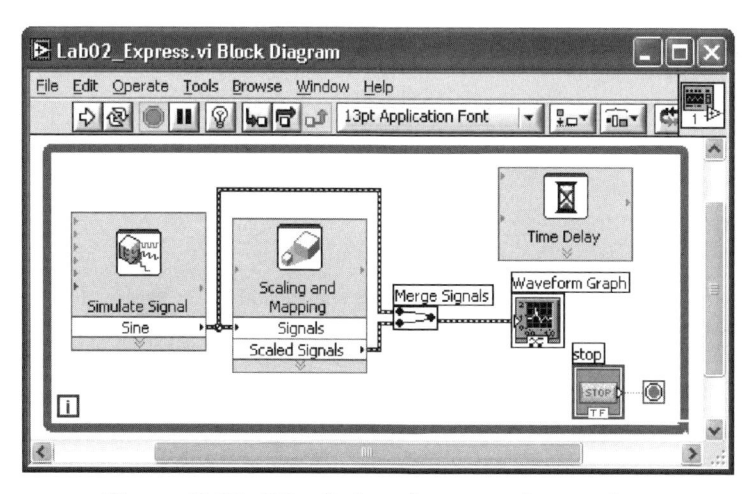

Figure 2-31: BD of signal generation and amplification system using Express VIs [1].

Lab 2

To build this BD, locate the `Simulate Signal` Express VI (**Functions → Input → Simulate Signal**) to generate a signal source. This brings up a configuration dialog as shown in Figure 2-32. Different types of signals including sine, square, triangle, sawtooth, or DC can be generated with this VI. Enter and adjust the parameters as indicated in Figure 2-32 to simulate a sinewave having a frequency of 200 Hz and an amplitude swinging between –100 and 100. Set the sampling frequency to 8000 Hz. A total of 128 samples spanning a time duration of 15.875 milliseconds (ms) are generated. Note that when the parameters are changed, the modified signal is displayed instantly in the **Result Preview** graph window.

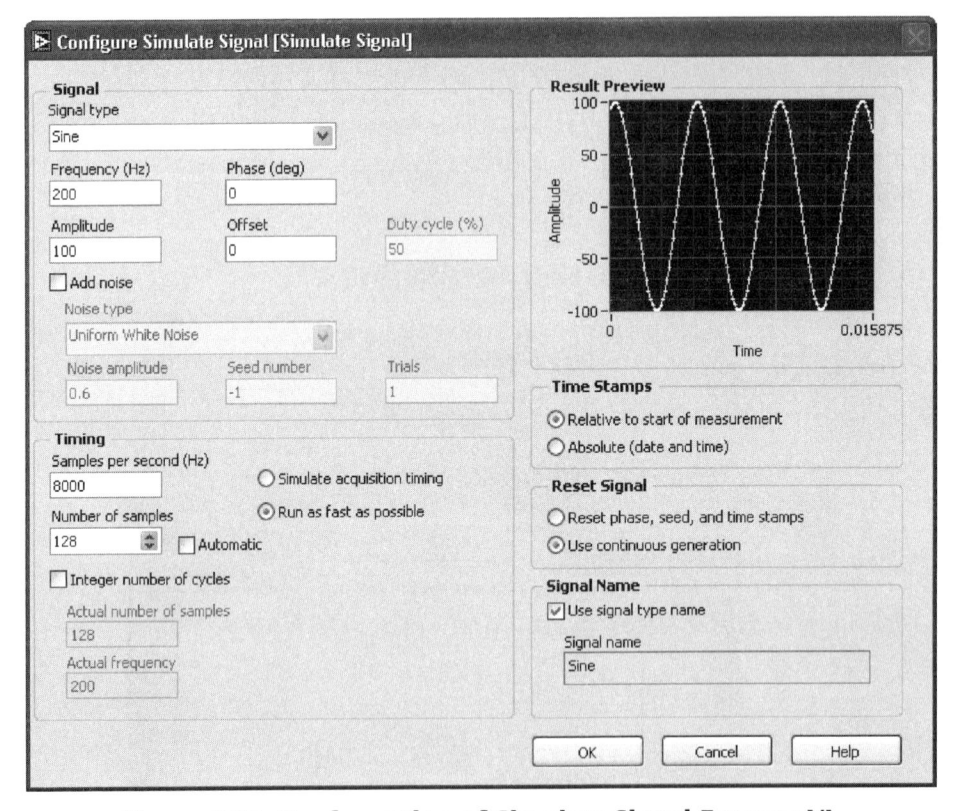

Figure 2-32: Configuration of Simulate Signal Express VI.

Next, place a `Scaling and Mapping` Express VI (**Functions → Arithmetic & Comparison → Scaling and Mapping**) to amplify or scale this simulated signal. When its configuration dialog is brought up, see Figure 2-33, choose **Linear (Y=mx+b)** and enter 5 in **Slope (m)** to scale the input signal 5 times.

Wire the `Sine` terminal of the `Simulate Signal` Express VI to the `Signals` terminal of the `Scaling and Mapping` Express VI. Note that a wire having a dynamic data type is created.

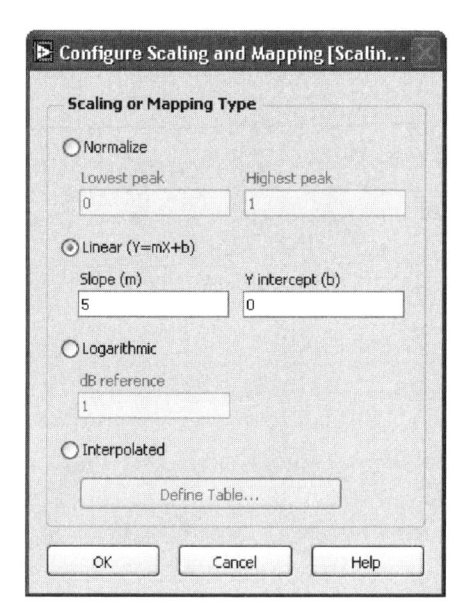

Figure 2-33: Configuration of Scaling and Mapping Express VI.

To display the output signal, place a `Waveform Graph` (**Controls → Graph Indicators → Waveform Graph**) on the FP. The `Waveform Graph` can also be created by right-clicking on the `Scaled Signals` terminal and choosing **Create → Graph Indicator** from the shortcut menu.

Now, in order to observe the original and the scaled signal together in the same graph, wire the `Sine` terminal of the `Simulate Signal` Express VI to the `Waveform Graph`. This inserts a `Merge Signals` function on the wire automatically. An automatic insertion of the `Merge Signals` function occurs when a signal having a dynamic data type is wired to other signals having the same or other data types. The `Merge Signals` function combines multiple inputs, thus allowing two signals, consisting of the original and scaled signals, to be handled by one wire. Since both the original and scaled signals are displayed in the same graph, resize the plot legend to display the two labels and markers. The use of the dynamic data type sets the signal labels automatically.

To run the VI continuously, place a `While Loop`. Position the `While Loop` to enclose all the Express VIs and the graph. Now the VI is ready to be run.

Lab 2

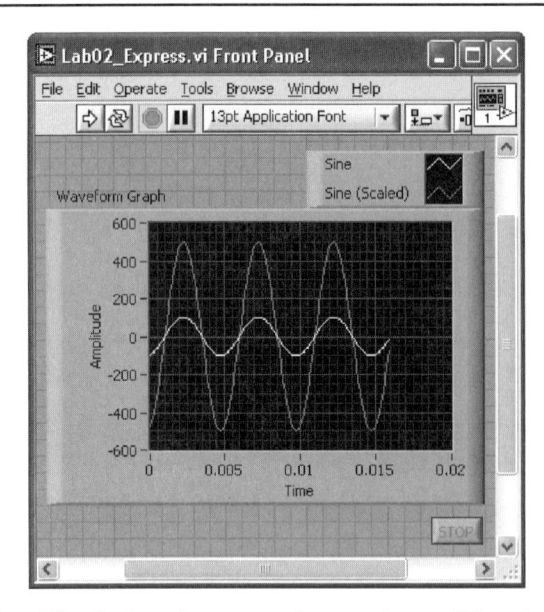

Figure 2-34: FP of signal generation and amplification system.

Run the VI and observe the Waveform Graph. The output should appear as shown in Figure 2-34. To extend the plot to the right end of the plotting area, right-click on the Waveform Graph and choose **X Scale**, then uncheck **Loose Fit** from the shortcut menu. The graph shown in Figure 2-35 should appear.

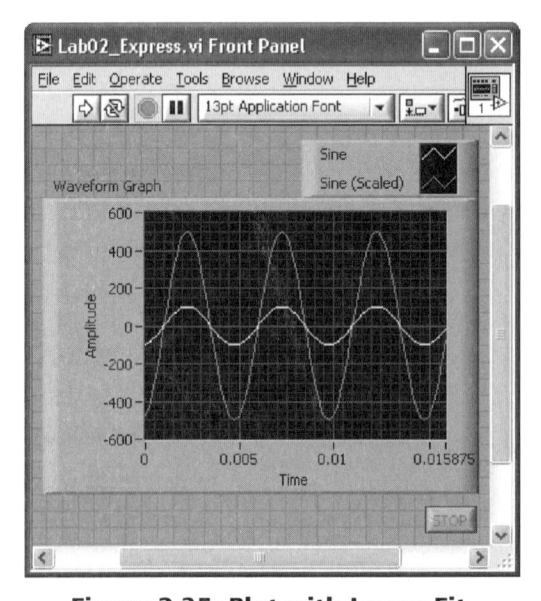

Figure 2-35: Plot with Loose Fit.

Getting Familiar with LabVIEW: Part II

If the plot runs too fast, a delay can be placed in the `While Loop`. To do this, place a `Time Delay` Express VI (**Functions → Execution Control → Time Delay**) and set the delay time to 0.2 in the configuration window. This way, the loop execution is delayed by 0.2 seconds in the BD shown in Figure 2-31.

Although this system runs successfully, no control of the signal frequency and gain is available during its execution since all the parameters are set in the configuration dialogs of the Express VIs. To gain such flexibility, some modifications need to be made.

To change the frequency at run time, place a `Vertical Pointer Slide` control (**Controls → Numeric Controls → Vertical Pointer Slide**) on the FP and wire it to the `Frequency` terminal of the `Simulate Signal` Express VI. The control is labeled as `Frequency`. The Express VI can be resized to show more terminals at the bottom of the expandable node. Resize the VI to show an additional terminal below the `Sine` terminal. Then, click on this new terminal, `error out` by default, to select `Frequency` from the list of the displayed terminals.

Next, replace the `Scaling and Mapping` Express VI with a `Multiply` function (**Functions → Arithmetic & Comparison → Express Numeric → Multiply**). Place another `Vertical Pointer Slide` control and wire it to the y terminal of the `Multiply` function to adjust the gain. This control is labeled as `Gain`. These modifications are illustrated in Figure 2-36.

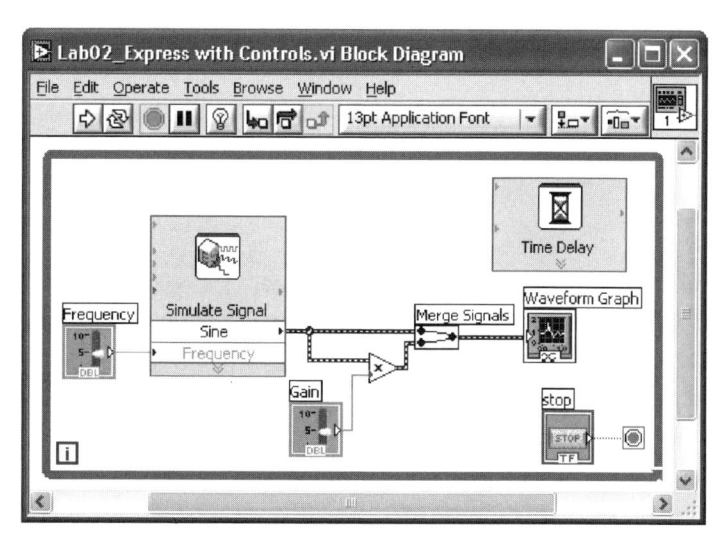

Figure 2-36: BD of signal generation and amplification system with controls.

Lab 2

Now on the FP, set the maximum range of each slide control to 1000 for the Frequency control and 5 for the Gain control, respectively. Also, set the default values for these controls to 200 and 2, respectively.

By running this modified VI, it can be observed that the two signals are displayed with the same label since the source of these signals, that is, the Sine terminal of the Simulate Signal Express VI, is the same. Also, due to the autoscale feature of the Waveform Graph, the scaled signal appears unchanged while the Y axis of the Waveform Graph changes appropriately. This is illustrated in Figure 2-37.

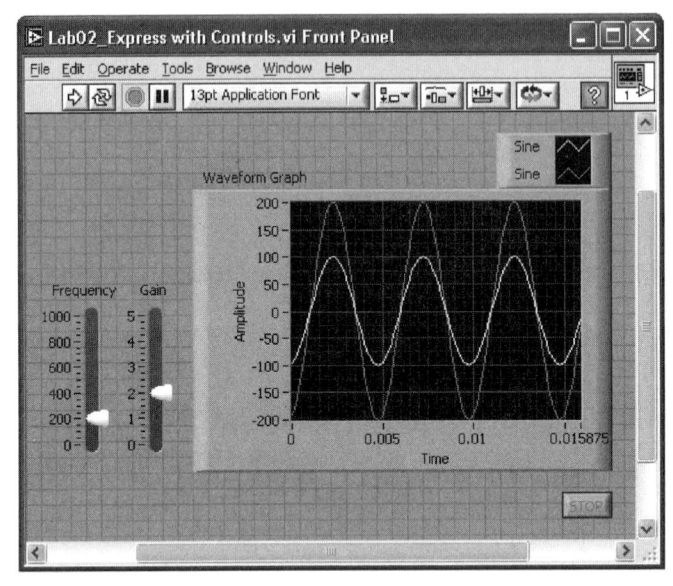

Figure 2-37: Autoscaled graph of two signals shown together.

Let us now modify the properties of the Waveform Graph. In order to disable the autoscale feature, right-click on the Waveform Graph and uncheck **Y Axis → AutoScale Y**. The maximum and minimum scale can also be adjusted. In this example –600 and 600 are used as the minimum and maximum values, respectively. This is done by modifying the maximum and minimum scale values of the Y axis with the Labeling tool. If the automatic tool selection mode is enabled, just click on the maximum or minimum scale of the Y axis to enter any desired scale value. To modify the labels displayed in the plot legend, right-click and choose **Ignore Attributes**. Then, edit the labels to read Original and Scaled using the Labeling tool. Changing the properties of the Waveform Graph can also be accomplished by using its properties dialog box. This box is brought up by right-clicking on the Waveform Graph and choosing **Properties** from the shortcut menu.

Getting Familiar with LabVIEW: Part II

The completed FP is shown in Figure 2-38. With this version of the VI, the frequency of the input signal and the gain of the output signal can be controlled using the controls on the FP.

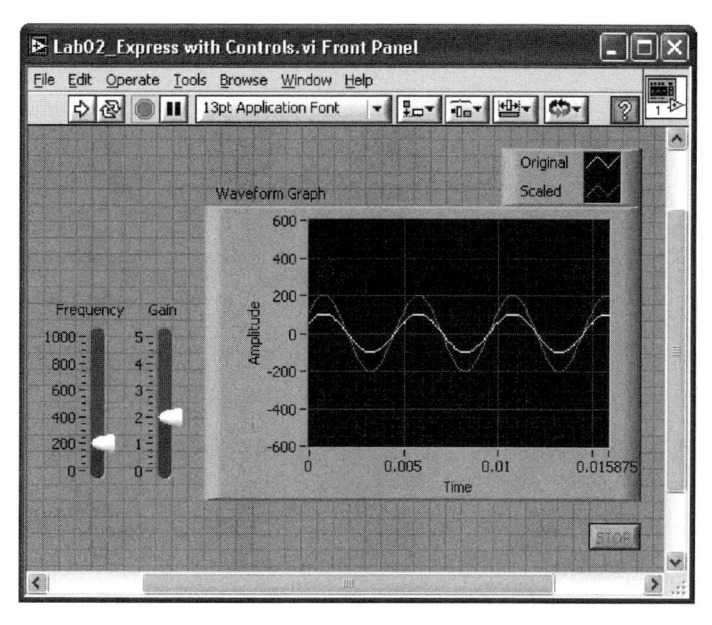

Figure 2-38: FP of signal generation and amplification system with controls.

L2.2 Building a System with Regular VIs

In this section, the implementation of the same system discussed above is achieved by using regular VIs.

After creating a blank VI, place a `While Loop` (**Functions → Execution Control → While Loop**) on the BD, which may need to be resized later. To provide the signal source of the system, place a `Basic Function Generator` VI (**Functions → All Functions → Analyze → Waveform Generation → Basic Function Generator**) inside the `While Loop`. To configure the parameters of the signal, appropriate controls and constants need to be wired. To create a control for the signal type, right-click on the `signal type` terminal of the `Basic Function Generator` VI and choose **Create → Control** from the shortcut menu. Note that an enumerated (`Enum`) type control for the signal gets located on the FP. Four items including sine, triangle, square and sawtooth are listed in this control.

Lab 2

Next, right-click on the `amplitude` terminal, and choose **Create → Constant** from the shortcut menu to create an amplitude constant. Enter 100 in the numeric constant box to set the amplitude of the signal. In order to configure the sampling frequency and the number of samples, create a constant on the `sampling information` terminal by right-clicking and choosing **Create → Constant** from the shortcut menu. This creates a cluster constant which includes two numeric constants. The first element of the cluster shown in the upper box represents the sampling frequency and the second element shown in the lower box represents the number of samples. Enter 8000 for the sampling frequency and 128 for the number of samples. Note that the same parameters were used in the previous section.

Now, toggle to the FP by pressing <Ctrl + E> and place two `Vertical Pointer Slide` controls on the FP by choosing **Controls → Numeric Controls → Vertical Pointer Slide**. Rename the controls `Frequency` and `Gain`, respectively. Set the maximum scale values to 1000 for the `Frequency` control and 5 for the `Gain` control. The `Vertical Pointer Slide` controls create corresponding icons on the BD. Make sure that the icons are located inside the `While Loop`. If not, select the icons and drag them inside the `While Loop`. The `Frequency` control should be wired to the `frequency` terminal of the `Basic Function Generator` VI in order to be able to adjust the frequency at run time. The `Gain` control is used at a later stage.

The output of the `Basic Function Generator` VI appears in the waveform data type. The waveform data type is a special cluster which bundles three components (`t0`, `dt`, and `Y`) together. The component `t0` represents the trigger time of the waveform, `dt` the time interval between two samples, and `Y` data values of the waveform.

Next, the generated signal needs to be scaled based on a gain factor. This is done by using a `Multiply` function (**Functions → Arithmetic & Comparison → Express Numeric → Multiply**) and a second `Vertical Pointer Slide` control, named `Gain`. Wire the generated waveform out of the `signal out` terminal of the `Basic Function Generator` VI to the x terminal of the `Multiply` function. Also, wire the `Gain` control to the y terminal of the `Multiply` function.

Recall that the `Merge Signals` function is used to combine two signals having dynamic data types into the same wire. To achieve the same outcome with regular VIs, place a `Build Array` function (**Functions → All Functions → Array → Build Array**) to build a 2D array, that is, two rows (or columns) of one dimensional signal. Resize the `Build Array` function to have two input terminals. Wire the original signal to the upper terminal of the `Build Array` function, and the output of the `Multiply` function to the lower terminal. Remember that the `Build Array`

Getting Familiar with LabVIEW: Part II

function is used to concatenate arrays or build n-dimensional arrays. Since the `Build Array` function is used for comparing the two signals, make sure that the **Concatenate Inputs** option is unchecked from the shortcut menu. More details on the use of the `Build Array` function can be found in [2].

A `Waveform Graph` (**Controls → Graph Indicators → Waveform Graph**) is then placed on the FP. Wire the output of the `Build Array` function to the input of the `Waveform Graph`. Resize the plot legend to display the labels and edit them. Similar to the example in the previous section, the **AutoScale** feature of the Y axis should be disabled and the **Loose Fit** option should be unchecked along the X axis.

Place a `Wait (ms)` function (**Functions → All Functions → Time & Dialog → Wait**) inside the `While Loop` to delay the execution in case the VI runs too fast. Right-click on the `milliseconds to wait` terminal and choose **Create → Constant** from the shortcut menu to create and wire a `Numeric Constant`. Enter 200 in the box created.

Figure 2-39 and Figure 2-40 illustrate the BD and FP of the designed signal generation system, respectively. Save the VI as *Lab02_Regular_Waveform.vi* and run it. Change the signal type, gain and frequency values to see the original and scaled signal in the `Waveform Graph`.

The waveform data type is not accepted by all of the functions or subVIs. To cope with this issue, the Y component (data value) of the waveform data type is extracted

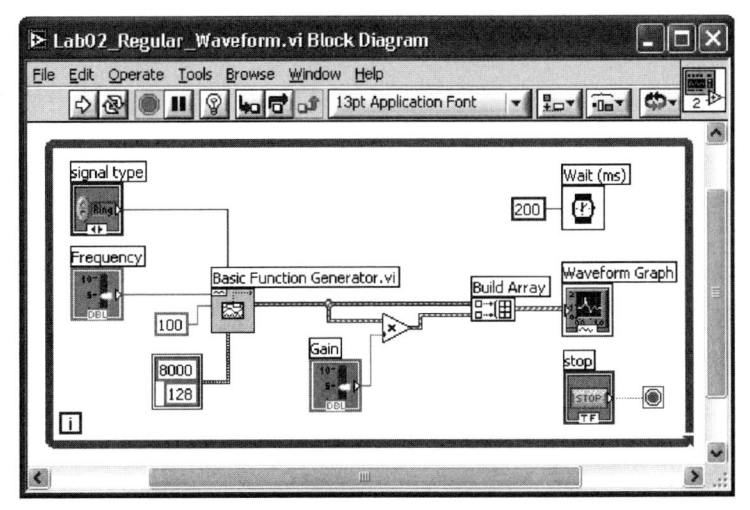

Figure 2-39: BD of signal generation and amplification system using regular VIs.

Lab 2

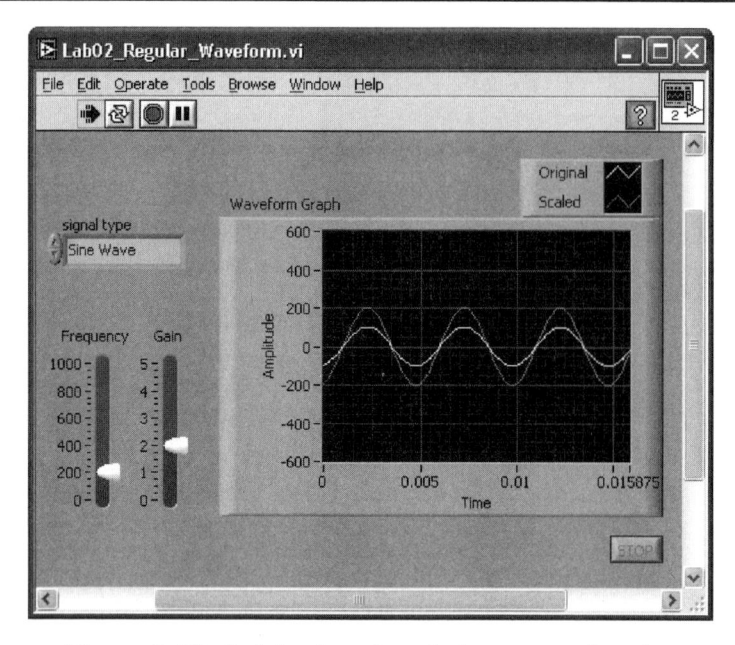

Figure 2-40: Original and scaled output signals.

to obtain the output signal as an array of data samples. This is done by placing a Get Waveform Components function (**Functions → All Functions → Waveform → Get Waveform Components**). Then, wire the signal out terminal of the Basic Function Generator VI to the waveform terminal of the Get Waveform Components function. Click on t0, the default terminal, of the Get Waveform Components function and choose Y as the output to extract data values from the waveform data type, see Figure 2-41. The remaining steps are the same as those for the version shown in Figure 2-39. In this version, however, the processed signal is an array of double precision samples.

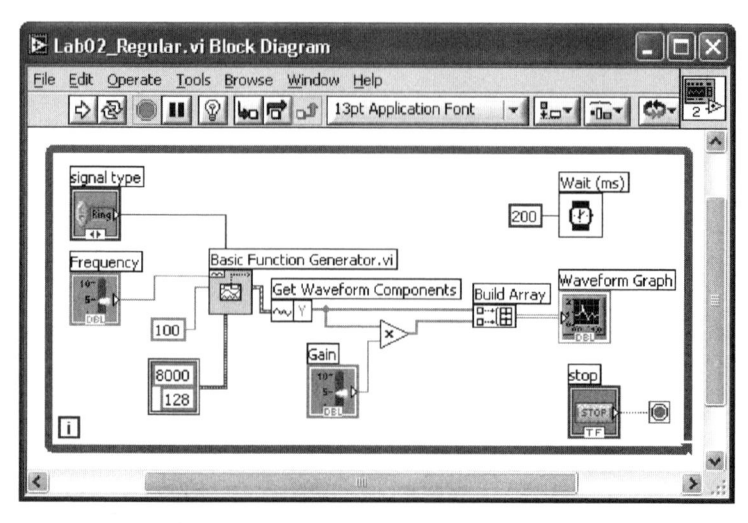

Figure 2-41: Matching data types.

40

L2.3 Profile VI

The Profile tool is used to gather timing and memory usage information. Make sure the VI is stopped before setting up a Profile window. Select **Tool** → **Advanced** → **Profile VIs…** to bring up a Profile window.

Place a checkmark in the **Timing Statistics** checkbox to display timing statistics of the VI. The **Timing Details** option provides more detailed statistics of the VI such as drawing time. To profile memory usage as well as timing, check the **Memory Usage** checkbox after checking the **Profile Memory Usage** checkbox. Note that this option can slow down the execution of the VI. Start profiling by clicking the Start button on the profiler, then run the VI. A snapshot of the profiler information can be obtained by clicking on the Snapshot button. After viewing the timing information, click the Stop button. The profile statistics can be stored into a text file by clicking the Save button.

An outcome of the profiler is exhibited in Figure 2-42 after running the Lab02_Regular VI. More details on the use of the Profile tool can be found in [3].

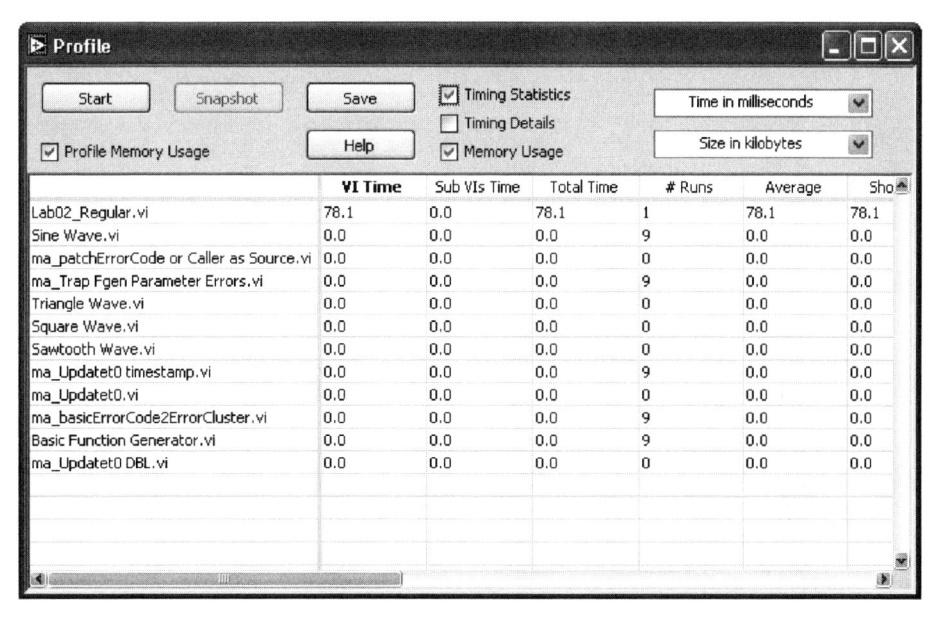

Figure 2-42: Profile window after running Lab02_Regular VI.

L2.4 Bibliography

[1] National Instruments, *Getting Started with LabVIEW*, Part Number 323427A-01, 2003.

[2] National Instruments, *LabVIEW User Manual*, Part Number 320999E-01, 2003.

[3] National Instruments, *LabVIEW Performance and Memory Management*, Application Note 168, Part Number 342078B-01, 2004.

CHAPTER 3

Analog-to-Digital Signal Conversion

The process of analog-to-digital signal conversion consists of converting a continuous time and amplitude signal into discrete time and amplitude values. Sampling and quantization constitute the steps needed to achieve analog-to-digital signal conversion. To minimize any loss of information that may occur as a result of this conversion, it is important to understand the underlying principles behind sampling and quantization.

3.1 Sampling

Sampling is the process of generating discrete time samples from an analog signal. First, it is helpful to mention the relationship between analog and digital frequencies. Let us consider an analog sinusoidal signal $x(t) = A\cos(\omega t + \phi)$. Sampling this signal at $t = nT_s$, with the sampling time interval of T_s, generates the discrete time signal

$$x[n] = A\cos(\omega n T_s + \phi) = A\cos(\theta n + \phi), \quad n = 0,1,2,\ldots, \tag{3.1}$$

where $\theta = \omega T_s = \dfrac{2\pi f}{f_s}$ denotes digital frequency with units being radians (as compared to analog frequency ω with units being radians/sec).

The difference between analog and digital frequencies is more evident by observing that the same discrete time signal is obtained from different continuous time signals if the product ωT_s remains the same. (An example is shown in Figure 3-1.) Likewise, different discrete time signals are obtained from the same analog or continuous time signal when the sampling frequency is changed. (An example is shown in Figure 3-2.) In other words, both the frequency of an analog signal f and the sampling frequency f_s define the frequency of the corresponding digital signal θ.

Chapter 3

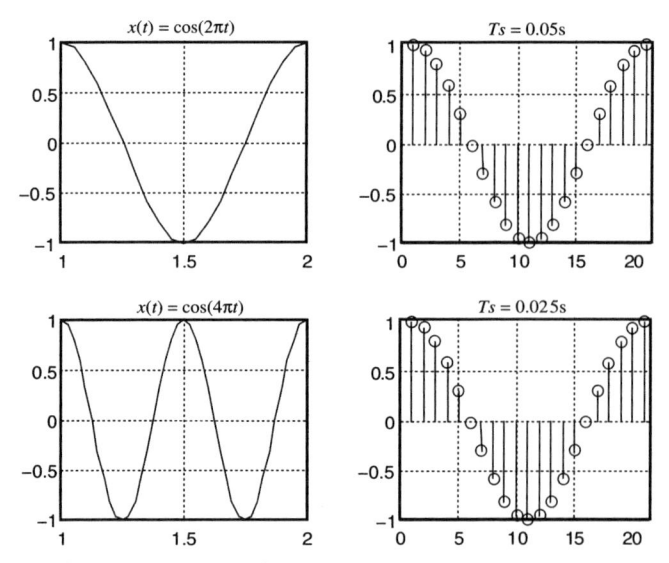

Figure 3-1: Sampling of two different analog signals leading to the same digital signal.

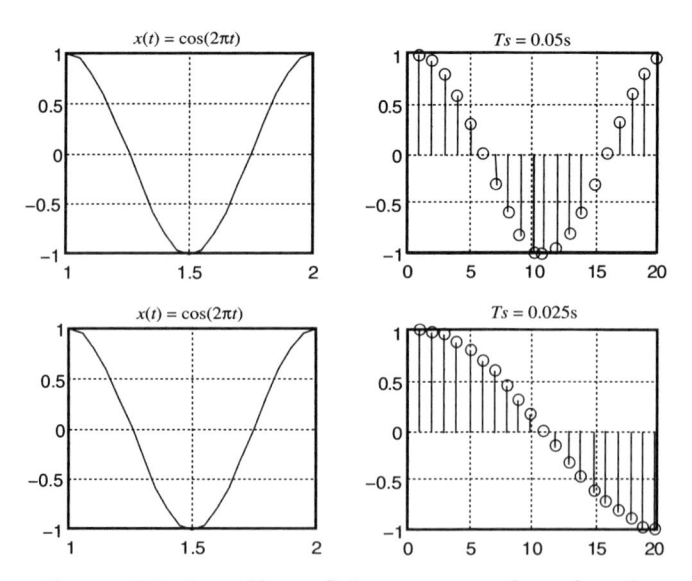

Figure 3-2: Sampling of the same analog signal leading to two different digital signals.

Analog-to-Digital Signal Conversion

It helps to understand the constraints associated with the above sampling process by examining signals in the frequency domain. The Fourier transform pairs in the analog and digital domains are given by

Fourier transform pair for analog signals

$$\begin{cases} X(j\omega) = \int_{-\infty}^{\infty} x(t) e^{-j\omega t} dt \\ x(t) = \dfrac{1}{2\pi} \int_{-\infty}^{\infty} X(j\omega) e^{j\omega t} d\omega \end{cases} \quad (3.2)$$

Fourier transform pair for discrete signals

$$\begin{cases} X(e^{j\theta}) = \sum_{n=-\infty}^{\infty} x[n] e^{-jn\theta}, \; \theta = \omega T_s \\ x[n] = \dfrac{1}{2\pi} \int_{-\pi}^{\pi} X(e^{j\theta}) e^{jn\theta} d\theta \end{cases} \quad (3.3)$$

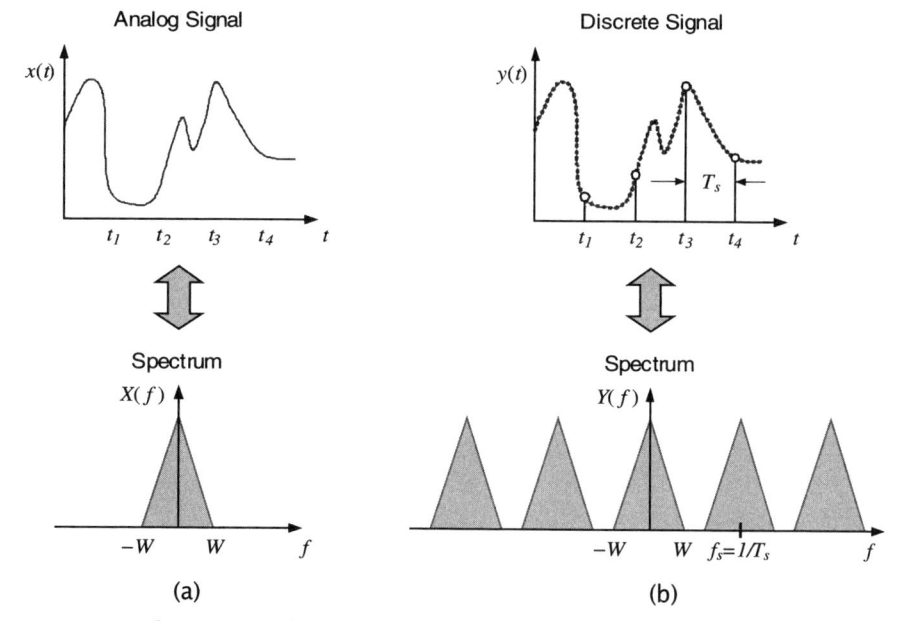

Figure 3-3: (a) Fourier transform of a continuous-time signal, and (b) its discrete time version.

As illustrated in Figure 3-3, when an analog signal with a maximum bandwidth of W (or a maximum frequency of f_{max}) is sampled at a rate of $T_s = \dfrac{1}{f_s}$, its corresponding frequency response is repeated every 2π radians, or f_s. In other words, the Fourier transform in the digital domain becomes a periodic version of the Fourier transform

Chapter 3

in the analog domain. That is why, for discrete signals, we are only interested in the frequency range $[0, f_s/2]$.

Therefore, in order to avoid any aliasing or distortion of the frequency content of the discrete signal, and hence to be able to recover or reconstruct the frequency content of the original analog signal, we must have $f_s \geq 2f_{max}$. This is known as the Nyquist rate; that is, the sampling frequency should be at least twice the highest frequency in the signal. Normally, before any digital manipulation, a front-end antialiasing low-pass analog filter is used to limit the highest frequency of the analog signal.

The aliasing problem can be further illustrated by considering an undersampled sinusoid as depicted in Figure 3-4. In this figure, a 1 kHz sinusoid is sampled at $f_s = 0.8$ kHz, which is less than the Nyquist rate of 2 kHz. The dashed-line signal is a 200 Hz sinusoid passing through the same sample points. Thus, at the sampling frequency of 0.8 kHz, the output of an A/D converter would be the same if either of the 1 kHz or 200 Hz sinusoids was the input signal. On the other hand, oversampling a signal provides a richer description than that of the signal sampled at the Nyquist rate.

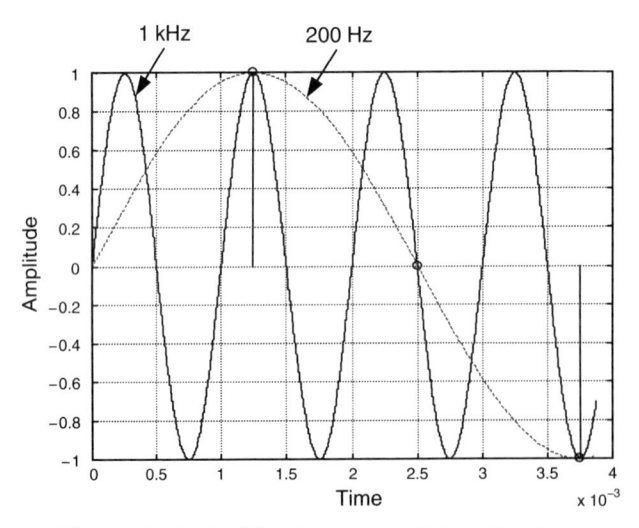

Figure 3-4: Ambiguity caused by aliasing.

3.1.1 Fast Fourier Transform

The Fourier transform of discrete signals is continuous over the frequency range $[0, f_s/2]$. Thus, from a computational standpoint, this transform is not suitable to use. In practice, the discrete Fourier transform (DFT) is used in place of the Fourier transform. DFT is the equivalent of the Fourier series in the analog domain. Detailed

Analog-to-Digital Signal Conversion

descriptions of signal transforms can be found in various textbooks on digital signal processing, for example [1], [2]. Fourier series and DFT transform pairs are expressed as

Fourier series for periodic analog signals
$$\begin{cases} X_k = \dfrac{1}{T}\int_{-T/2}^{T/2} x(t)e^{-j\omega_0 kt}dt \\ x(t) = \sum_{k=-\infty}^{\infty} X_k e^{j\omega_0 kt} \end{cases} \quad (3.4)$$

where T denotes period and ω_0 fundamental frequency.

Discrete Fourier transform (DFT) for periodic discrete signals
$$\begin{cases} X[k] = \sum_{n=0}^{N-1} x[n]e^{-j\frac{2\pi}{N}nk}, \; k = 0,1,...,N-1 \\ x[n] = \dfrac{1}{N}\sum_{k=0}^{N-1} X[k]e^{j\frac{2\pi}{N}nk}, \; n = 0,1,...,N-1 \end{cases} \quad (3.5)$$

It should be noted that DFT and Fourier series pairs are defined for periodic signals. Hence, when computing DFT, it is required to assume periodicity with a period of N samples. Figure 3-5 illustrates a sampled sinusoid which is no longer periodic. In order to make sure that the sampled version remains periodic, the analog frequency should satisfy this condition [3]

$$f = \frac{m}{N}f_s \quad (3.6)$$

where m denotes the number of cycles over which DFT is computed.

Figure 3-5: Periodicity condition of sampling.

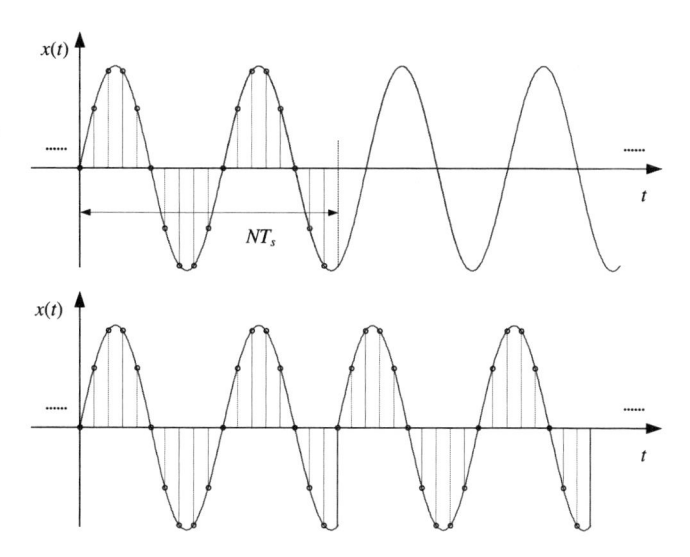

Chapter 3

The computational complexity (number of additions and multiplications) of DFT is reduced from N^2 to $N \log N$ by using fast Fourier transform (FFT) algorithms. In these algorithms, N is normally considered to be a power of two. Figure 3-6 shows the effect of the periodicity constraint on the FFT computation. In this figure, the FFTs of two sinusoids with frequencies of 250 Hz and 251 Hz are shown. The amplitudes of the sinusoids are unity. Although there is only a 1 Hz difference between the sinusoids, the FFT outcomes are significantly different due to the improper sampling.

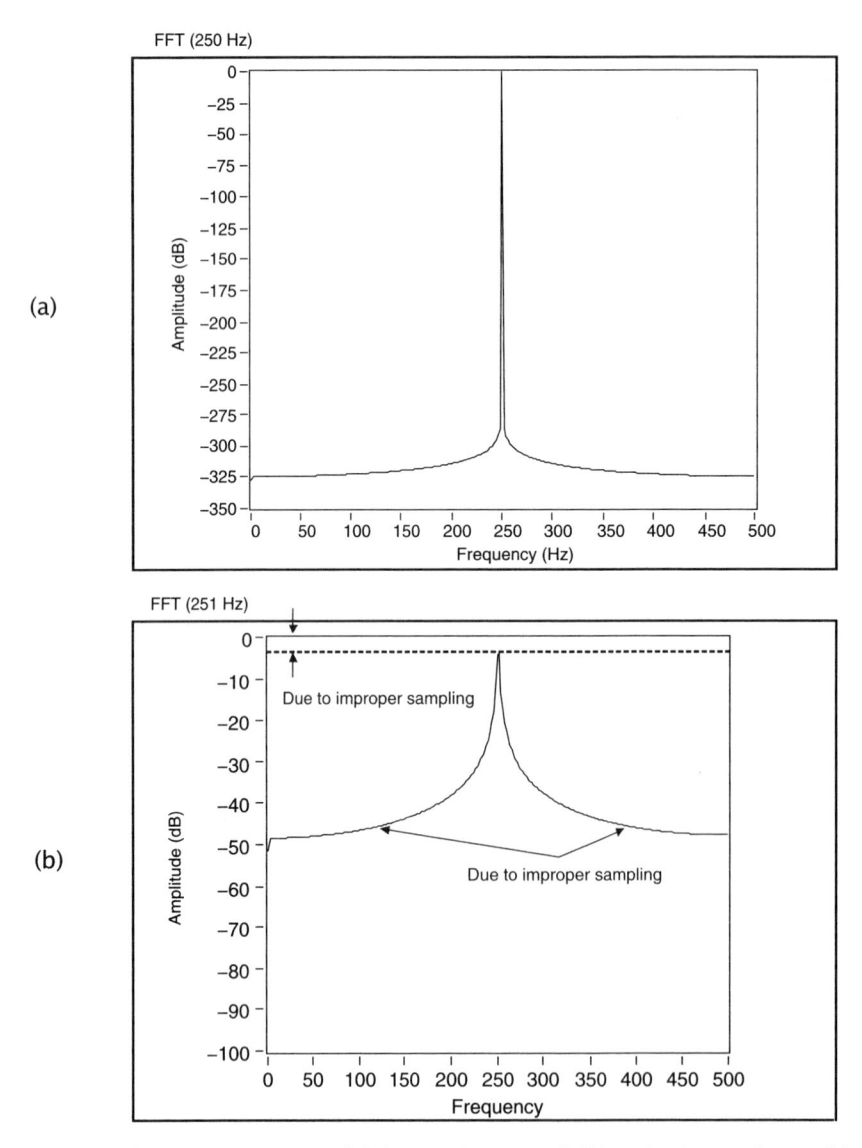

Figure 3-6: FFTs of (a) a 250 Hz and (b) a 251 Hz sinusoid.

Analog-to-Digital Signal Conversion

3.2 Quantization

An A/D converter has a finite number of bits (or resolution). As a result, continuous amplitude values are represented or approximated by discrete amplitude levels. The process of converting continuous into discrete amplitude levels is called quantization. This approximation leads to errors called quantization noise. The input/output characteristic of a 3-bit A/D converter is shown in Figure 3-7 to see how analog voltage values are approximated by discrete voltage levels.

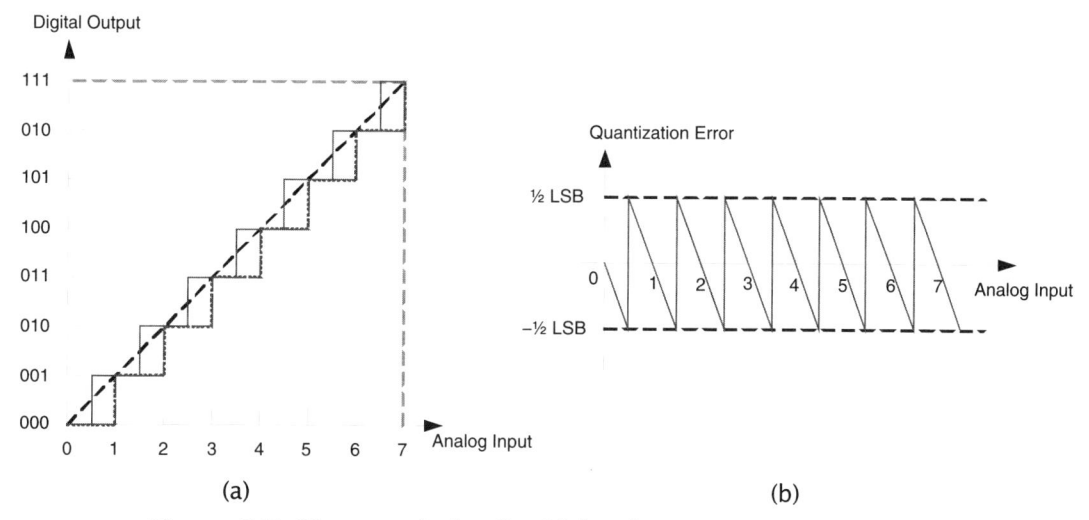

Figure 3-7: Characteristic of a 3-bit A/D converter: (a) input/output transfer function, and (b) additive quantization noise.

A quantization interval depends on the number of quantization or resolution levels, as illustrated in Figure 3-8. Clearly the amount of quantization noise generated by an A/D converter depends on the size of the quantization interval. More quantization bits translate into a narrower quantization interval and hence into a lower amount of quantization noise.

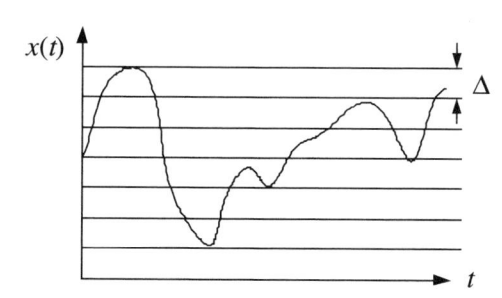

Figure 3-8: Quantization levels.

Chapter 3

In Figure 3-8, the spacing Δ between two consecutive quantization levels corresponds to one least significant bit (LSB). Usually, it is assumed that quantization noise is signal independent and is uniformly distributed over −0.5 LSB and 0.5 LSB. Figure 3-9 shows the quantization noise of an analog signal quantized by a 3-bit A/D converter.

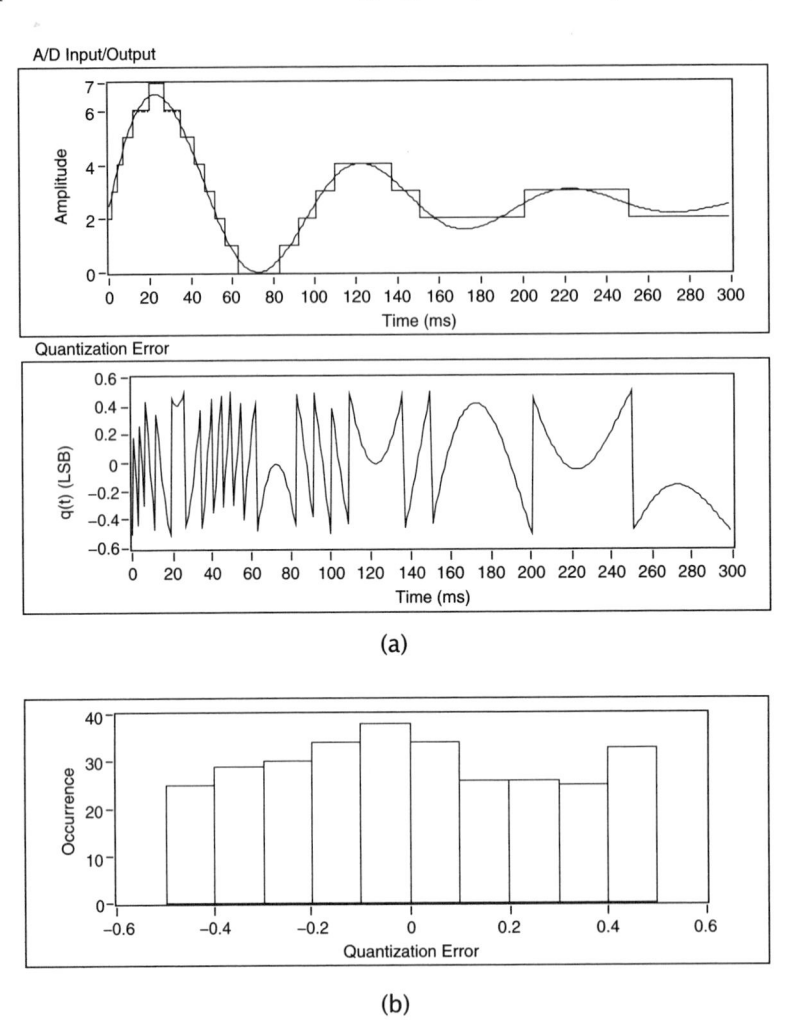

(a)

(b)

Analog-to-Digital Signal Conversion

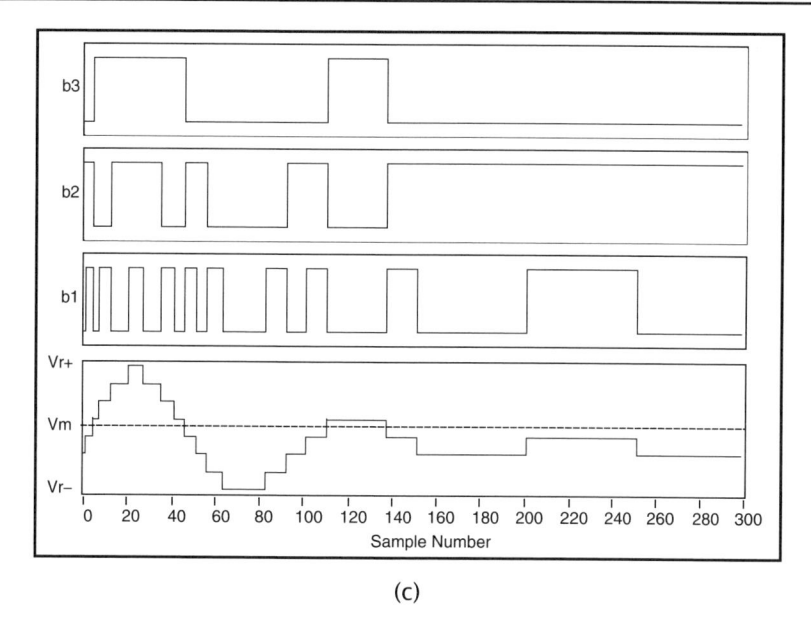

(c)

Figure 3-9: Quantization of an analog signal by a 3-bit A/D converter: (a) output signal and quantization error, (b) histogram of quantization error, and (c) bit stream.

3.3 Signal Reconstruction

So far, we have examined the forward process of sampling. It is also important to understand the inverse process of signal reconstruction from samples. According to the Nyquist theorem, an analog signal v_a can be reconstructed from its samples by using the following equation:

$$v_a(t) = \sum_{k=-\infty}^{\infty} v_a[kT_s] \left[\text{sinc}\left(\frac{t - kT_s}{T_s}\right) \right] \qquad (3.7)$$

One can see that the reconstruction is based on the summations of shifted sinc functions. Figure 3-10 illustrates the reconstruction of a sinewave from its samples.

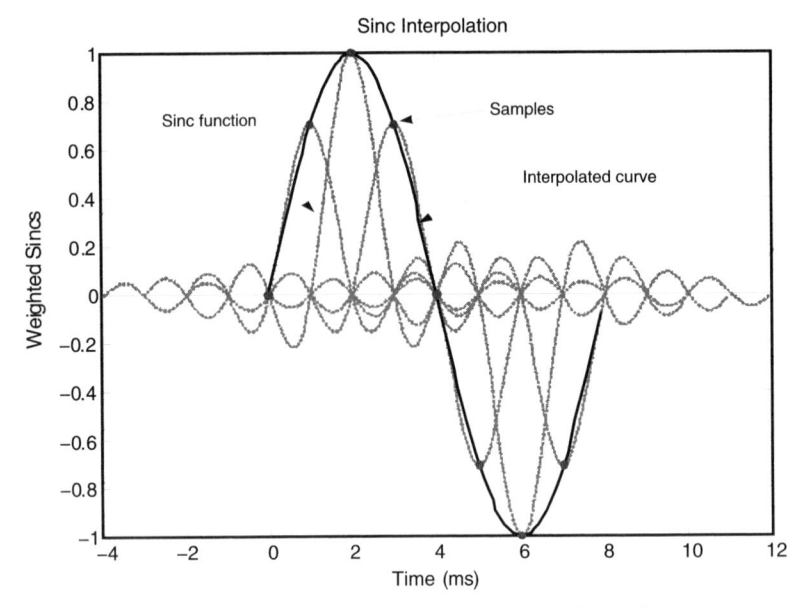

Figure 3-10: Reconstruction of an analog sinewave based on its samples, $f = 125$ Hz, and $f_s = 1$ kHz.

It is very difficult to generate sinc functions by electronic circuitry. That is why, in practice, an approximation of a sinc function is used. Figure 3-11 shows an approximation of a sinc function by a pulse, which is easy to realize in electronic circuitry. In fact, the well-known sample and hold circuit performs this approximation [3].

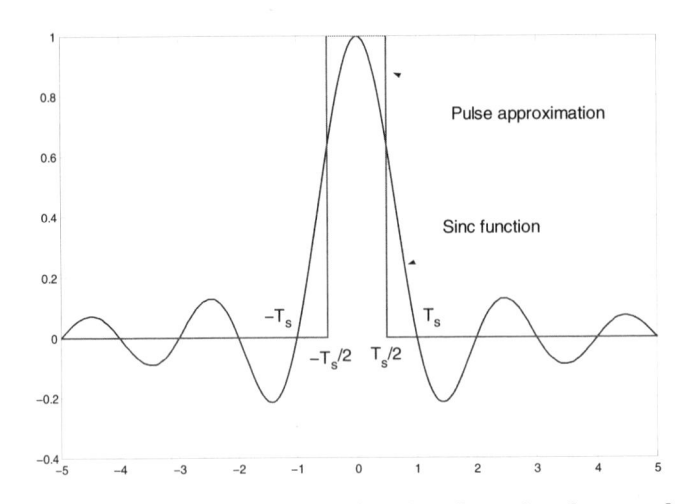

Figure 3-11: Approximation of a sinc function by a pulse.

Bibliography

[1] J. Proakis and D. Manolakis, *Digital Signal Processing: Principles, Algorithms, and Applications*, Prentice-Hall, 1996.

[2] S. Mitra, *Digital Signal Processing: A Computer-Based Approach*, McGraw-Hill, 2001.

[3] B. Razavi, *Principles of Data Conversion System Design*, IEEE Press, 1995.

Lab 3: Sampling, Quantization and Reconstruction

This lab covers several examples to further convey sampling, quantization, and reconstruction aspects of analog-to-digital and digital-to-analog signal conversion processes.

L3.1 Aliasing

In this example, a discrete signal is generated by sampling a sinusoidal signal. When the normalized frequency f/f_s of the discrete signal becomes greater than 0.5, or the Nyquist frequency, the aliasing effect becomes evident.

A sampling process is done by setting the sampling frequency f_s to 1 kHz, and the number of samples N to 10. This results in a 10 ms sampled signal. The signal frequency is arranged to vary between 0 to 1000 Hz using a FP control. Figure 3-12 shows a sinusoidal signal having a frequency of 300 Hz which is sampled at 1 kHz for 10 ms producing 10 samples, which are displayed in a Waveform Graph. In this

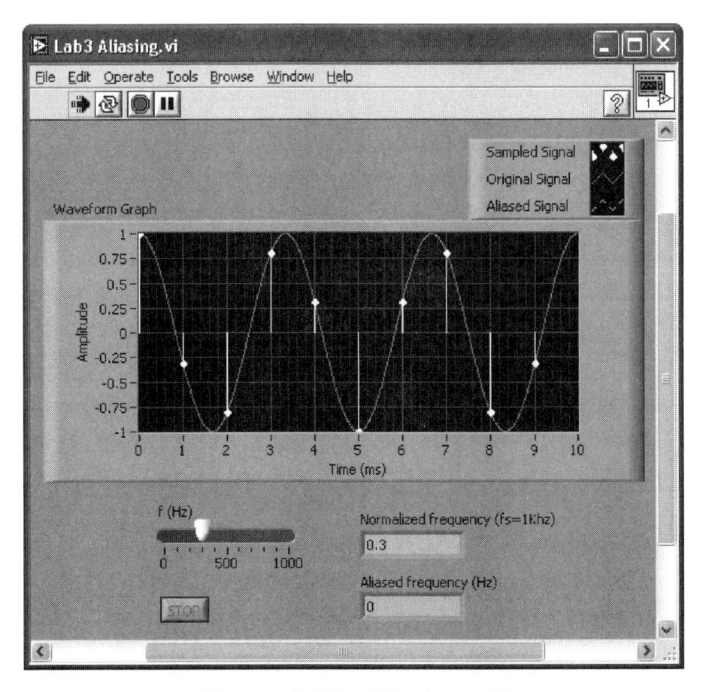

Figure 3-12: Aliasing effect.

Lab 3

graph, an analog signal representation is also made by oversampling the sinusoidal signal 100 times faster. In other words, an analog signal representation is obtained by considering a sampling frequency of 100 kHz generating 1000 samples.

The FP of the VI includes a `Horizontal Slide` control for the signal frequency, and two `Numeric Indicators` for the normalized frequency and aliased frequency. A `Stop Button` associated with a `While Loop` on the BD is located on the FP. This button is used to stop the execution of the VI.

Figure 3-13 shows the BD for this sampling system. To generate the analog and discrete sinusoids, three `Sine Wave` VIs (**Functions → All Functions → Analyze → Signal Processing → Signal Generation → Sine Wave**) are used. These VIs are arranged vertically in the middle of the BD. The inputs to these VIs consist of the number of samples, amplitude, frequency, and phase offset. Amplitude is set to 1 by default in the absence of any wiring to the `amplitude` terminal. The `f` terminal requires frequency to be specified in cycles per sample, which is the reciprocal of number of samples per period. For phase, the numeric constant 90 is wired to the `phase in` terminal.

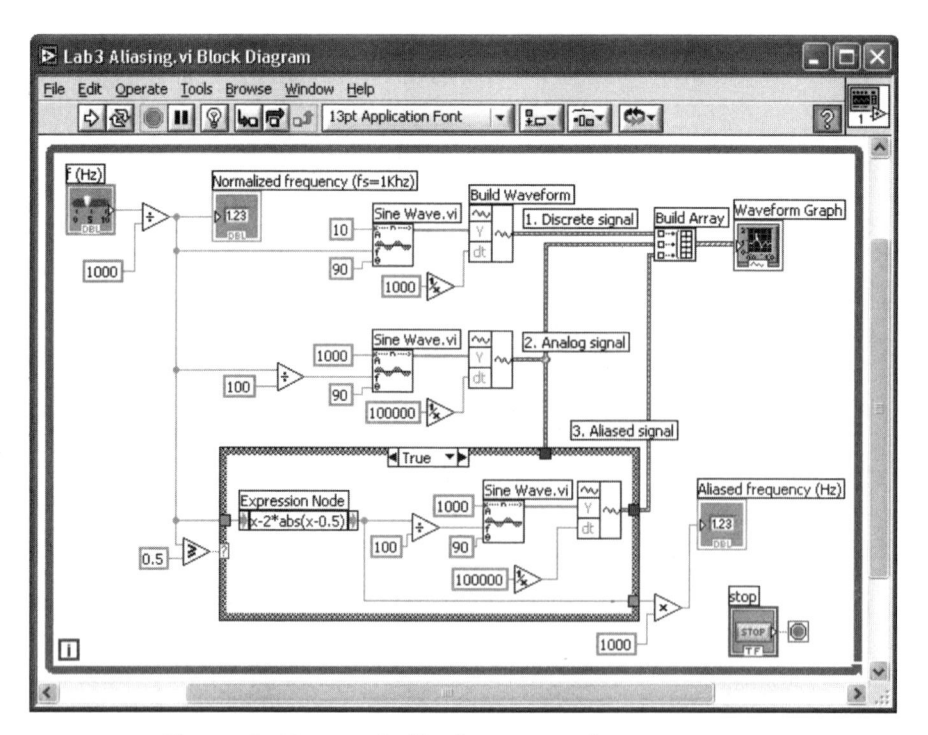

Figure 3-13: BD of aliasing example—True case.

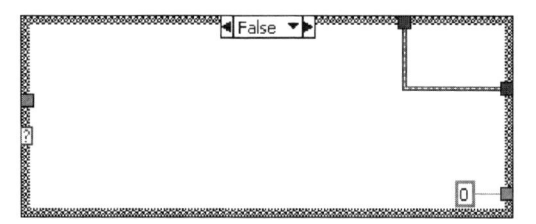

Figure 3-14: False case.

The output of the VI consists of an array of its sinewave samples. The Build Waveform function (**Functions → All Functions → Waveform → Build Waveform**) is used to build a waveform by combining the samples into the Y terminal, and the time duration between samples, $T_s = 1/1000$, into the dt terminal. As discussed earlier, the number of samples for the analog representation of the signal is set to 100 times that of the discrete signal. Thus, to keep the ratio of frequency to samples the same as that of the discrete signal, the value wired to the f terminal is divided by 100. Also, the time interval of the analog signal is set to one hundredth of that of the discrete signal.

Among the three Sine Wave VIs shown in Figure 3-13, the top VI generates the discrete signal, the middle VI generates the analog signal, and the bottom VI generates the aliased signal when the signal frequency is higher than the Nyquist frequency.

A Case Structure is used to handle the sampling cases with aliasing and without aliasing. If the normalized frequency is greater than 0.5, corresponding to the True case, the third Sine Wave VI generates an aliased signal. All the inputs except for the aliased signal frequency are the same.

Note that an Expression Node (**Functions → All Functions → Numeric → Expression Node**) is used to obtained the aliased frequency. An Expression Node is usually used to calculate an expression of a single variable. Many built-in functions, such as abs (absolute), can be used in an Expression Node to evaluate an equation. More details on the use of Expression Node can be found in [1].

For the False case, that is, sampling without aliasing, there is no need to generate an aliased signal. Thus, the analog signal is connected to the output of the case structure so that the same signal is drawn on the waveform graph and the frequency of the aliased signal is set to 0. This is illustrated in Figure 3-14. It should be remembered that, when using a Case Structure, it is necessary to wire all the outputs for each case.

Lab 3

An aliasing outcome is illustrated in Figure 3-15, where samples of a 700 Hz sinusoid are shown. Note that these samples could have also been obtained from a 300 Hz sinusoid, shown by the dotted line in Figure 3-15.

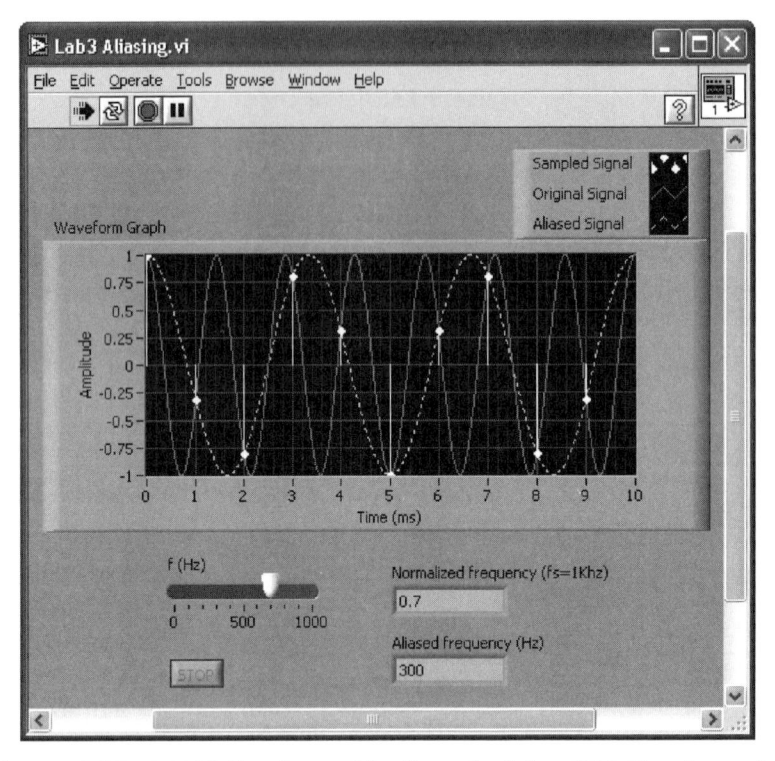

Figure 3-15: A 700 Hz sinusoid aliased with a 300 Hz sinusoid.

All the three waveforms are bundled together by using the `Build Array` function and displayed in the same graph. The properties of the `Waveform Graph` should be configured as shown in Figure 3-15. This is done by expanding the plot legend vertically to display the three entries and renaming the labels appropriately. Right-click on the `Waveform Graph` and choose **Properties** from the shortcut menu. A Waveform Graph Properties dialog box will be brought up. Select the **Plots** tab to modify the plot style. Choose `Sampled Signal` in the **Plot** drop down menu, see Figure 3-16. Also, choose the options for **Point Style**, **Plot Interpolation**, and **Fill to** as indicated in this figure. Adjust the line style of the aliased signal to dotted line.

Rename all the controls and indicators, and modify the maximum scale of the `Horizontal Pointer Slide` control to 1000 to complete the VI.

Sampling, Quantization and Reconstruction

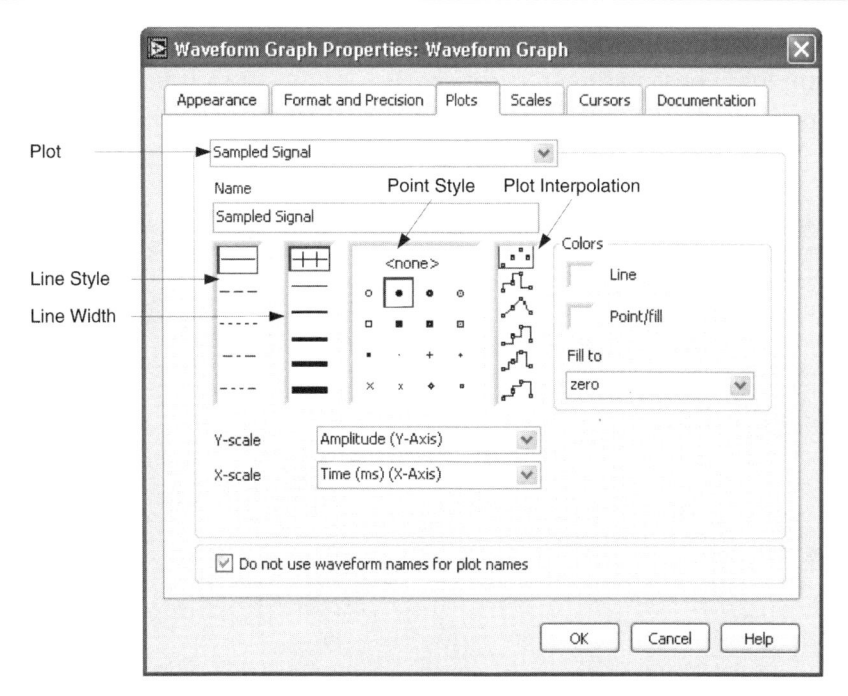

Figure 3-16: Waveform Graph Properties dialog box.

L3.2 Fast Fourier Transform

The analog frequency should satisfy the condition in Equation (3.6) to avoid any discontinuity in DFT. Let us build the example shown in Figure 3-17 using Express VIs to demonstrate the required periodicity of DFT.

Use two `Simulate Signal` Express VIs (**Functions → Input → Simulate Signal**) to simulate the signals. Placing a `Simulate Signal` Express VI brings up a configuration dialog for setting up the parameters including signal type, frequency, amplitude, and sampling frequency, as shown in Figure 3-18. Choose **Sine** for the signal type, set the frequency to 250, the amplitude to 1,

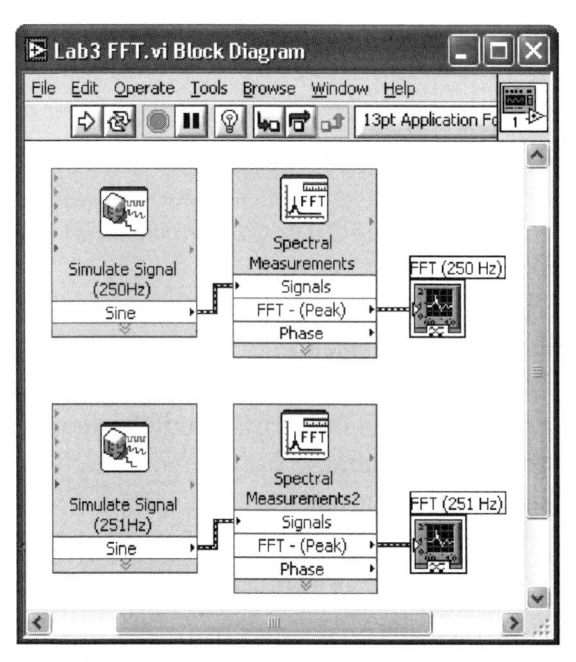

Figure 3-17: BD of Express VI FFTs.

Lab 3

and the phase to 90. Furthermore, enter 1000 as the sampling frequency and 512 as the number of samples. These parameters satisfy the condition in Equation (3.6). As for the 251 Hz sinusoid, use the same parameters except for the frequency, which is to be set to 251.

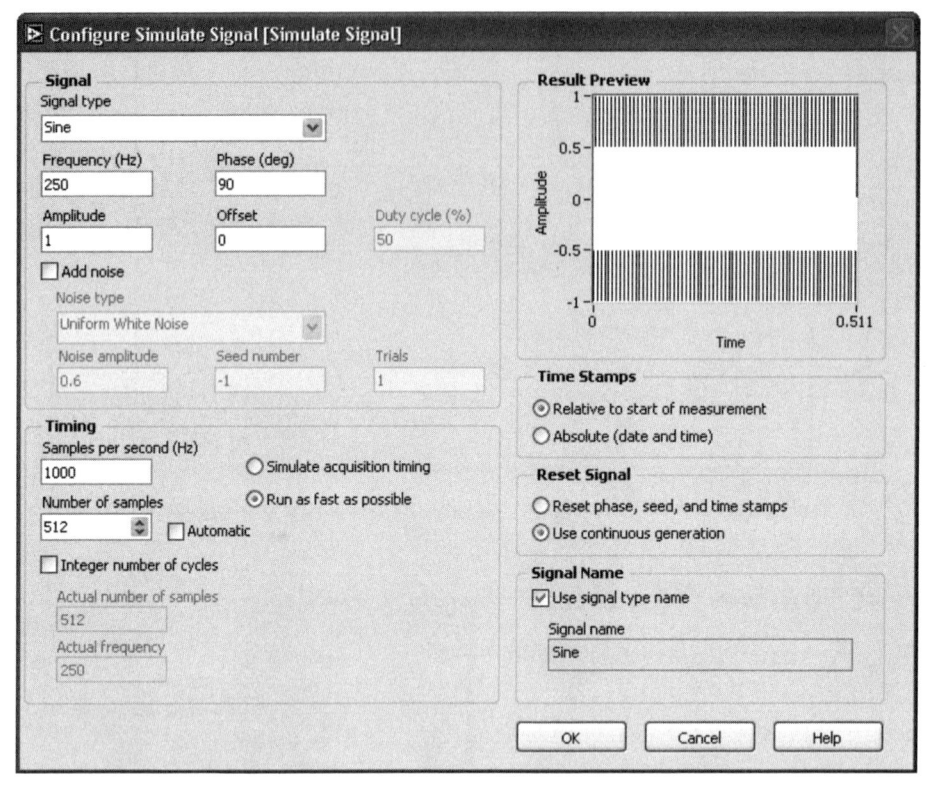

Figure 3-18: Configuration dialog of Simulate Signal Express VI.

Now, place two `Spectral Measurements` Express VIs (**Functions → Analysis → Spectral Measurements**) to compute the FFTs of the signals. The configuration dialog entries need to be adjusted as shown in Figure 3-19. The adjustments shown in this figure provide the spectrum in dB scale without using a spectral leakage window. Notice that when the parameters are adjusted, the preview windows are updated based on the current setting.

The spectra of the two signals are shown in Figure 3-20. As seen from this figure, the spectrum of the 251 Hz signal is spread over a wide range due to the improper sampling. Also, its peak drops by nearly 4 dB.

Sampling, Quantization and Reconstruction

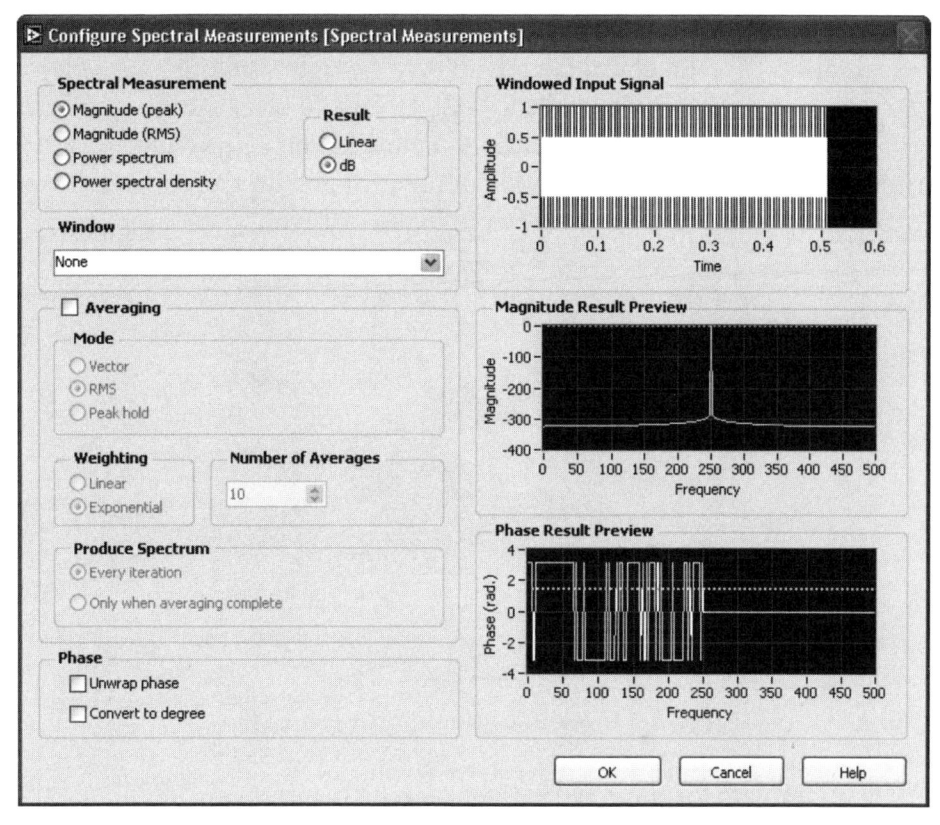

Figure 3-19: Configuration dialog of Spectral Measurements Express VI.

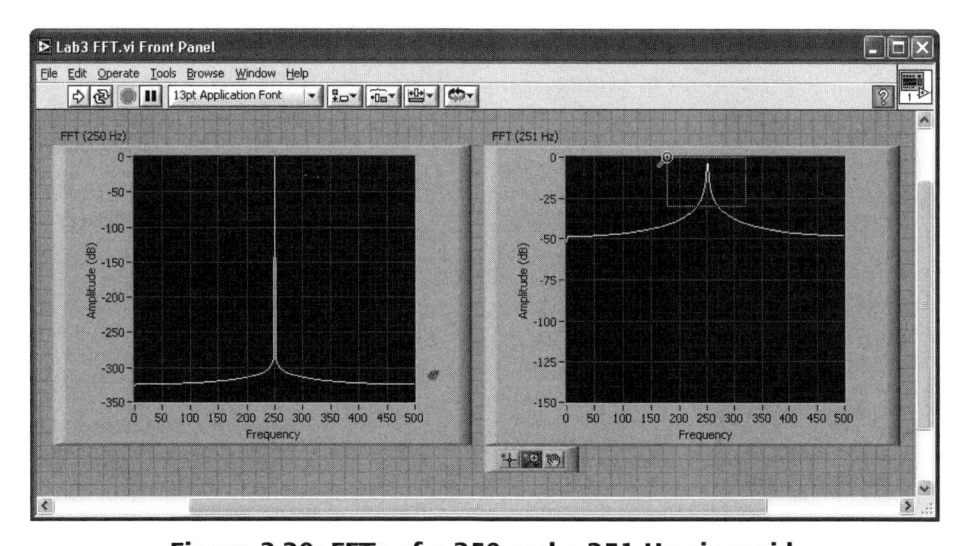

Figure 3-20: FFTs of a 250 and a 251 Hz sinusoid.

Lab 3

The plot in the `Waveform Graph` can be magnified using the **Graph Palette** for better visualization. The **Graph Palette** is displayed by right-clicking on the `Waveform Graph` and choosing **Visible Item → Graph Palette** from the shortcut menu. The options **Cursor Movement Tool**, **Zoom**, and **Panning Tool** are provided in the palette. More specific options for zooming in and out are available in the expanded menu when the **Zoom** option is chosen as shown in Figure 3-21.

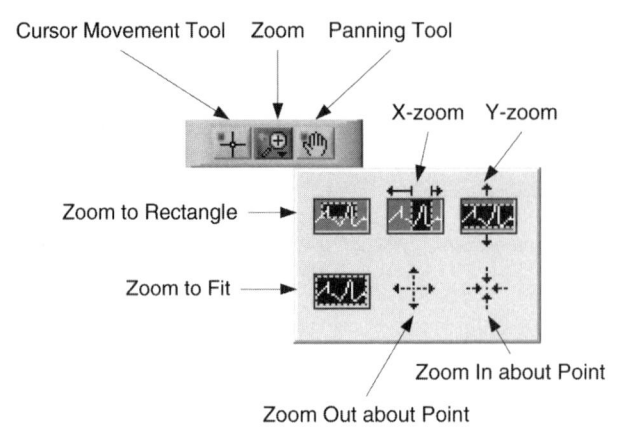

Figure 3-21: Menu options of Graph palette.

The improper sampling for the 251 Hz signal can be corrected by modifying the sampling parameters. The configuration dialog of the `Simulate Signal` Express VI provides a useful option, **Integer number of cycles**, to satisfy the sampling condition. This is illustrated in Figure 3-22.

Checking the **Integer number of cycles** option alters the number of samples and frequency to 502 and 250.996, respectively. As a result, a proper sampling condition is established. The spectrum of this resampled signal is shown in Figure 3-23. As seen from this figure, the frequency leakage is considerably reduced in this case.

Sampling, Quantization and Reconstruction

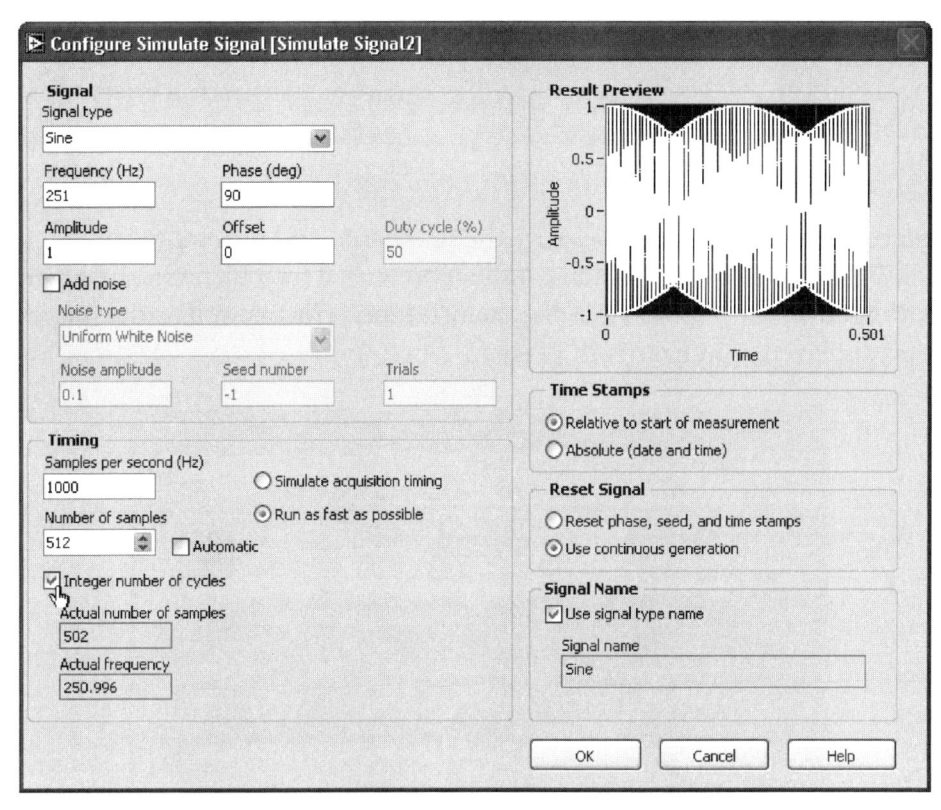

Figure 3-22: Modifying sampling parameters.

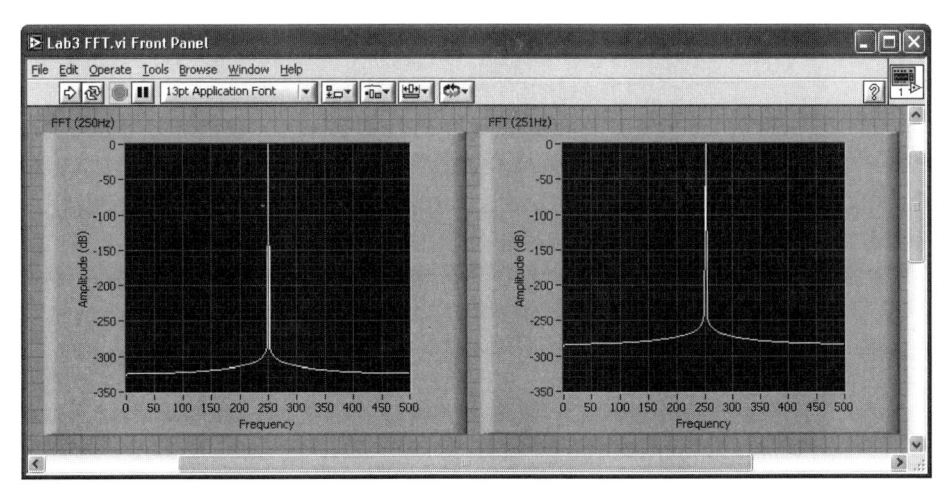

Figure 3-23: FFTs of a 250 and a 251 Hz sinusoids (modified sampling condition).

Lab 3

L3.3 Quantization

Let us now build an A/D converter VI to illustrate the quantization effect. An analog signal given by

$$y(t) = 5.2\exp(-10t)\sin(20\pi t) + 2.5$$

is considered for this purpose. Note that the maximum and minimum values of the signal fall in the range 0 to 7, which can be represented by 3 bits. On the FP, the quantization error, the histogram of the quantization error, as well as the quantized output are displayed as indicated in Figure 3-24.

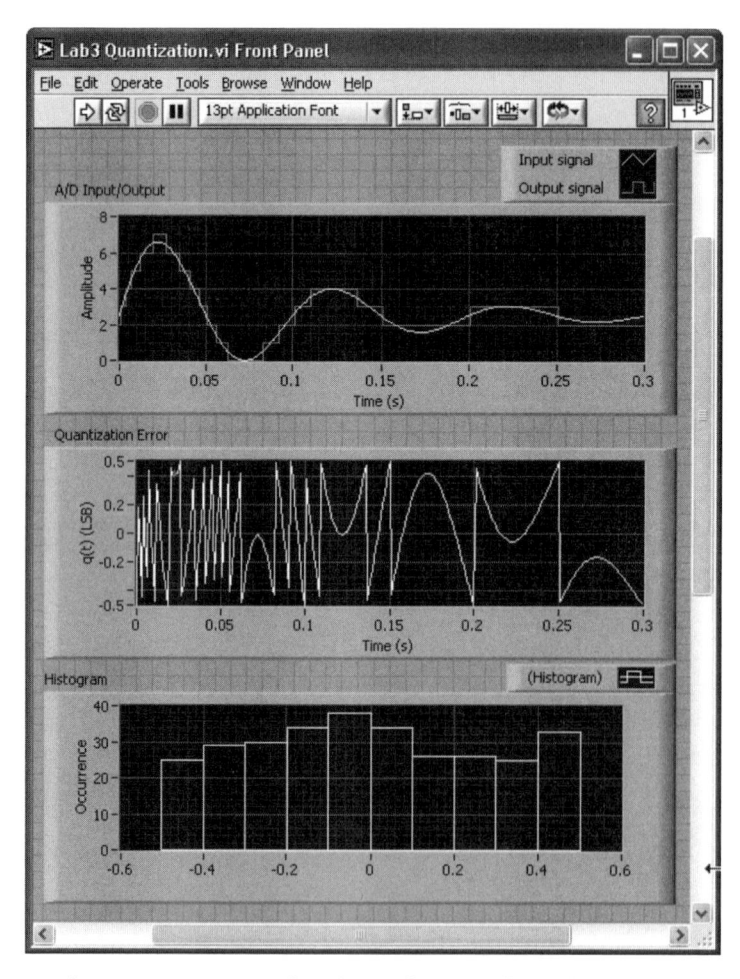

Figure 3-24: Quantization of an analog signal by a 3-bit A/D converter: output signal, quantization error, and histogram of quantization error.

Sampling, Quantization and Reconstruction

To build the converter BD, see Figure 3-25, the `Formula Waveform` VI (**Functions → All Functions → Waveform → Analog Waveform → Waveform Generation → Formula Waveform**) is used. The inputs to this VI comprise a string constant specifying the formula, amplitude, frequency, and sampling information. The values of the output waveform, Y component, are extracted with the `Get Waveform Components` function.

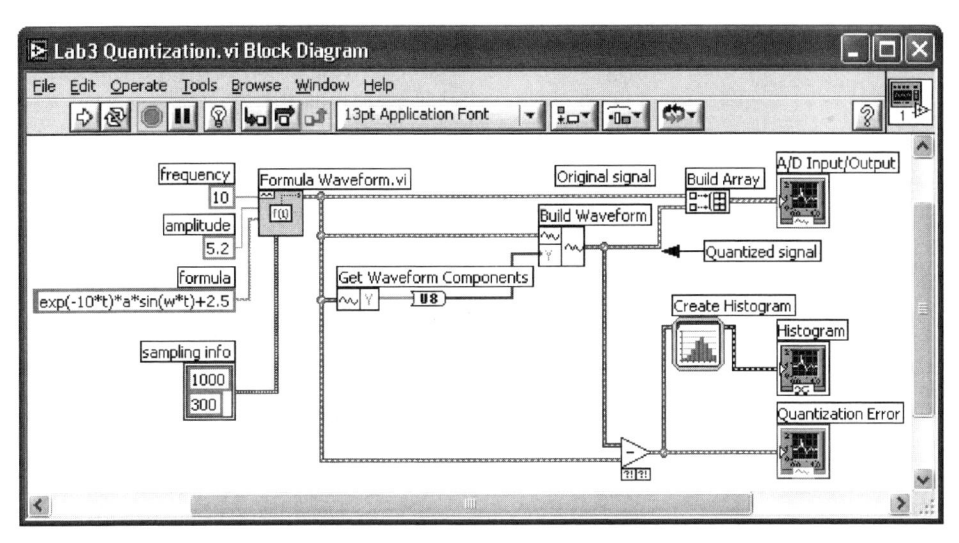

Figure 3-25: Quantization of an analog signal by a 3-bit A/D converter.

To exhibit the quantization process, the double precision signal is converted into an unsigned integer signal by using the `To Unsigned Byte Integer` function (**Functions → All Functions → Numeric → Conversion → To Unsigned Byte Integer**). The resolution of quantization is assumed to be 3 bits, noting that the amplitude of the signal remains between 0 and 7. Values of the analog waveform are replaced by quantized values forming a discretized waveform. This is done by wiring the quantized values to a `Build Waveform` function while the other properties are kept the same as the analog waveform.

Now the difference between the input and quantized output values is found by using a `Subtract` function. This difference represents the quantization error. Also, the histogram of the quantization error is obtained by using the `Create Histogram` Express VI (**Functions → Signal Analysis → Create Histogram**). Placing this VI brings up a configuration dialog as shown in Figure 3-26. The maximum and minimum quantization errors are 0.5 and −0.5, respectively. Hence, the number of bins is set to 10 in order to divide the errors between -0.5 and 0.5 into 10 levels. In addition, for the **Amplitude Representation** option, choose **Sample count** to generate the histogram.

Lab 3

A `Waveform Graph` is created by right-clicking on the `Histogram` node of the `Create Histogram` Express VI and choosing **Create → Graph Indicator**.

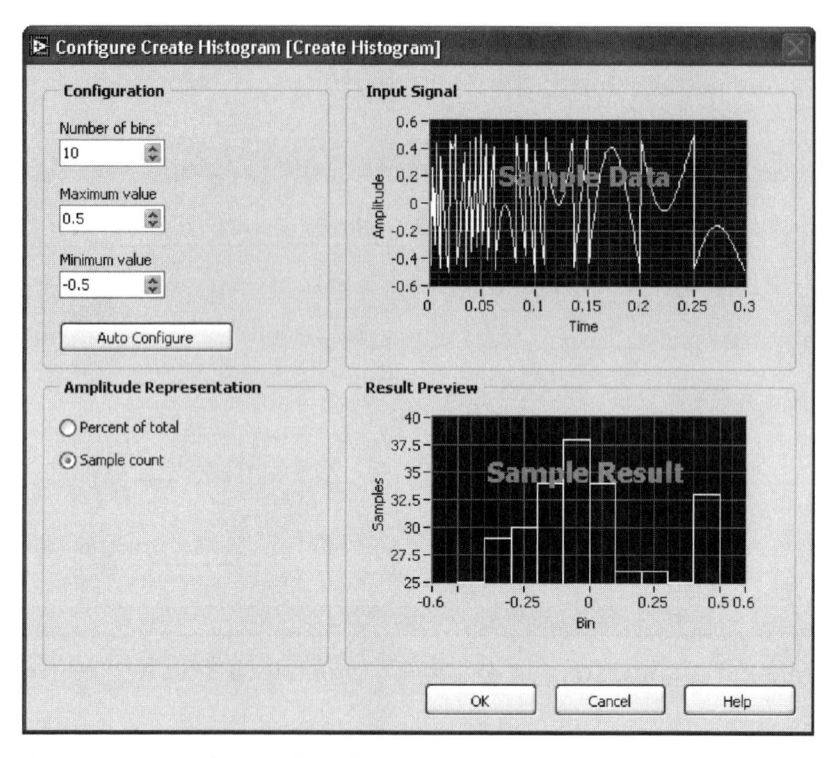

Figure 3-26: Configuration dialog of Create Histogram Express VI.

Return to the FP and change the property of the graph for a more understandable display of the discrete signal. Add the plot legend to the waveform graph and resize it to display the two signals. Rename the analog signal as `Input Signal` and the discrete signal as `Output Signal`.

To display the discrete signal, bring up the properties dialog by right-clicking and choosing **Properties** from the shortcut menu. Click the **Plots** tab and choose the signal plot `Output Signal`. Then, choose **stepwise horizontal**, indicated by the icon, from the **Plot Interpolation** option as the interpolation method. Now the VI is complete, see Figure 3-25.

Next, let us build a VI which can analyze the quantized discrete waveform into a bitstream resembling a logic analyzer. For a 3-bit A/D converter, the bitstream is represented as $b_3 b_2 b_1$ in binary format. The discrete waveform and its bit decomposition are shown in Figure 3-27.

Sampling, Quantization and Reconstruction

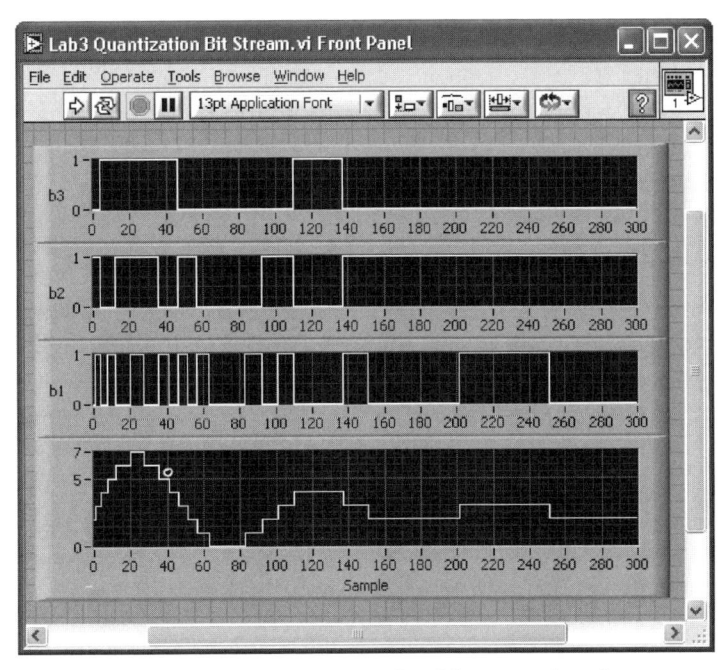

Figure 3-27: Bitstream of 3-bit quantization.

The same analog signal used in the previous example is considered here. The analog signal is generated by a `Formula Waveform` VI, and quantized by using a `To Unsigned Byte Integer` function. Locate a `For Loop` to repeat the quantization as many as the number of samples. This number is obtained by using the `Array Size` function (**Functions → All Functions → Array → Array Size**). Wire this number to the `Count` terminal of the `For Loop`.

Wiring the input array to the `For Loop` places a `Loop Tunnel` on the loop border. Note that auto indexing is enabled by default when inputting an array into a For Loop. With auto indexing enabled, each element of the input array is passed into the loop one at a time per loop iteration.

In order to obtain a binary bitstream, each value passed into the `For Loop` is converted into a Boolean array via a `Number To Boolean Array` function (**Functions → All Functions → Boolean → Number To Boolean Array**). The elements of the Boolean array represent the decomposed bits of the 8-bit integer. The value of a specific bit can be accessed by passing the Boolean array into an `Index Array` function (**Functions → All Functions → Array → Index Array**) and specifying the bit location with a `Numeric Constant`. Since the values stored in the array are Boolean, that is, False or True, they are then converted into 0 and 1, respectively,

Lab 3

using the `Boolean To (0,1)` function (**Functions → All Functions → Boolean → Boolean To (0,1)**). Data from each bit location are wired out of the `For Loop`. Note that an array output is created with the auto-indexing being enabled.

As configured in the previous example, the **stepwise horizontal** interpolation method is used for the waveform graph of the discrete signal. The completed VI is shown in Figure 3-28.

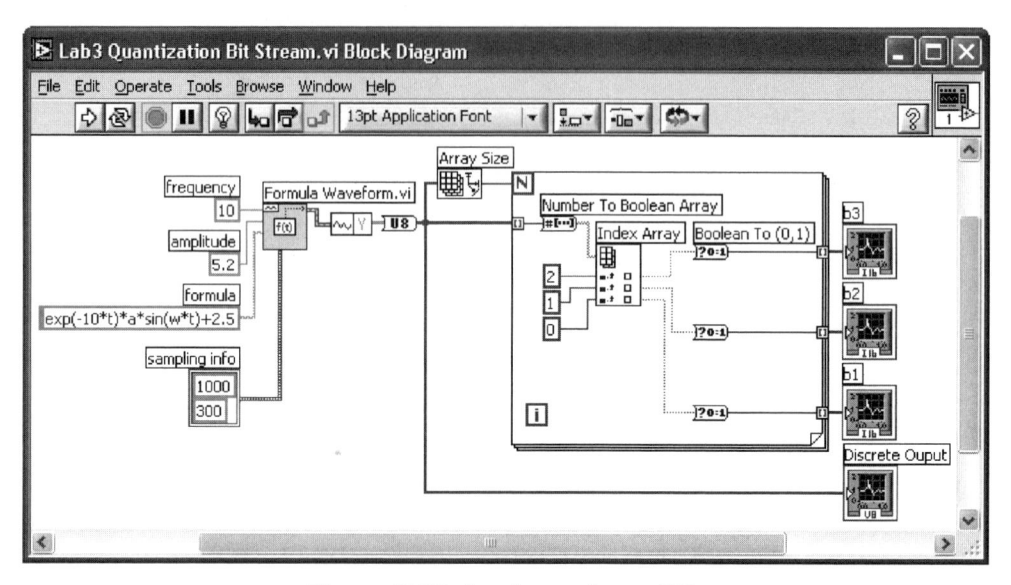

Figure 3-28: Logic analyzer BD.

L3.4 Signal Reconstruction

As the final example in this lab, a signal reconstruction VI is presented. Let us examine the FP shown in Figure 3-29 exhibiting a sampled signal and its reconstructed version. The reconstruction kernel is also shown in this FP.

The sampled signal is shown via bars in the top `Waveform Graph`. In order to reconstruct an analog signal from the sampled signal, a convolution operation with a sinc function is carried out as specified by Equation (3.7).

Sampling, Quantization and Reconstruction

Let us now build the VI. It is assumed that a unity amplitude sinusoid of 10 Hz is sampled at 80 Hz. To display the reconstructed analog signal, the sampling frequency and number of samples are set to 100 times those of the discrete signal. The two waveforms are merged and displayed in the same `Waveform Graph` as shown in Figure 3-29.

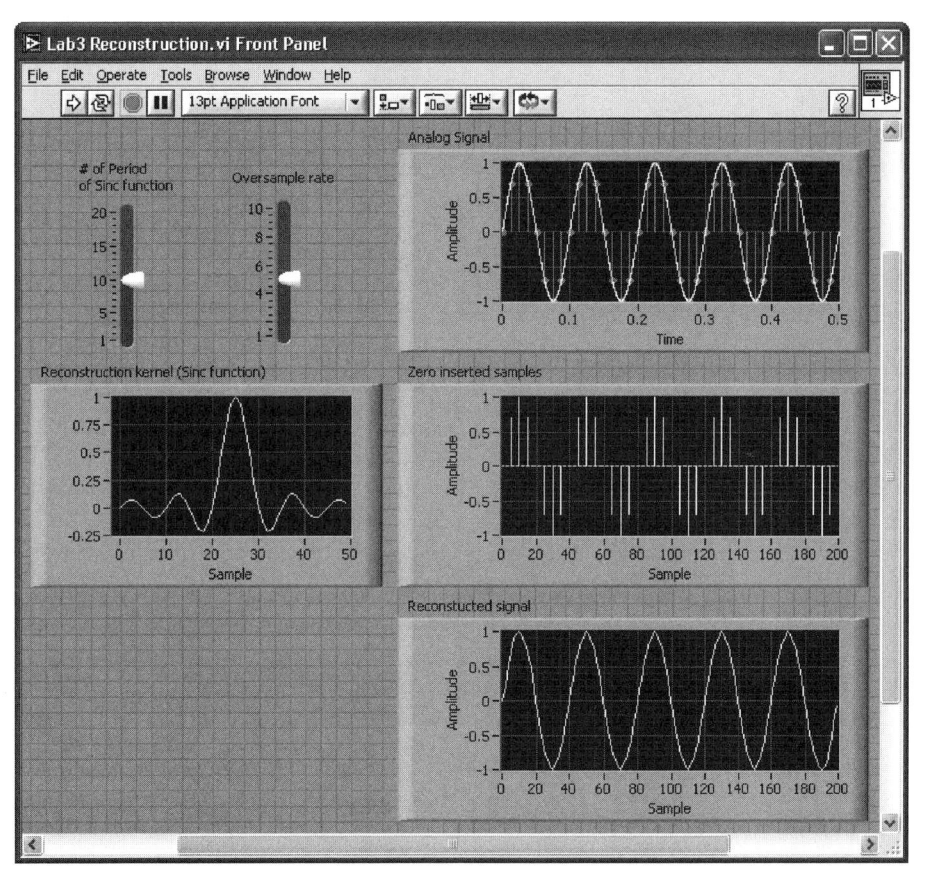

Figure 3-29: FP of a reconstructed sinewave from its samples.

Lab 3

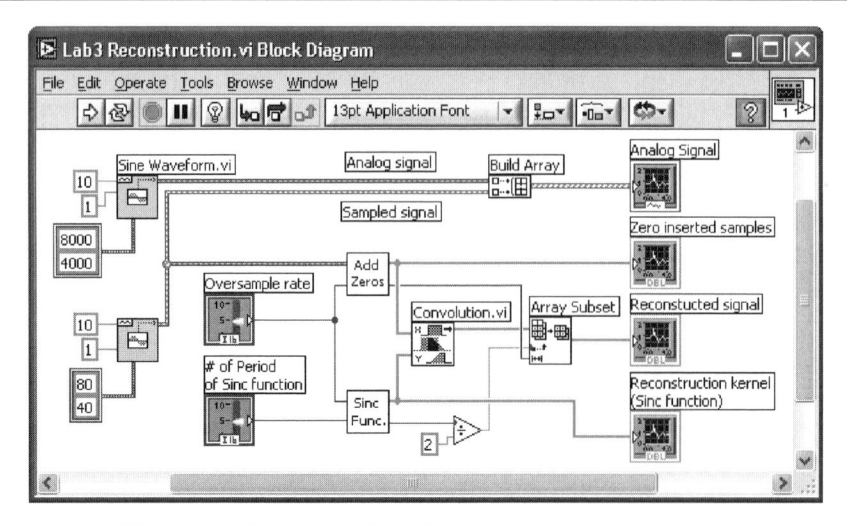

Figure 3-30: BD of signal reconstruction system.

The BD of the signal reconstruction system is shown in Figure 3-30. Two custom subVIs are shown on the BD. The `Add Zeros` VI is used to insert zeros between consecutive samples to simulate oversampling, and the `Sinc Function` VI is used to generate samples of a sinc function with a specified number of zero-crossings.

The BD of each subVI is briefly explained here. In the `Add Zeros` VI, see Figure 3-31, zero rows are concatenated to the 1D signal array. The augmented 2D array is then transposed and reshaped to 1D so that the zeros are located between the samples. The number of zeros inserted between the samples can be controlled by wiring a numeric control. The output waveform shown in the BD takes its input from the other VI and is created by right-clicking on the `Get Waveform Components`

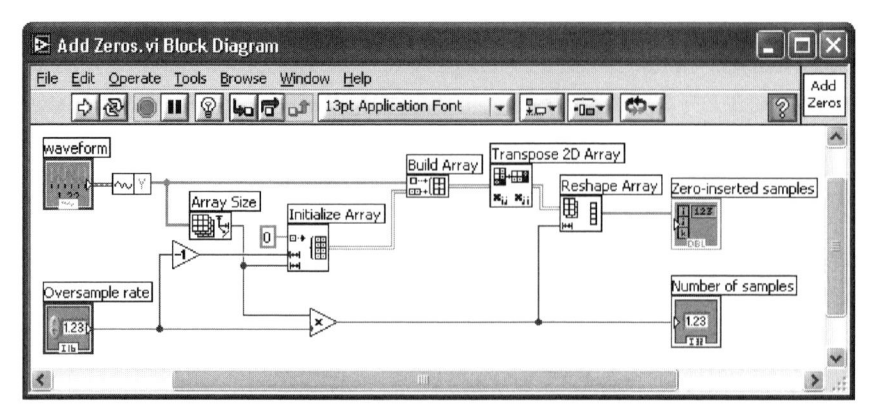

Figure 3-31: Add Zeros subVI.

Sampling, Quantization and Reconstruction

function and choosing **Create → Control**. The outputs of the VI comprise the array of zero-inserted samples and the total number of samples. The connector pane of the VI consists of two input terminals and two output terminals. The input terminals are wired to the controls and the output terminals to the indicators, respectively.

The Sinc Function VI, see Figure 3-32, generates samples of a sinc function based on the number of samples, delay, and sampling interval parameters.

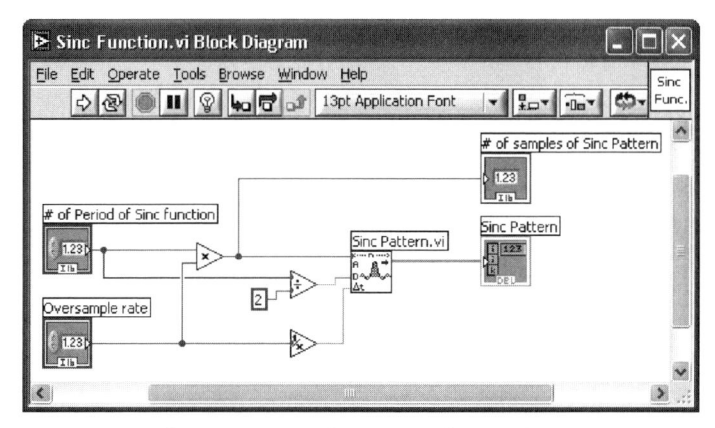

Figure 3-32: Sinc Function subVI.

Finally, let us return to the BD shown in Figure 3-30. The two signals generated by the subVIs, i.e., the zero-inserted signal and sinc signal, are convolved using the Convolution VI (**Functions → All Functions → Analyze → Signal Processing → Time Domain → Convolution**). Note that the length of the convolved array obtained from the Convolution VI is one less than the sum of the samples in the two signals, that is, 249 for the example shown in Figure 3-29. Since the number of the input samples is 200, only a 200 sample portion (samples indices between 25 and 224) of the convolved output is displayed for better visualization.

L3.5 Bibliography

[1] National Instruments, *LabVIEW User Manual*, Part Number 320999E-01, 2003.

CHAPTER 4

Digital Filtering

Filtering of digital signals is a fundamental concept in digital signal processing. Here, it is assumed that the reader has already taken a course in digital signal processing or is already familiar with finite impulse response (FIR) and infinite impulse response (IIR) filter design methods.

In this chapter, the structure of digital filters is briefly mentioned followed by a discussion on the LabVIEW Digital Filter Design (DFD) toolkit. This toolkit provides various tools for the design, analysis and simulation of digital filters.

4.1 Digital Filtering

4.1.1 Difference Equations

As a difference equation, an FIR filter is expressed as

$$y[n] = \sum_{k=0}^{N} b_k x[n-k] \qquad (4.1)$$

where b's denote the filter coefficients and N the number of zeros or filter order. As described by this equation, an FIR filter operates on a current input $x[n]$ and a number of previous inputs $x[n-k]$ to generate a current output $y[n]$.

The equi-ripple method, also known as the Remez algorithm, is normally used to produce an optimal FIR filter [1]. Figure 4-1 shows the filter responses using the available design methods consisting of equi-ripple, Kaiser window and Dolph-Chebyshev window. Among these methods, the equi-ripple method generates a response whose deviation from the desired response is evenly distributed across the passband and stopband [2].

Chapter 4

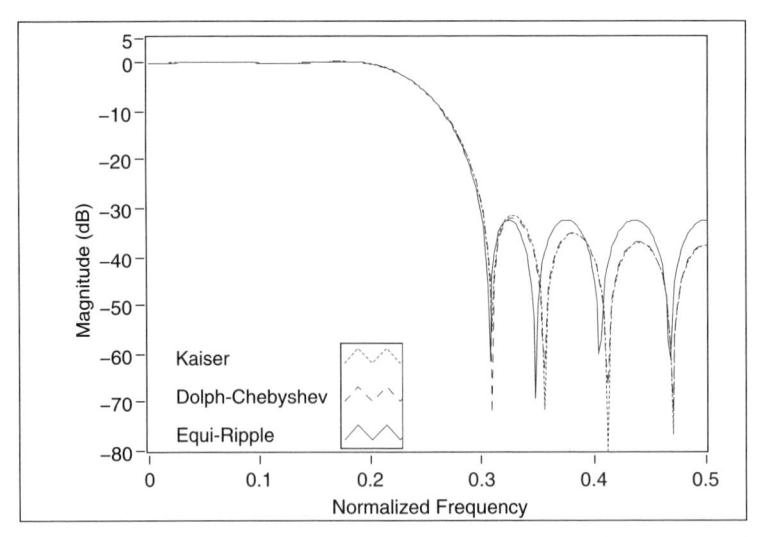

Figure 4-1: Responses of different FIR filter design methods.

The difference equation of an IIR filter is given by

$$y[n] = \sum_{k=0}^{N} b_k x[n-k] - \sum_{k=1}^{M} a_k y[n-k] \qquad (4.2)$$

where b's and a's denote the filter coefficients and N and M the number of zeros and poles, respectively. As indicated by Equation (4.2), an IIR filter uses a number of previous outputs $y[n-k]$ as well as a current and a number of previous inputs to generate a current output $y[n]$.

Several methods are widely used to design IIR filters. They include Butterworth, Chebyshev, Inverse Chebyshev, and Elliptic methods. In Figure 4-2, the magnitude response of an IIR filter designed by these methods having the same order are shown for comparison purposes. For example, the elliptic method generates a relatively narrower transition band and more ripples in passband and stopband while the Butterworth method generates a monotonic type of response [2]. Table 4-1 summarizes the characteristics of these design methods.

Digital Filtering

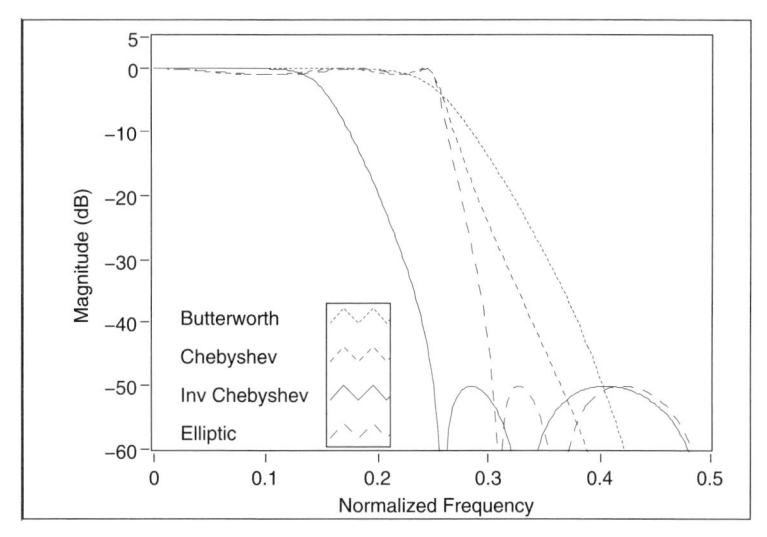

Figure 4-2: Responses of different IIR filter design methods.

Table 4-1: Comparison of different IIR filter design methods [1].

IIR filter	Ripple in passband?	Ripple in stopband?	Transition bandwidth for a fixed order	Order for given filter specifications
Butterworth	No	No	Widest	Highest
Chebyshev	Yes	No	Narrower	Lower
Inverse Chebyshev	No	Yes	Narrower	Lower
Elliptic	Yes	Yes	Narrowest	Lowest

4.1.2 Stability and Structure

In general, as compared to IIR filters, FIR filters require less precision and are computationally more stable. The stability of an IIR filter depends on whether its poles are located inside the unit circle in the complex plane. Consequently, when an IIR filter is implemented on a fixed-point processor, its stability can be affected. Furthermore, the phase response of an FIR filter is always linear. In Table 4-2, a summary of the differences between the attributes of FIR and IIR filters is listed.

Chapter 4

Table 4-2: FIR filter attributes versus IIR filter attributes [1].

Attribute	FIR filter	IIR filter
Linear phase response	Possible	Not possible
Stability	Always stable	Conditionally stable
Fixed-point implementation	Easy to perform	Can be complicated
Computational complexity	More operations	Fewer operations
Datapath precision	Less precision required	Higher precision required
Limit cycles[1]	Does not produce	May produce

[1] A limit cycle is an oscillation in the output caused by rounding errors.

Let us now discuss the stability and structure of IIR filters. The transfer function of an IIR filter is expressed as

$$H(z) = \frac{b_0 + b_1 z^{-1} + \ldots + b_N z^{-N}}{1 + a_1 z^{-1} + \ldots + a_M z^{-M}} \quad (4.3)$$

It is well known that as far as stability is concerned, the direct-form implementation is sensitive to coefficient quantization errors. Noting that the second-order cascade form produces a more robust response to quantization noise [2], the above transfer function can be rewritten as

$$H(z) = \prod_{k=1}^{N_s} \frac{b_{0k} + b_{1k} z^{-1} + B_{2k} z^{-2}}{1 + a_{1k} z^{-1} + a_{2k} z^{-2}} \quad (4.4)$$

where $N_s = \lfloor N/2 \rfloor$, $\lfloor \cdot \rfloor$ represents the largest integer less than or equal to the inside value. This serial or cascaded structure is illustrated in Figure 4-3.

Figure 4-3: Cascaded filter stages.

It is worth mentioning that each second-order filter is considered to be of direct-form II, see Figure 4-4, in order to have a more memory efficient implementation.

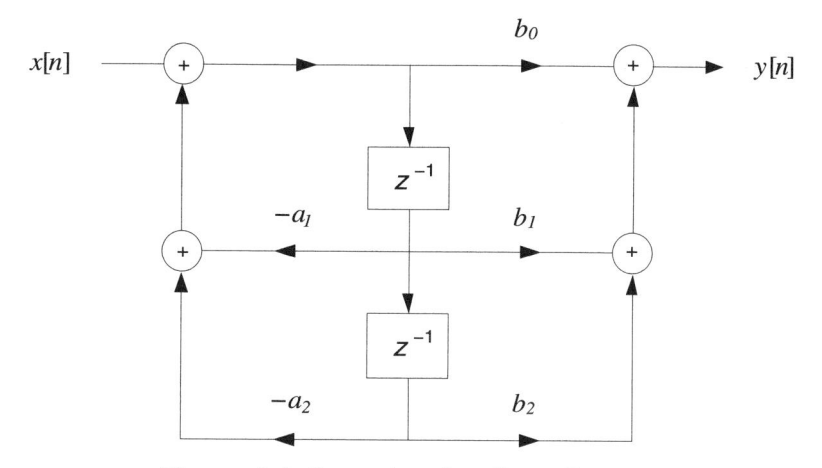

Figure 4-4: Second order direct-form II.

4.2 LabVIEW Digital Filter Design Toolkit

There exist various software tools for designing digital filters. Here, we have used the LabVIEW Digital Filter Design (DFD) toolkit. Any other filter design tool may be used to obtain the coefficients of a digital filter. The DFD toolkit provides various tools to design, analyze, and simulate floating-point and fixed-point implementations of digital filters [1].

4.2.1 Filter Design

The Filter Design VIs of the DFD toolkit allow one to design a digital filter with ease by specifying its parameters. For example, the `DFD Classical Filter Design` Express VI provides a graphical user interface to design and analyze digital filters, and the `DFD Pole-Zero Placement` Express VI can be used to alter the locations of poles and zeros in the complex plane easily.

The filter design methods provided in the DFD toolkits include Kaiser Window, Dolph-Chebyshev Window, and Equi-Ripple for FIR filters, and Butterworth, Chebyshev, Inverse Chebyshev, and Elliptic for IIR filters.

In addition, the DFD toolkit has some Special Filter Design VIs. These VIs are used to design special filters such as notch/peak filter, comb filter, maximally flat filter, narrowband filter, and group delay compensator.

4.2.2 Analysis of Filter Design

A comprehensive analysis of a digital filter can be achieved by using the Analysis VIs of the DFD toolkit. These VIs provide magnitude response, phase response, impulse response, step response, and zero/pole plot.

4.2.3 Fixed-Point Filter Design

The Fixed-Point Tools VIs of the DFD toolkit can be used to examine the outcome of a fixed-point implementation. Note that when changing a filter structure from the direct-form to the cascade-form or any other form, a different filter response is obtained, in particular when a fixed-point implementation is realized.

4.2.4 Multirate Digital Filter Design

The DFD toolkit also provides a group of VIs, named Multirate Filter Design VIs, for the design, analysis, and implementation of multirate filters. These multirate filters include single-stage, multistage, halfband, Nyquist, raised cosine, and cascaded integrator comb (CIC) filters [1].

4.3 Bibliography

[1] National Instruments, *Digital Filter Design Toolkit User Manual*, Part Number 371353A-01, 2005.

[2] J. Proakis and D. Manolakis, Digital Signal Processing: Principles, Algorithms, and Applications, Prentice-Hall, 1995.

Lab 4: FIR/IIR Filtering System Design

In this lab, an FIR and an IIR filter are designed using the VIs as part of the Digital Filter Design (DFD) toolkit.

L4.1 FIR Filtering System

An FIR lowpass filtering system is designed and built in this section.

L4.1.1 Design FIR Filter with DFD Toolkit

Let us design a lowpass filter having the following specifications: passband response = 0.1 dB, passband frequency = 1200 Hz, stopband attenuation = 30 dB, stopband frequency = 2200 Hz and sampling rate = 8000 Hz. In order to design this filter using the DFD toolkit, place the DFD Classical Filter Design Express VI (**Functions → All Functions → Digital Filter Design → Filter Design → DFD Classical Filter Design**) on the BD. Enter the specifications of the filter in the configuration dialog box which appears when placing this Express VI. The magnitude response of the filter and the zero/pole plot are displayed based on the filter specifications in the configuration dialog box, see Figure 4-5. Here, the equi-ripple method is chosen as the design method.

Once this Express VI is configured, its label is changed based on the filter type specified, such as the Equi-Ripple FIR Lowpass Filter in this example. The filter type gets displayed on the BD as shown in Figure 4-6.

Additional information on the designed filter, such as phase, group delay, impulse response, unit response, frequency response, and zero/pole plot can be seen by using the DFD Filter Analysis Express VI (**Functions → All Functions → Digital Filter Design → Filter Analysis → DFD Filter Analysis**). As indicated in Figure 4-6, wire five Waveform Graphs to the output terminals of the DFD Filter Analysis Express VI except for the Z Plane terminal. The DFD Pole-Zero Plot control (**Controls → All Controls → Digital Filter Design → DFD Pole-Zero Plot**) needs to be placed on the FP to obtain the zero/pole plot. This locates a terminal icon on the BD. Then, wire the Z Plane terminal of the DFD Filter Analysis Express VI to the DFD Pole-Zero Plot control.

Lab 4

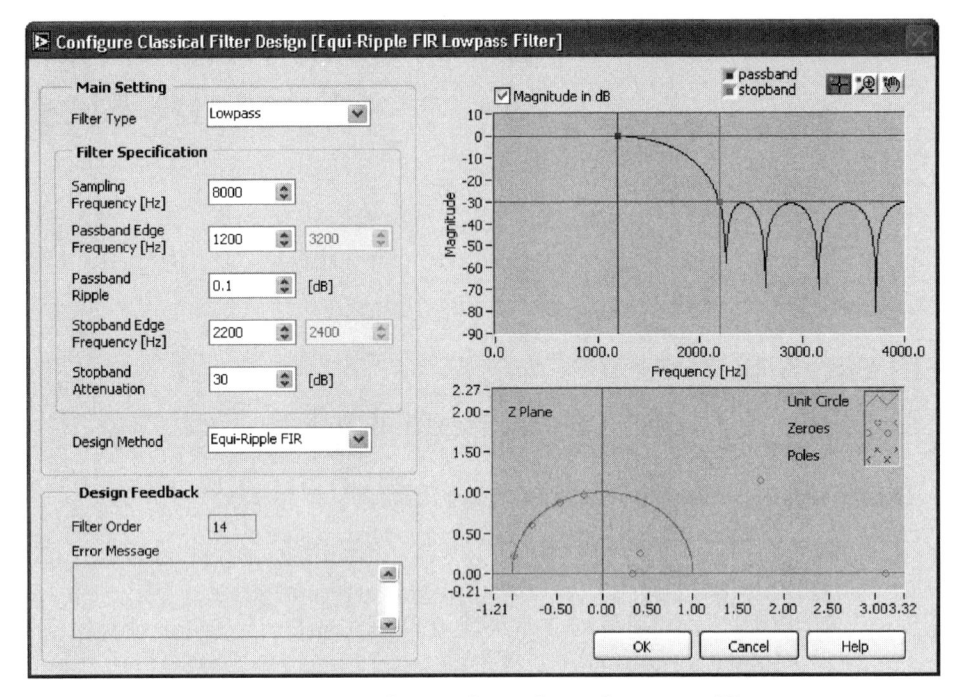

Figure 4-5: Configuration of FIR lowpass filter.

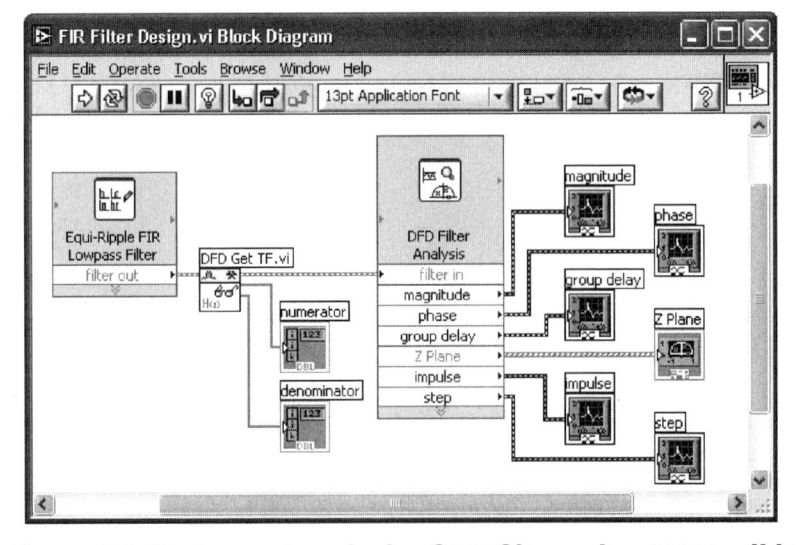

Figure 4-6: Design and analysis of FIR filter using DFD toolkit.

FIR/IIR Filtering System Design

The coefficients of the filter are obtained by wiring the DFD Get TF VI (**Functions → All Functions → Digital Filter Design → Utilities → DFD Get TF**) to the filter cluster, i.e., the output of the DFD Classical Filter Design Express VI. The DFD Get TF VI retrieves the transfer function of the filter designed by the DFD Classical Filter Design Express VI. For FIR filters, the numerator values of the transfer function correspond to the b coefficients of the filter and the denominator to unity. The transfer function of the designed filter can be observed by creating two Numeric Indicators. To do this, right-click on the numerator terminal of the DFD Get TF VI and choose **Create → Indicator** from the shortcut menu. The second indicator is created and wired to the denominator terminal of the VI.

Save the VI as *FIR Filter Design.vi* and then run it. The response of the designed FIR filter is illustrated in Figure 4-7. Notice that the indicator array on the FP needs to be resized to display all the elements of the coefficient set.

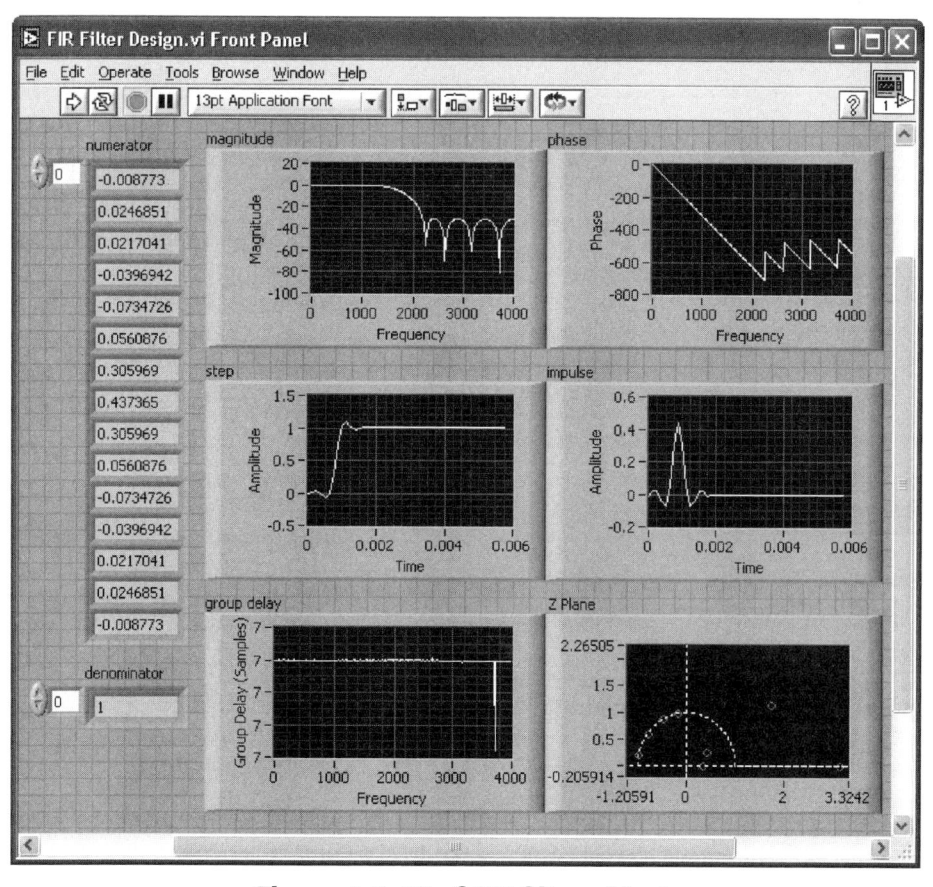

Figure 4-7: FP of FIR filter object.

Lab 4

L4.1.2 Creating a Filtering System VI

The VI of the filtering system built here consists of signal generation, filtering, and graphical output components. Three sinusoidal signals are summed and passed through the designed FIR filter and the filtered signal is then displayed and verified.

Let us build the FP of the filtering system. Place three `Horizontal Pointer Slide` controls (**Controls → Numeric Controls → Horizontal Pointer Slide**) to adjust the frequency of the signals. Place three `Waveform Graphs` to display the input signal and filtered signal in the time and frequency domains. See that the corresponding terminal icons for the `Horizontal Pointer Slide` controls and `Waveform Graphs` are created on the BD, see Figure 4-8.

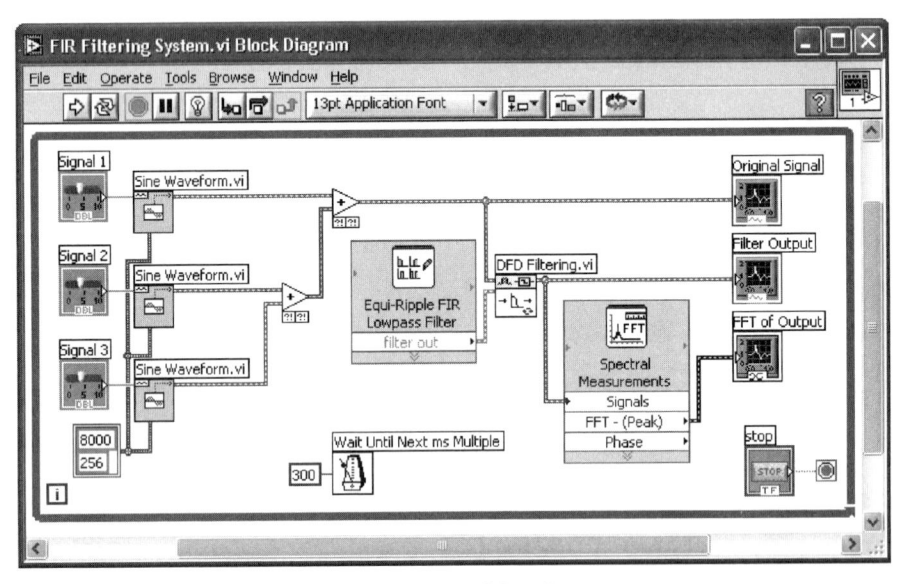

Figure 4-8: BD of FIR filtering system.

Next, switch to the BD. To provide the signal source of the system, place three `Sine Waveform` VIs (**Functions → All Functions → Analyze → Waveform Generation → Sine Waveform**) on the BD. The amplitude of the output sinusoid is configured to be its default value of unity in the absence of an input. The icons of the `Horizontal Pointer Slide` controls are wired to the `frequency` terminal of each `Sine Waveform` VI.

Create a cluster constant to incorporate the sampling information. This is done by right-clicking on the `sampling info` terminal of the `Sine Waveform` VI and choosing **Create → Constant**. Enter 8000 as the sampling rate and 256 as the number of samples. Wire the cluster constant to all three VIs so that all the signals have the same sampling rate and length. The three signal arrays are summed together to construct the input signal of the filtering system. This is done by using two `Add` functions (**Functions** → **Arithmetic & Comparison** → **Express Numeric** → **Add**) as shown in Figure 4-8.

Now the filtering component is described. The filter is designed by using the `DFD Classical Filter Design` Express VI (**Functions** → **All Functions** → **Digital Filter Design** → **Filter Design** → **DFD Classical Filter Design**) as described earlier. This VI creates a filter object in the form of a cluster based on the configured filter specifications. The filter object is wired to the `filter in` terminal of the `DFD Filtering` VI (**Functions** → **All Functions** → **Digital Filter Design** → **Processing** → **DFD Filtering**) in order to filter the input signal, which is wired from the cascaded `Add` function.

The input signal and the output of the `DFD Filtering` VI are wired to two `Waveform Graphs` to observe the filtering effect in the time domain. To have a spectral measurement of the signal, place a `Spectral Measurements` Express VI on the BD. On the configuration dialog box of the Express VI, configure the **Spectral Measurement** field as Magnitude (peak), the **Result** field as dB, and the **Window** field as None. Wire the FFT output in dB to a `Waveform Graph`. Place a `While Loop` on the BD to enclose all the sections of the code on the BD. The completed BD is shown in Figure 4-8.

Now, return to the FP to change the properties of the FP objects. Rename the labels of the controls and Waveform Graphs as shown in Figure 4-9. First, let us change the properties of the three `Horizontal Pointer Slide` controls. Right-click on each control and choose **Properties** from the shortcut menu. This brings up a properties dialog box. Change the maximum scale of all the three controls to the Nyquist frequency, 4000 Hz, in the **Scale** tab, and set the default frequency values to 750 Hz, 2500 Hz and 3000 Hz, respectively, in the **Data Range** tab.

Next, let us modify the properties of the `Waveform Graph` labeled as `FFT of Output` in Figure 4-9. Right-click on the `Waveform Graph` and choose **Properties** from the shortcut menu to bring up a properties dialog box. Uncheck **Autoscale** of the Y axis and change the minimum scale to –80 in the **Scales** tab to observe peaks of the waveform more closely. In the other two graphs corresponding to the time domain signal, uncheck **X Scale** → **Loose Fit** from the shortcut menu to fit the plot into the entire plotting area.

Lab 4

Save the VI as *FIR Filtering System.vi* and then run it. Note that among the three signals 750 Hz, 2500 Hz, and 3000 Hz, the 2500 Hz and 3000 Hz signals should be filtered out and only the 750 Hz signal should be seen at the output. The waveform result on the FP during run time is shown in Figure 4-9.

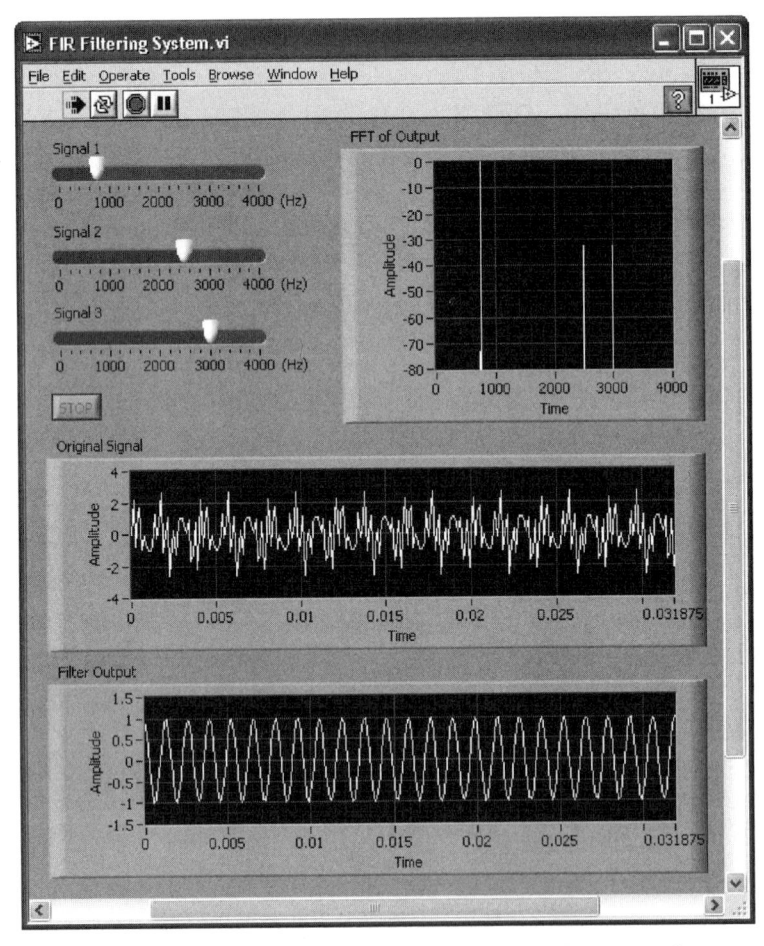

Figure 4-9: FP of FIR filtering system during run time.

FIR/IIR Filtering System Design

L4.2 IIR Filtering System

An IIR bandpass filter is designed and built in this section.

L4.2.1 IIR Filter Design

Let us consider a bandpass filter with the following specifications: passband response = 0.5 dB, passband frequency = 1333 to 2666 Hz, stopband attenuation = 20 dB, stopband frequency = 1000 to 3000 Hz and sampling frequency = 8000 Hz. The design of an IIR filter is achieved by using the `DFD Classical Filter Design` Express VI described earlier. Enter the specifications of the filter in the configuration dialog box which is brought up by placing this Express VI on the BD, see Figure 4-10. The elliptic method is chosen here as the design method to achieve a narrow transition band.

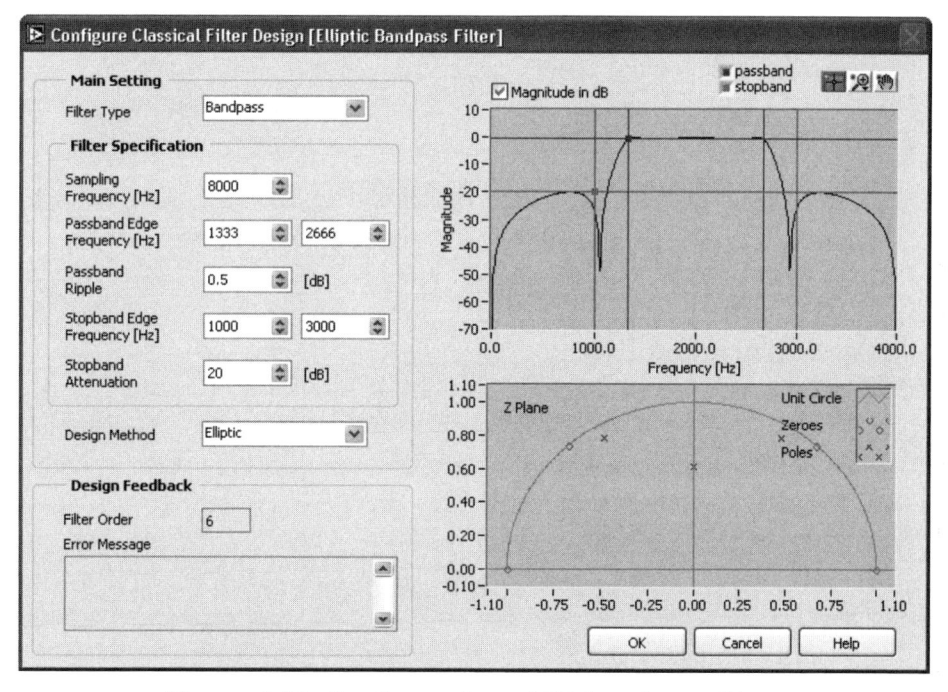

Figure 4-10: Configuration of IIR bandpass filter.

Lab 4

Similar to FIR filtering, the label of the Express VI is changed to `Elliptic Bandpass Filter` by altering the configuration as shown in Figure 4-11. The response of the designed filter can be obtained by using the `DFD Filter Analysis` Express VI and wiring five `Waveform Graphs` and a `DFD Pole-Zero Plot` control. These steps are similar to those mentioned for FIR filtering.

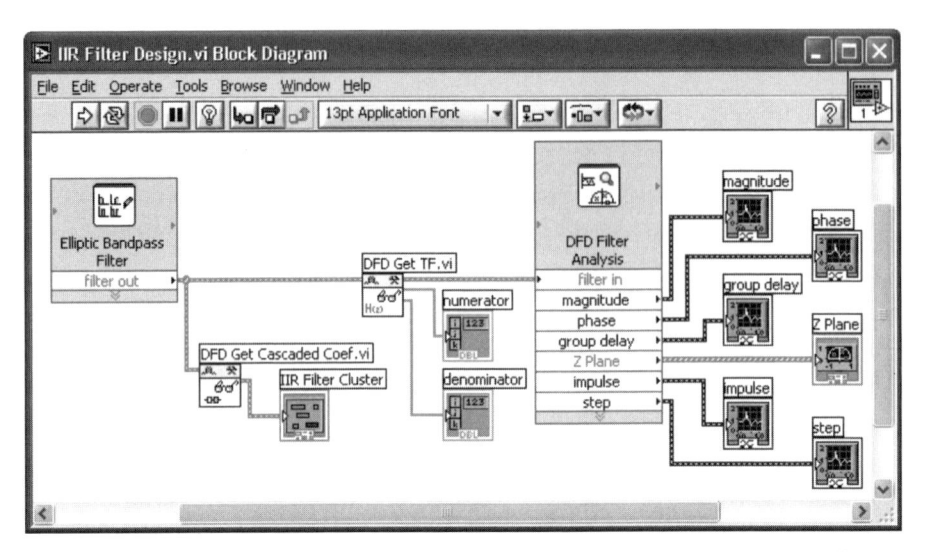

Figure 4-11: Design and analysis of IIR filter using DFD toolkit.

The filter coefficients provided by the `DFD Classical Filter Design` Express VI correspond to the 'IIR cascaded second order sections form II' structure by default. To observe the cascaded coefficients, the filter cluster is wired to the `DFD Get Cascaded Coef` VI. A cluster of indicators is created by right-clicking on the `IIR Filter Cluster` terminal of the VI and choosing **Create → Indicator**. The filter coefficients corresponding to the 'IIR direct form II' structure are obtained by using the `DFD Get TF` VI similar to FIR filtering.

Save the VI as *IIR Filtering Design.vi* and then run it. The response of the IIR bandpass filter is illustrated in Figure 4-12.

FIR/IIR Filtering System Design

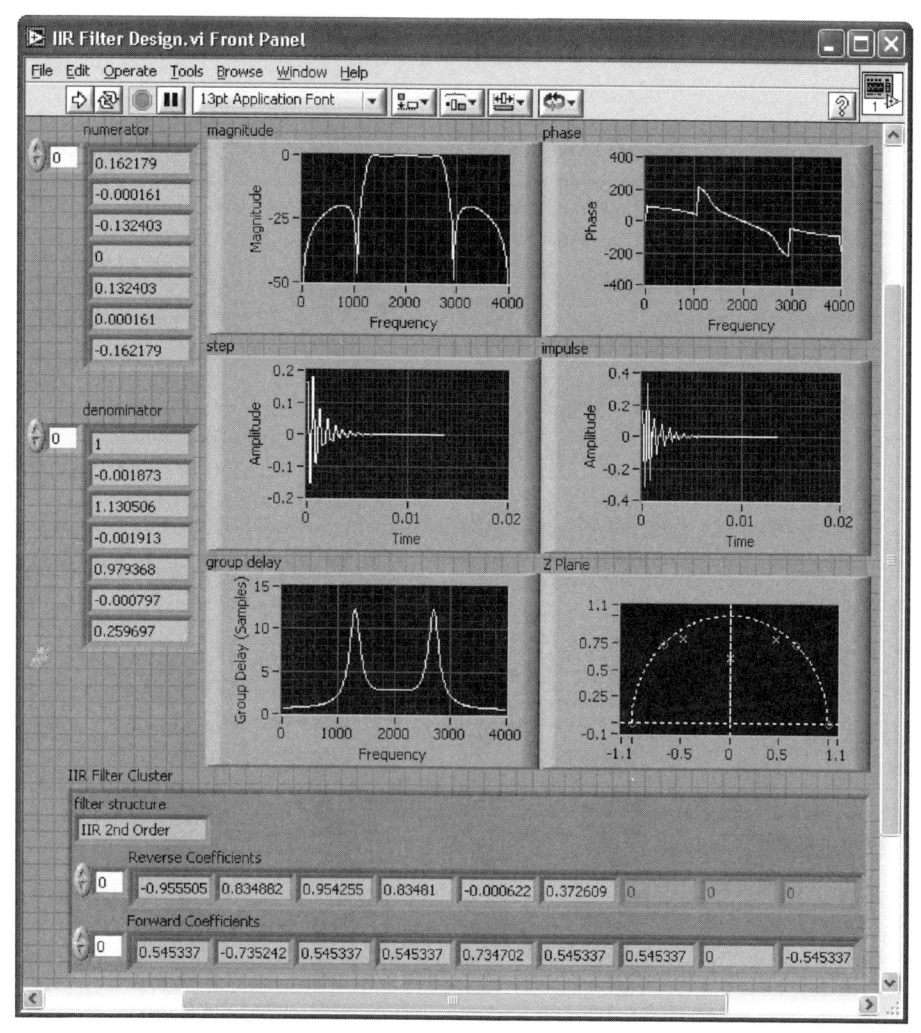

Figure 4-12: FP of IIR filter object.

Notice that the filter coefficients are displayed as truncated values in Figure 4-12. The format of the numeric indicators is configured to be floating-point with six digits of precision. This is done by right-clicking on the numeric indicators on the FP and choosing **Format & Precision...** from the shortcut menu. A properties dialog box is brought up, see Figure 4-13. Configure the representation to **Floating point** and the precision to **6 Digits of precision** as shown in Figure 4-13.

Lab 4

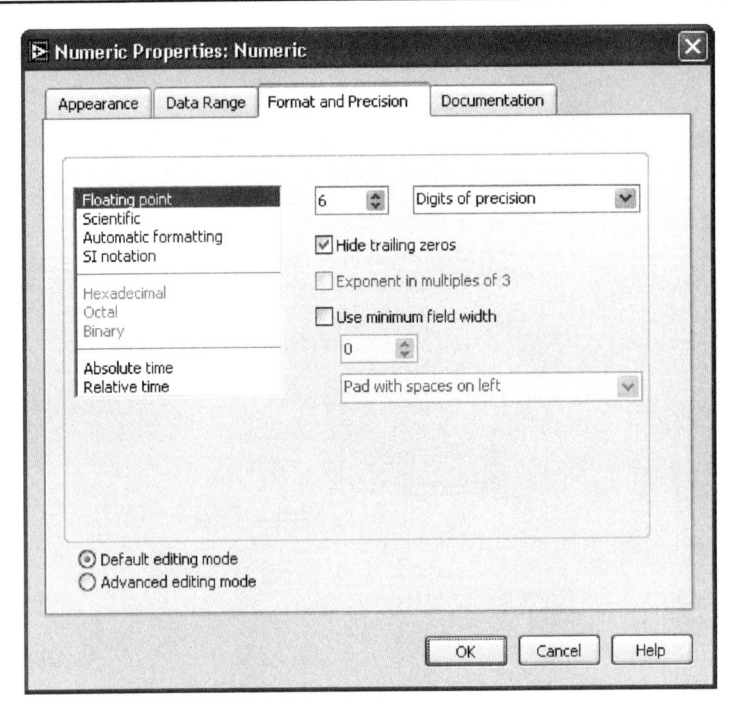

Figure 4-13: Changing properties of numeric indicator format and precision.

From the coefficient set, the transfer function of the IIR filter is given by

$$H[z] = \frac{0.162179 - 0.000161z^{-1} - 0.132403z^{-2} + 0.132403z^{-4} + 0.000161z^{-5} - 0.162179z^{-6}}{1 - 0.001873z^{-1} + 1.130506z^{-2} - 0.001913z^{-3} + 0.979368z^{-4} - 0.000797z^{-5} + 0.259697z^{-6}}$$

$$= H_1[z] \cdot H_2[z] \cdot H_3[z]$$

where $H_1[z]$, $H_2[z]$, and $H_3[z]$ denote the transfer functions of the three second-order sections. From the cascade coefficient cluster, the three transfer functions are

$$H_1[z] = \frac{0.545337 - 0.735242z^{-1} + 0.545337z^{-2}}{1 - 0.955505z^{-1} + 0.834882z^{-2}}$$

$$H_2[z] = \frac{0.545337 + 0.734702z^{-1} + 0.545337z^{-2}}{1 + 0.954255z^{-1} + 0.834810z^{-2}}$$

$$H_3[z] = \frac{0.545337 - 0.545337z^{-2}}{1 - 0.000622z^{-1} + 0.372609z^{-2}}$$

L4.2.2 Filtering System

Using the FIR Filtering System VI created in the previous section, the filter portion is replaced with the IIR bandpass filter just designed, see Figure 4-14. The VI is then saved as *IIR Filtering System.vi*.

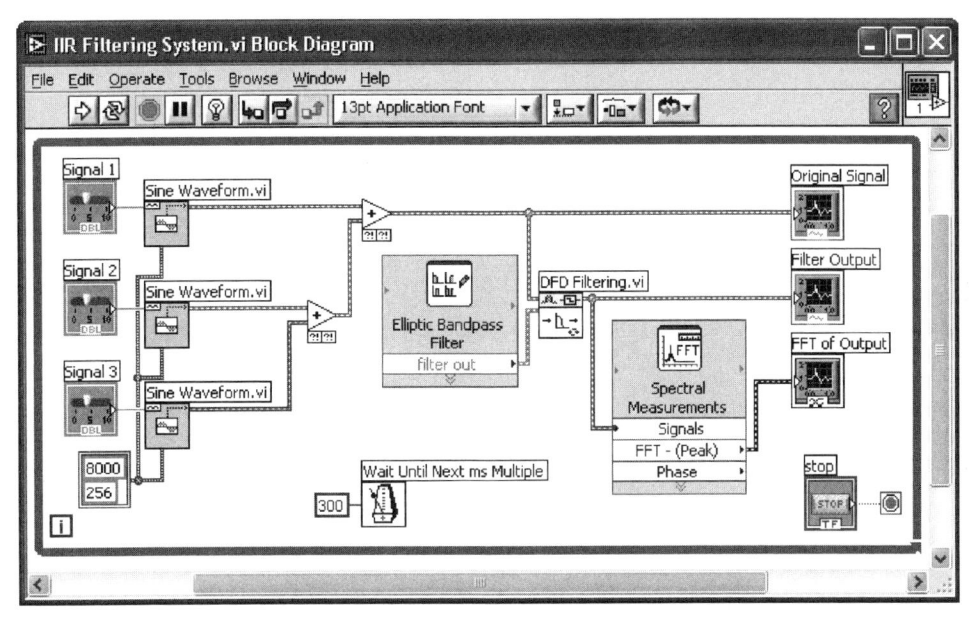

Figure 4-14: BD of IIR filtering system.

Let us change the default values of the three frequency controls on the FP to 1000 Hz, 2000 Hz, and 3000 Hz to see if the IIR filter is functioning properly. The signals having the frequencies 1000 Hz and 3000 Hz should be filtered out while only the signal having the frequency 2000 Hz should remain and be seen in the output. The output waveform as seen on the FP is shown in Figure 4-15. From the FFT of the output, it can be seen that the desired stopband attenuation of 20 dB is obtained.

Lab 4

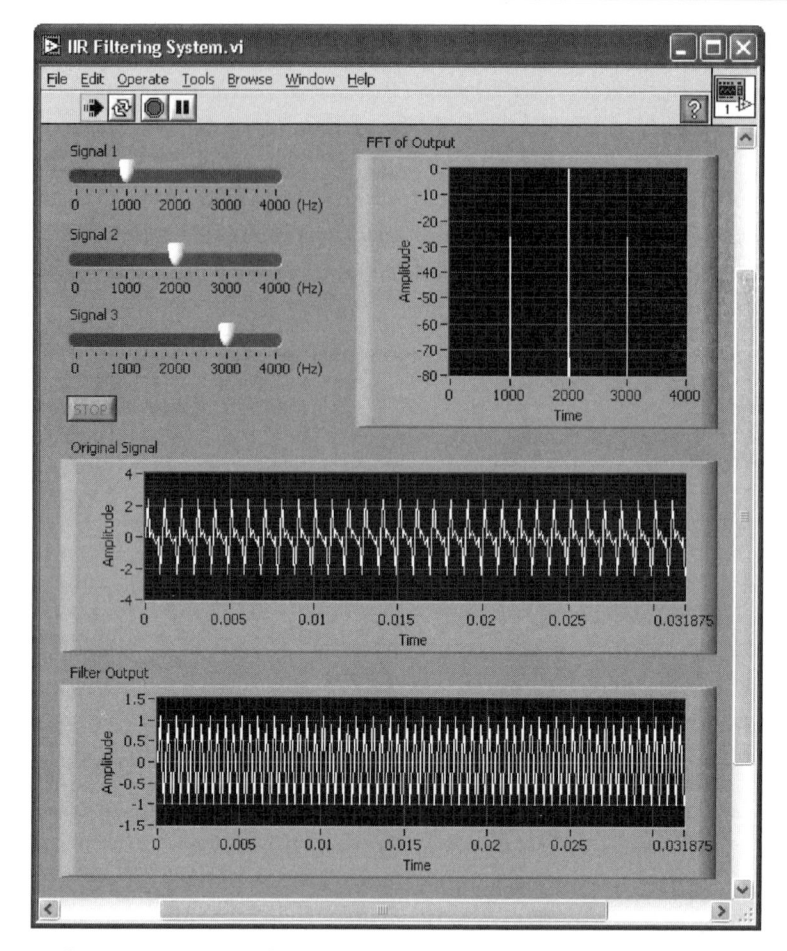

Figure 4-15: FP of IIR filtering system during run time.

L4.3 Building a Filtering System Using Filter Coefficients

There are various tools which one can use to compute coefficient sets of digital filters based on their specifications. In this section, the creation of a filter object is discussed when using different tools for obtaining its coefficients.

Figure 4-16 illustrates two ways to build a filter object using arrays of numeric constants containing filter coefficients. The DFD Build Filter from TF VI (**Functions → All Functions → Digital Filter Design → Utilities → DFD Build Filter from TF**) can be used to build a filter object if the direct-form coefficients of the filter are available, see Figure 4-16(a). For an IIR filter in the second-order cascade form, the DFD Build Filter from Cascaded Coef VI (**Functions → All Functions → Digital Filter Design → Utilities → DFD Build Filter from Cascaded Coef**) can be used

FIR/IIR Filtering System Design

to build a filter object, see Figure 4-16(b). The input cluster to this VI consists of a numeric constant for the filter structure and two arrays of numeric constants, labeled as Reverse Coefficients and Forward Coefficients. Each filter section consists of two reverse coefficients in the denominator, and three forward coefficients in the numerator, considering that the first coefficient of the denominator is regarded as 1.

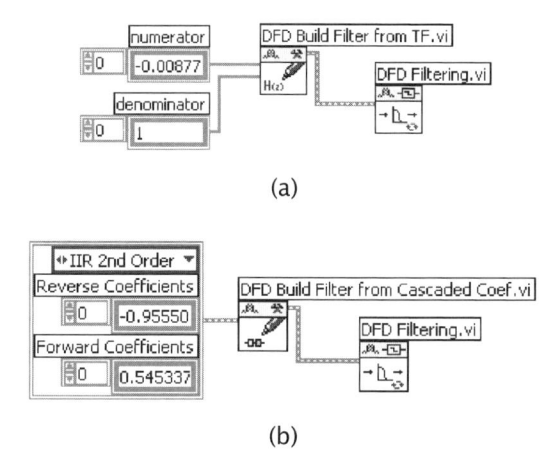

(a)

(b)

Figure 4-16: DFD Build Filter: (a) using direct-form coefficients, and (b) using cascade-form coefficients.

L4.4 Filter Design Without Using DFD Toolkit

The examples explained in the preceding sections can be implemented without using the DFD toolkit. This can be achieved by using the Digital FIR Filter VI (**Functions → All Functions → Analyze → Waveforms Conditioning → Digital FIR Filter**) and Digital IIR Filter VI (**Functions → All Functions → Analyze → Waveforms Conditioning → Digital IIR Filter**).

Similar to the Classic Filter Design Express VI of the DFD toolkit, the Digital FIR Filter VI is configured based on the filter specifications, thus there is no need to obtain the filter coefficients before building the filtering system. As a result, the specifications can be adjusted on the fly. The BD corresponding to this approach is shown in Figure 4-17.

For the Digital FIR Filter VI, the filter specifications are defined via two inputs in the form of a cluster constant. A cluster constant is created by right-clicking on the FIR filter specifications terminal and choosing **Create → Constant**. This cluster specifies the filter type, number of taps, and lower/upper

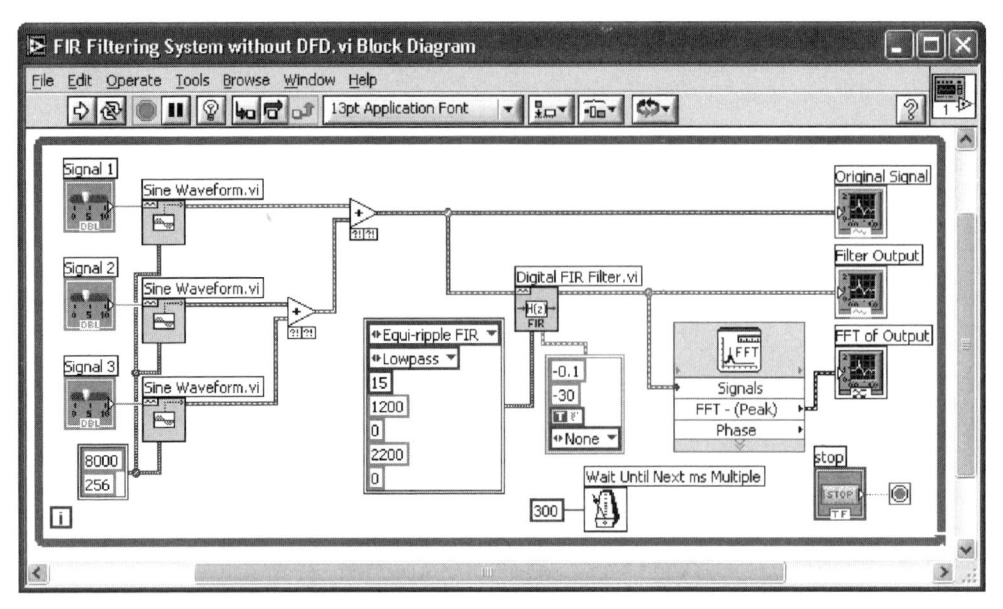

Figure 4-17: BD of FIR filtering system without using DFD VI.

passband or stopband. Another cluster constant specifying the passband gain, stopband gain, and window type can be wired to the Optional FIR filter specifications terminal. More details on the use of cluster constants as related to the Digital FIR Filter VI can be found in [1].

Rename the FP objects and set the maximum and default values of the controls. Save the VI as *FIR Filtering System without DFD.vi* and then run it. The FP of the VI during run time is shown in Figure 4-18. Observe that the 750 Hz signal falling in the passband remains while the 2500 Hz and 3000 Hz signals falling in the stopband, that is, greater than 2200 Hz, are attenuated by 30 dB.

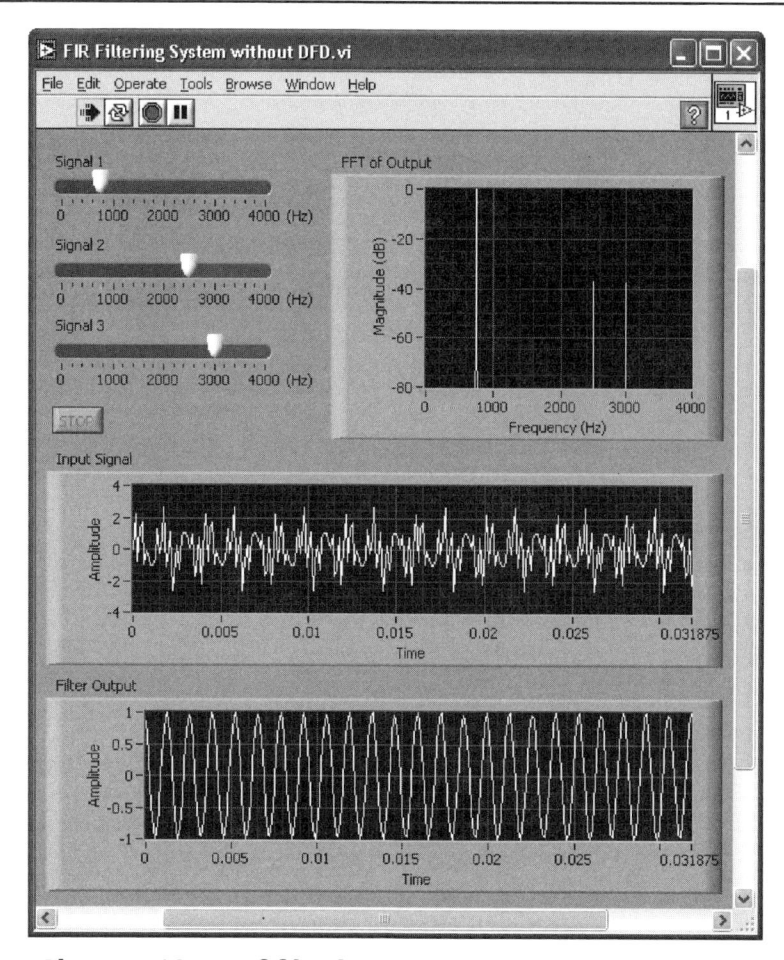

Figure 4-18: FP of filtering system using specifications.

L4.5 Bibliography

[1] National Instruments, *Signal Processing Toolset User Manual*, Part Number 322142C-01, 2002.

CHAPTER 5

Fixed-Point versus Floating-Point

From an arithmetic point of view, there are two ways a DSP system can be implemented in LabVIEW to match its hardware implementation on a processor. These are fixed-point and floating-point implementations. In this chapter, we discuss the issues related to these two hardware implementations.

In a fixed-point processor, numbers are represented and manipulated in integer format. In a floating-point processor, in addition to integer arithmetic, floating-point arithmetic can be handled. This means that numbers are represented by the combination of a mantissa (or a fractional part) and an exponent part, and the processor possesses the necessary hardware for manipulating both of these parts. As a result, in general, floating-point processors are slower than fixed-point ones.

In a fixed-point processor, one needs to be concerned with the dynamic range of numbers, since a much narrower range of numbers can be represented in integer format as compared to floating-point format. For most applications, such a concern can be virtually ignored when using a floating-point processor. Consequently, fixed-point processors usually demand more coding effort than do their floating-point counterparts.

5.1 Q-format Number Representation

The decimal value of an N-bit 2's-complement number, $B = b_{N-1}b_{N-2}...b_1b_0, b_i \in \{0,1\}$, is given by

$$D(B) = -b_{N-1}2^{N-1} + b_{N-2}2^{N-2} + ... + b_1 2^1 + b_0 2^0 \qquad (5.1)$$

The 2's-complement representation allows a processor to perform integer addition and subtraction by using the same hardware. When using the unsigned integer representation, the sign bit is treated as an extra bit. This way only positive numbers can be represented.

Chapter 5

There is a limitation of the dynamic range of the foregoing integer representation scheme. For example, in a 16-bit system, it is not possible to represent numbers larger than $2^{15} - 1 = 32767$ and smaller than $-2^{15} = -32768$. To cope with this limitation, numbers are often normalized between -1 and 1. In other words, they are represented as fractions. This normalization is achieved by the programmer moving the implied or imaginary binary point (note that there is no physical memory allocated to this point) as indicated in Figure 5-1. This way, the fractional value is given by

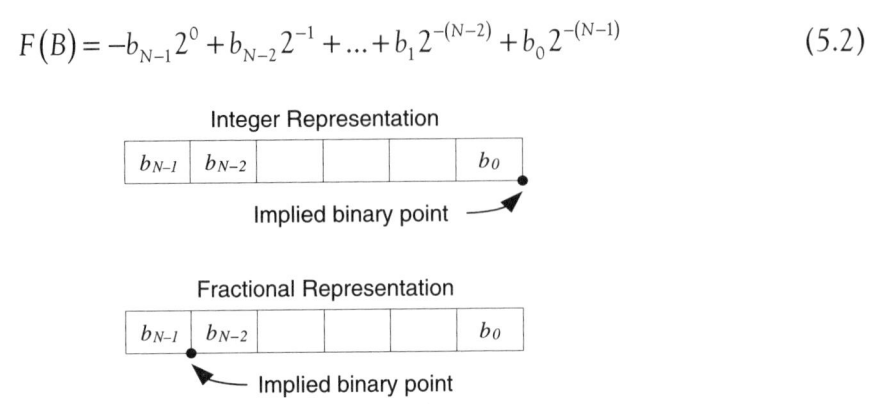

$$F(B) = -b_{N-1}2^0 + b_{N-2}2^{-1} + \ldots + b_1 2^{-(N-2)} + b_0 2^{-(N-1)} \tag{5.2}$$

Figure 5-1: Number representations.

This representation scheme is referred to as the Q-format or fractional representation. The programmer needs to keep track of the implied binary point when manipulating Q-format numbers. For instance, let us consider two Q15 format numbers. Each number consists of 1 sign bit plus 15 fractional bits. When these numbers are multiplied, a Q30 format number is generated (the product of two fractions is still a fraction), with bit 31 being the sign bit and bit 32 another sign bit (called an extended sign bit). Assuming a 16-bit wide memory, not enough bits are available to store all 32 bits, and only 16 bits can be stored. It makes sense to store the sixteen most significant bits. This requires storing the upper portion of the 32-bit product by doing a 15-bit right shift. In this manner, the product would be stored in Q15 format. (See Figure 5-2.)

Based on the 2's-complement representation, a dynamic range of $-2^{N-1} \leq D(B) \leq 2^{N-1} - 1$ can be covered, where N denotes the number of bits. For illustration purposes, let us consider a 4-bit system where the most negative number is -8 and the most positive number 7. The decimal representations of the numbers are shown in Figure 5-3. Notice how the numbers change from most positive to most negative with the sign bit. Since only the integer numbers falling within

Fixed-Point versus Floating-Point

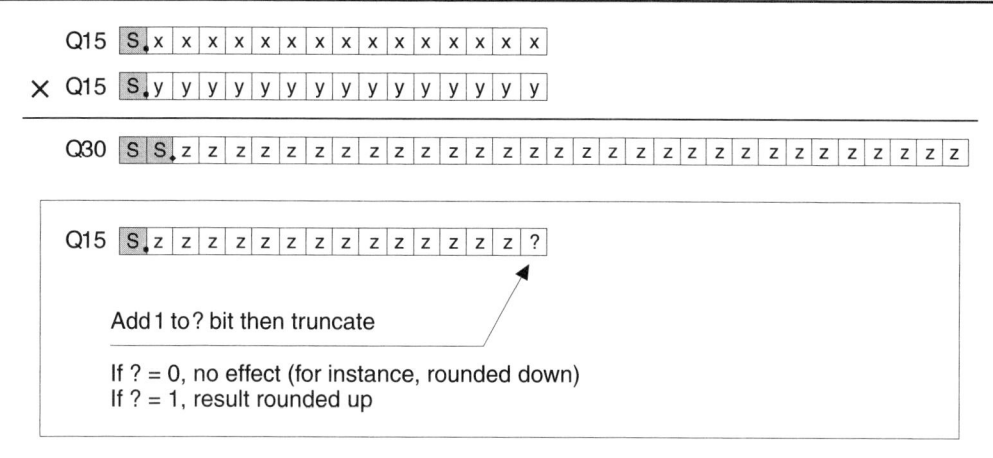

```
Q15  S x x x x x x x x x x x x x x x
×Q15  S y y y y y y y y y y y y y y y
─────────────────────────────────────
Q30  S S z z z z z z z z z z z z z z z z z z z z z z z z z z z z z z

Q15  S z z z z z z z z z z z z z z z ?
```

Add 1 to ? bit then truncate

If ? = 0, no effect (for instance, rounded down)
If ? = 1, result rounded up

Figure 5-2: Multiplying and storing Q15 numbers.

the limits −8 and 7 can be represented, it is easy to see that any multiplication or addition resulting in a number larger than 7 or smaller than −8 will cause overflow. For example, when 6 is multiplied by 2, we get 12. Hence, the result is greater than the representation limits and will be wrapped around the circle to 1100, which is −4.

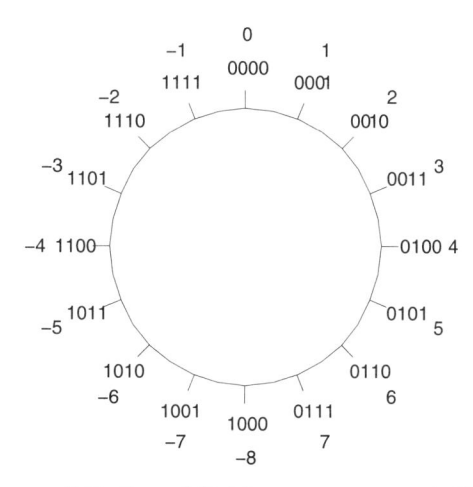

Figure 5-3: Four-bit binary representation.

The Q-format representation solves this problem by normalizing the dynamic range between −1 and 1. This way, any resulting multiplication will be within these limits. Using the Q-format representation, the dynamic range is divided into 2^N sections, where $2^{-(N-1)}$ is the size of a section. The most negative number is always −1 and the most positive number is $1 - 2^{-(N-1)}$.

Chapter 5

The following example helps one to see the difference in the two representation schemes. As shown in Figure 5-4, the multiplication of 0110 by 1110 in binary is equivalent to multiplying 6 by –2 in decimal, giving an outcome of –12, a number exceeding the dynamic range of the 4-bit system. Based on the Q3 representation, these numbers correspond to 0.75 and –0.25, respectively. The result is –0.1875, which falls within the fractional range. Notice that the hardware generates the same 1's and 0's; what is different is the interpretation of the bits.

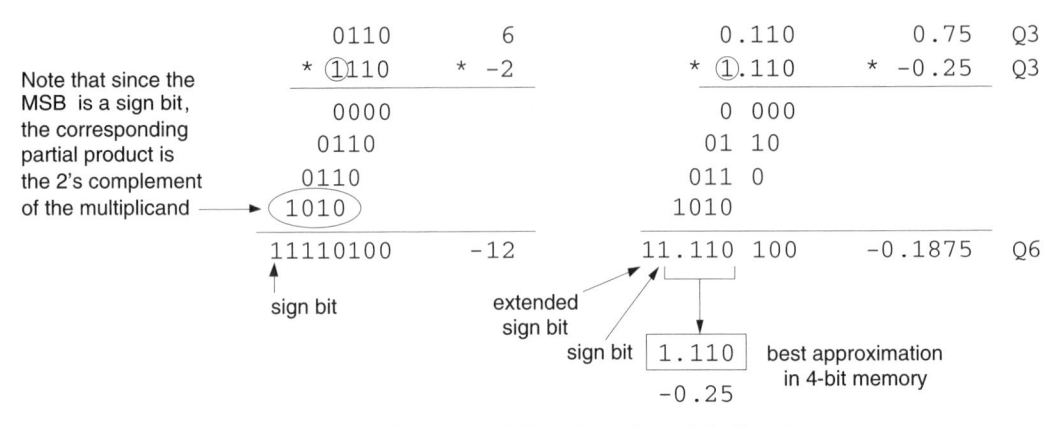

```
                              0110              6                0.110           0.75    Q3
                           *  ①110           *  -2             *  ①.110       * -0.25    Q3
Note that since the           ————              ——                ————            ————
MSB is a sign bit,            0000                                0 000
the corresponding             0110                                01 10
partial product is            0110                                011 0
the 2's complement            1010                                1010
of the multiplicand ─────►  ( 1010 )                              ————
                           ————————            ———             ————————        ————————
                           11110100             -12            11.110 100      -0.1875   Q6
                            ↑                                                
                           sign bit          extended                         
                                             sign bit                          
                                                         sign bit  1.110   best approximation
                                                                  -0.25    in 4-bit memory
```

Figure 5-4: Binary and fractional multiplication.

When multiplying Q-N numbers, it should be remembered that the result will consist of 2N fractional bits, one sign bit, and one or more extended sign bits. Based on the data type used, the result has to be shifted accordingly. If two Q15 numbers are multiplied, the result will be 32 bits wide, with the most significant bit being the extended sign bit followed by the sign bit. The imaginary decimal point will be after the 30th bit. So a right shift of 15 is required to store the result in a 16-bit memory location as a Q15 number. It should be realized that some precision is lost, of course, as a result of discarding the smaller fractional bits. Since only 16 bits can be stored, the shifting allows one to retain the higher precision fractional bits. If a 32-bit storage capability is available, a left shift of 1 can be done to remove the extended sign bit and store the result as a Q31 number.

To further understand a possible precision loss when manipulating Q-format numbers, let us consider another example where two Q12 numbers corresponding to 7.5 and 7.25 are multiplied. As can be seen from Figure 5-5, the resulting product must be left shifted by 4 bits to store all the fractional bits corresponding to Q12 format.

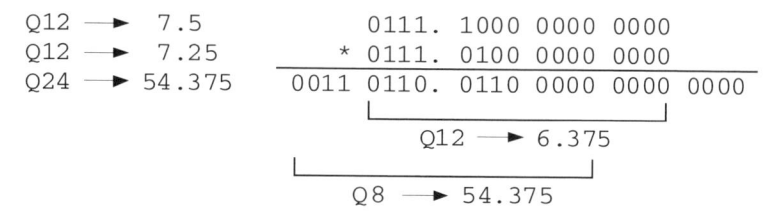

```
Q12  →   7.5              0111. 1000 0000 0000
Q12  →   7.25           * 0111. 0100 0000 0000
Q24  →  54.375        0011 0110. 0110 0000 0000 0000
                           |_____|
                              Q12  →  6.375
                      |_____|
                           Q8  →  54.375
```

Figure 5-5: Q-format precision loss example.

However, doing so results in a product value of 6.375, which is different from the correct value of 54.375. If the product is stored in a lower precision Q-format—say, in Q8 format—then the correct product value can be obtained and stored.

Although Q-format solves the problem of overflow during multiplications, addition and subtraction still pose a problem. When adding two Q15 numbers, the sum could exceed the range of the Q15 representation. To solve this problem, the scaling approach, discussed later in this chapter, needs to be employed.

5.2 Finite Word Length Effects

Due to the fact that the memory or registers of a processor have a finite number of bits, there could be a noticeable error between desired and actual outcomes on a fixed-point processor. The so-called finite word length quantization effect is similar to the input data quantization effect introduced by an A/D converter.

Consider fractional numbers quantized by a $b + 1$ bit converter. When these numbers are manipulated and stored in a $M + 1$ bit memory, with $M < b$, there is going to be an error (simply because $b - M$ of the least significant fractional bits are discarded or truncated). This finite word length error could alter the behavior of a DSP system by an unacceptable degree. The range of the magnitude of truncation error ε_t is given by $0 \leq |\varepsilon_t| \leq 2^{-M} - 2^{-b}$. The lowest level of truncation error corresponds to the situation when all the thrown-away bits are zeros, and the highest level to the situation when all the thrown-away bits are ones.

This effect has been extensively studied for FIR and IIR filters, for example see [1]. Since the coefficients of such filters are represented by a finite number of bits, the roots of their transfer function polynomials, or the positions of their zeros and poles, shift in the complex plane. The amount of shift in the positions of poles and zeros can be related to the amount of quantization errors in the coefficients. For example,

Chapter 5

for an Nth-order IIR filter, the sensitivity of the ith pole p_i with respect to the kth coefficient a_k can be derived to be (see [1]),

$$\frac{\partial p_i}{\partial a_k} = \frac{-p_i^{N-k}}{\prod_{\substack{l=1 \\ l \neq i}}^{N}(p_i - p_l)} \qquad (5.3)$$

This means that a change in the position of a pole is influenced by the positions of all the other poles. That is the reason the implementation of an Nth order IIR filter is normally achieved by having a number of second-order IIR filters in cascade or series in order to decouple this dependency of poles.

Also, note that as a result of coefficient quantization, the actual frequency response $\hat{H}(e^{j\theta})$ is different than the desired frequency response $H(e^{j\theta})$. For example, for an FIR filter having N coefficients, it can be easily shown that the amount of error in the magnitude of the frequency response, $|\Delta H(e^{j\theta})|$, is bounded by

$$|\Delta H(e^{j\theta})| = |H(e^{j\theta}) - \hat{H}(e^{j\theta})| \leq N2^{-b} \qquad (5.4)$$

In addition to the above effects, coefficient quantization can lead to limit cycles. This means that in the absence of an input, the response of a supposedly stable system (poles inside the unit circle) to a unit sample is oscillatory instead of diminishing in magnitude.

5.3 Floating-Point Number Representation

Due to relatively limited dynamic ranges of fixed-point processors, when using such processors, one should be concerned with the scaling issue, or how big the numbers get in the manipulation of a signal. Scaling is not of concern when using floating-point processors, since the floating-point hardware provides a much wider dynamic range.

As an example, let us consider the C67x processor, which is the floating-point version of the TI family of TMS320C6000 DSP processors. There are two floating-point data representations on the C67x processor: single precision (SP) and double precision (DP). In the single precision format, a value is expressed as (see [2])

$$-1^s \times 2^{(exp-127)} \times 1.frac \qquad (5.5)$$

where s denotes the sign bit (bit 31), *exp* the exponent bits (bits 23 through 30), and *frac* the fractional or mantissa bits (bits 0 through 22). (See Figure 5-6.)

Fixed-Point versus Floating-Point

```
 31 30                23 22                           0
  s      exp                      frac
```

Figure 5-6: C67x floating-point data representation.

Consequently, numbers as big as 3.4×10^{38} and as small as 1.175×10^{-38} can be processed. In the double precision format, more fractional and exponent bits are used as indicated below

$$-1^s \times 2^{(exp-1023)} \times 1.frac \tag{5.6}$$

where the exponent bits are from bits 20 through 30 and the fractional bits are all the bits of a first word and bits 0 through 19 of a second word. (See Figure 5-7.) In this manner, numbers as big as 1.7×10^{308} and as small as 2.2×10^{-308} can be handled.

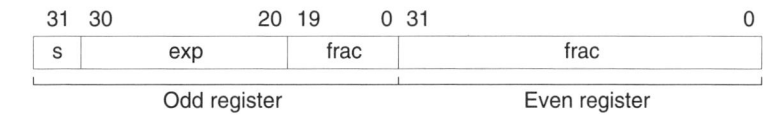

```
 31 30           20 19       0 31                     0
  s     exp          frac              frac
       Odd register                Even register
```

Figure 5-7: C67x double precision floating-point representation.

When using a floating-point processor, all the steps needed to perform floating-point arithmetic are done by the floating-point hardware. For example, consider adding two floating-point numbers represented by

$$a = a_{frac} \times 2^{a_{exp}}$$
$$b = b_{frac} \times 2^{b_{exp}} \tag{5.7}$$

The floating-point sum c has the following exponent and fractional parts:

$$\begin{aligned} c &= a + b \\ &= \left(a_{frac} + \left(b_{frac} \times 2^{-(a_{exp}-b_{exp})}\right)\right) \times 2^{a_{exp}} \quad \text{if } a_{exp} \geq b_{exp} \\ &= \left(\left(a_{frac} \times 2^{-(b_{exp}-a_{exp})}\right) + b_{frac}\right) \times 2^{b_{exp}} \quad \text{if } a_{exp} < b_{exp} \end{aligned} \tag{5.8}$$

These parts are computed by the floating-point hardware. This shows that, though possible, it is inefficient to perform floating-point arithmetic on fixed-point processors, since all the operations involved, such as those in Equation (5.8), must be done in software.

Chapter 5

5.4 Overflow and Scaling

As stated before, fixed-point processors have a much smaller dynamic range than their floating-point counterparts. It is due to this limitation that the Q15 representation of numbers is normally considered. For instance, a 16-bit multiplier can be used to multiply two Q15 numbers and produce a 32-bit product. Then the product can be stored in 32 bits or shifted back to 16 bits for storage or further processing.

When multiplying two Q15 numbers, which are in the range of −1 and 1, as discussed earlier, the product will be in the same range. However, when two Q15 numbers are added, the sum may fall outside this range, leading to an overflow. Overflows can cause major problems by generating erroneous results. When using a fixed-point processor, the range of numbers must be closely examined and if necessary adjusted to compensate for overflows. The simplest correction method for overflow is scaling.

The idea of scaling is to scale down the system input before performing any processing and then to scale up the resulting output to the original size. Scaling can be applied to most filtering and transform operations. An easy way to achieve scaling is by shifting. Since a right shift of 1 is equivalent to a division by 2, we can scale the input repeatedly by 0.5 until all overflows disappear. The output can then be rescaled back to the total scaling amount.

As far as FIR and IIR filters are concerned, it is possible to scale coefficients to avoid overflows. Let us consider the output of an FIR filter $y[n] = \sum_{k=0}^{N-1} h[k]x[n-k]$, where h's denote coefficients or unit sample response terms and x's input samples. In case of IIR filters, for a large enough N, the terms of the unit sample response become so small that they can be ignored. Let us suppose that x's are in Q15 format (i.e., $|x[n-k]| \leq 1$). Therefore, we can write $|y[n]| \leq \sum_{k=0}^{N-1} |h[k]|$. This means that, to ensure no output overflow (i.e., $|y[n]| \leq 1$), the condition $\sum_{k=0}^{N-1} |h[k]| \leq 1$ must be satisfied. This condition can be satisfied by repeatedly scaling (dividing by 2) the coefficients or unit sample response terms.

5.5 Data Types in LabVIEW

The numeric data types in LabVIEW together with their symbols and ranges are listed in Table 5-1.

Fixed-Point versus Floating-Point

Table 5-1: Numeric data types in LabVIEW [4].

Terminal symbol	Numeric data type	Bits of storage on disk
SGL	Single-precision, floating-point	32
DBL	Double-precision, floating-point	64
EXT	Extended-precision, floating-point	128
CSG	Complex single-precision, floating-point	64
CDB	Complex double-precision, floating-point	128
CXT	Complex extended-precision, floating-point	256
I8	Byte signed integer	8
I16	Word signed integer	16
I32	Long signed integer	32
U8	Byte unsigned integer	8
U16	Word unsigned integer	16
U32	Long unsigned integer	32
∑	128-bit time stamp	<64.64>

Note that, other than the numeric data types shown in Table 5-1, there exist other data types in LabVIEW, such as cluster, waveform, and dynamic data type, see Table 5-2. For more details on all the LabVIEW data types, refer to [3,4].

Table 5-2: Other data types in LabVIEW [4].

Terminal symbol	Data type
<>	Enumerated type
TF	Boolean
abc	String
[ ]	Array—Encloses the data type of its elements in square brackets and takes the color of that data type.
	Cluster—Encloses several data types. Cluster data types are brown if all elements in the cluster are numeric or pink if all the elements of the cluster are different types.
	Path
	Dynamic—(Express VIs) Includes data associated with a signal and the attributes that provide information about the signal, such as the name of the signal or the date and time the data was acquired.
	Waveform—Carries the data, start time, and dt of a waveform.
	Digital waveform—Carries start time, delta x, the digital data, and any attributes of a digital waveform.

Terminal symbol	Data type
0101	Digital—Encloses data associated with digital signals.
D	Reference number (refnum)
	Variant—Includes the control or indicator name, the data type information, and the data itself.
I/O	I/O name—Passes resources you configure to I/O VIs to communicate with an instrument or a measurement device.
	Picture—Includes a set of drawing instructions for displaying pictures that can contain lines, circles, text, and other types of graphic shapes.

5.6 Bibliography

[1] J. Proakis and D. Manolakis, *Digital Signal Processing: Principles, Algorithms, and Applications*, Prentice-Hall, 1996.

[2] Texas Instruments, *TMS320C6000 CPU and Instruction Set Reference Guide*, Literature ID# SPRU189F, 2000.

[3] National Instruments, *LabVIEW Data Storage*, Application Note 154, Part Number 342012C-01, 2004.

[4] National Instruments, *LabVIEW User Manual*, Part Number 320999E-01, 2003.

Lab 5: Data Type and Scaling

Fixed-point implementation of a DSP system requires one to examine permissible ranges of numbers so that necessary adjustments are made to avoid overflows. The most widely used approach to cope with overflows is scaling. The scaling approach is covered in this lab.

L5.1 Handling Data types in LabVIEW

In LabVIEW, the data type of data exchanged between two blocks is exhibited by the color of their connecting wires as well as their icons. A mismatched data type is represented by a coercion dot on a function or subVI input terminal, alerting that the input data type is being coerced into a different type. In general, a lower precision data value gets converted to a higher precision value. Coercion dots can lead to an increase in memory usage and run time [1]. Thus, it is recommended to resolve coercion dots in a VI.

An example exhibiting a mismatched data type is depicted in Figure 5-8. A double precision value and a 16-bit integer are wired to the input terminals of an Add function. As can be seen from this figure, a coercion dot appears at the y terminal of the Add function since the input to this terminal is of 16-bit integer type while the other input is of double precision type.

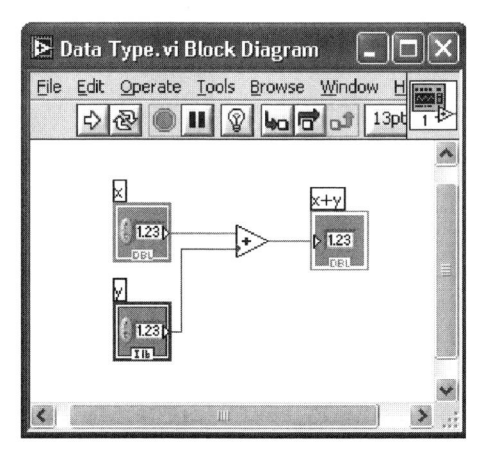

Figure 5-8: Data type mismatch.

Lab 5

Let us build the VI shown in Figure 5-8. Place an Add function and create two input controls by right-clicking and choosing **Create** → **Control** from the shortcut menu at each input terminal. By default, the data types of the two controls are set to double precision. In order to change the data type of the second Numeric Control, labeled y, right-click on the icon on the BD and select **Representation** → **Word**, which is represented by **I16**.

Create a Numeric Indicator by right-clicking on the x+y terminal of the Add function and choosing **Create** → **Indicator** from the shortcut menu. The data type of the newly created Numerical Indicator is double precision since the addition of two double precision values results in another double precision value.

Let us switch to the FP of the VI to demonstrate the importance of specifying the correct data type to the Numeric Control/Indicator. If the value entered on the FP control does not match the data type specified by the Numeric Control/Indicator, the input value is automatically converted into the data type specified by the Numeric Control. In the example shown in Figure 5-9, the value 1.5 is entered in both of the Numeric Controls on the FP. As can be seen, the entered value of the second Numeric Control, labeled y, automatically gets converted to a 16-bit integer or 2.

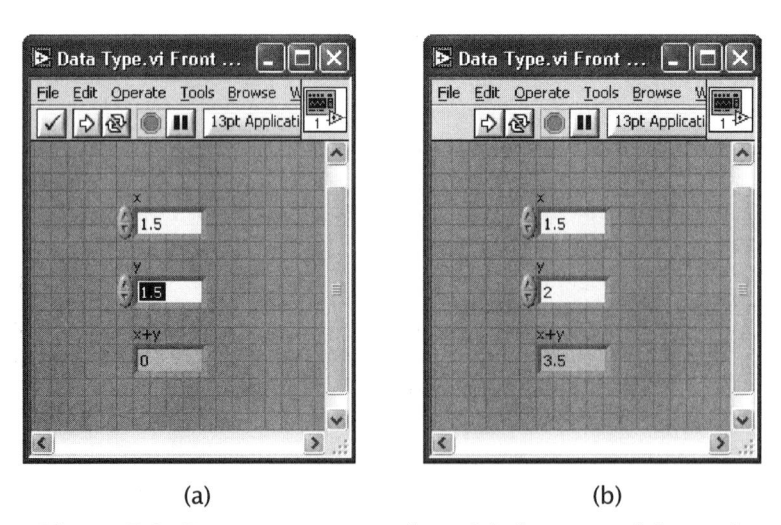

(a)   (b)

Figure 5-9: Data type conversion: (a) data typed in, and (b) data is converted to 16-bit integer by LabVIEW.

Data Type and Scaling

Coercion dots can be avoided if appropriate conversion functions are inserted in the BD. For example, as shown in Figure 5-10, the addition of a double precision and a 16-bit integer value is achieved without getting a coercion dot by inserting a To Double Precision Float function.

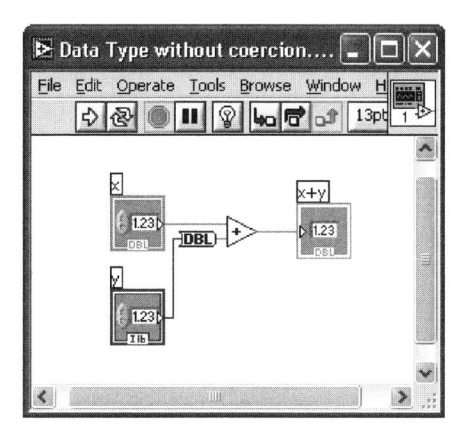

Figure 5-10: Data type conversion.

L5.2 Overflow Handling

An overflow occurs when the outcome of an operation is too large or too small for a processor to handle. In a 16-bit system, when manipulating integer numbers, they must remain in the range −32768 to 32767. Otherwise, any operation resulting in a number smaller than −32768 or larger than 32767 will cause overflow. For example, when 32767 is multiplied by 2, we get 65534, which is beyond the representation limit of a 16-bit system.

Consider the BD shown in Figure 5-11. In this BD, samples of a sinusoidal signal having an amplitude of 30000 are multiplied by 2. To illustrate the overflow problem, the input values generated by the Sine Waveform VI are converted to word signed integers or 16-bit integers (I16). This is done by inserting a To Word Integer function (**Functions → All Functions → Numeric → Conversion → To Word Integer**) at the output of the Get Waveform Component function. After the insertion of this function, the color of the wire connected to the output terminal of the function should appear blue indicating integer data type.

In addition, the multiplicand constant should also be converted to the I16 data type to avoid a coercion dot. To achieve this, right-click on the Numeric Constant and select **Representation → Word**. As a result, the data type of the multiplication outcome or product is automatically set to word signed integer (I16).

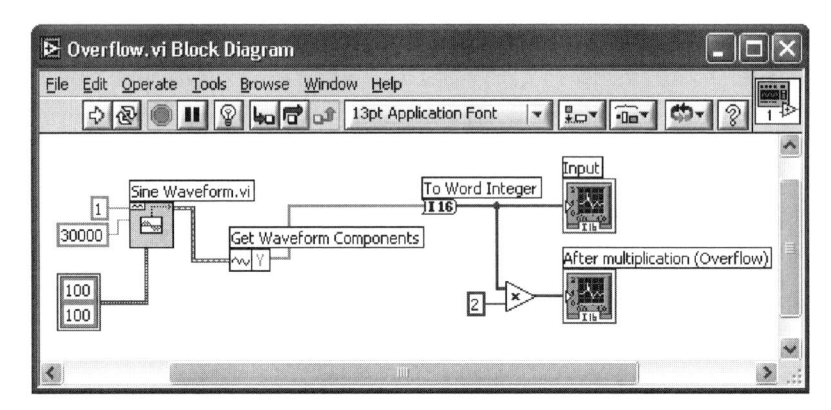

Figure 5-11: Signal multiplication data type conversion.

It should be noted that the definition of data types is software/hardware dependent. For example, the "word" data type in LabVIEW has a length of 16 bits, while it is 32 bits in the C6x DSP. That is, the "word" data type in LabVIEW is equivalent to the "short" data type in the C6x DSP [2].

The multiplication of a sinusoidal signal by 2 is expected to generate another sinusoidal signal with twice the amplitude. However, as seen from the FP in Figure 5-12, the result is distorted and clipped when the product is beyond the word integer (I16) range. Let us examine whether any overflow is caused by these multiplications. From Figure 5-12, it can be seen that the output signal includes wrong values due to overflows. For example, –5536 is shown to be the result of the multiplication of 30000 (the maximum input value) by 2, which is incorrect.

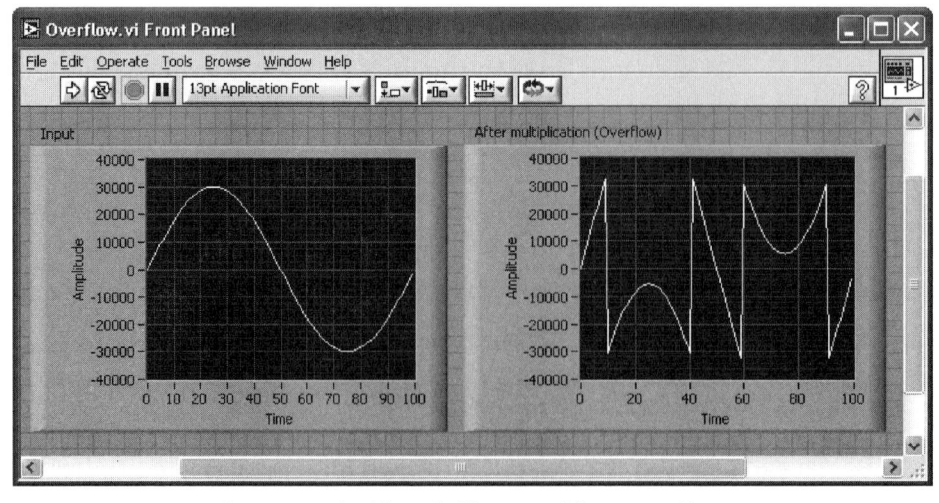

Figure 5-12: Signal distorted by overflow.

Data Type and Scaling

L5.2.1 Q-Format Conversion

Let us now consider the conversion of single or double precision values to Q-format. As shown in Figure 5-13, a double precision input value is first checked to see whether it is in the range of –1 and 1. This is done by using the In Range and Coerce function (**Functions → All Functions → Comparison → In Range and Coerce**). The input is scaled so that it falls within the range of 16-bit signed integer data type, or –32768 to 32767, by multiplying it with its maximum allowable value, that is 32768. Then, the product is converted to 16-bit signed integer data type by using a To Word Integer function. This ensures that the product falls within the range of the specified data type. In the worst case, the product gets clipped or saturated to the maximum or minimum allowable value.

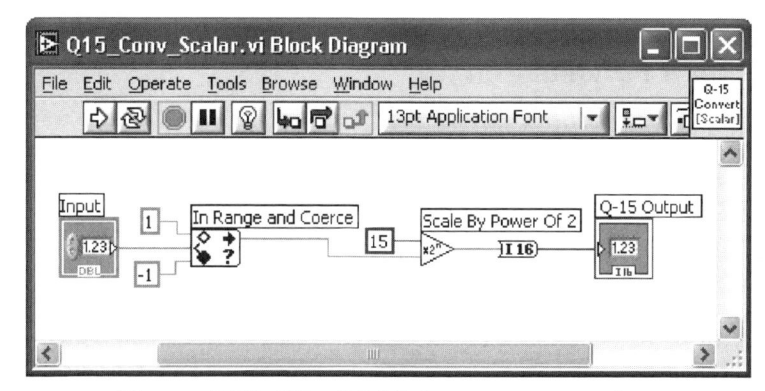

Figure 5-13: BD of Q15 format conversion.

Edit the icon of the VI as shown in Figure 5-13. The connector pane of the VI has one input and one output terminal. Assign the input terminal to the Numeric Control and the output terminal to the Numeric Indicator. Save the VI in a file named *Q15_Conv_Scalar.vi* and use it as a Q15 format converter for scalar inputs.

Next, the Q15_Conv_Scalar VI is modified to perform Q15 conversion for array type inputs/outputs as follows. As shown in Figure 5-14, replace the scalar numeric control and indicator with an array of controls and indicators, respectively. An array of controls or indicators can be created by first placing an Array shell (**Controls → All Controls → Array & Cluster → Array**) and then by dragging a control or indicator into it. The icon and the connector pane of the modified VI should be reconfigured accordingly. Save the modified VI in a file named *Q15_Conv_Array.vi*.

Lab 5

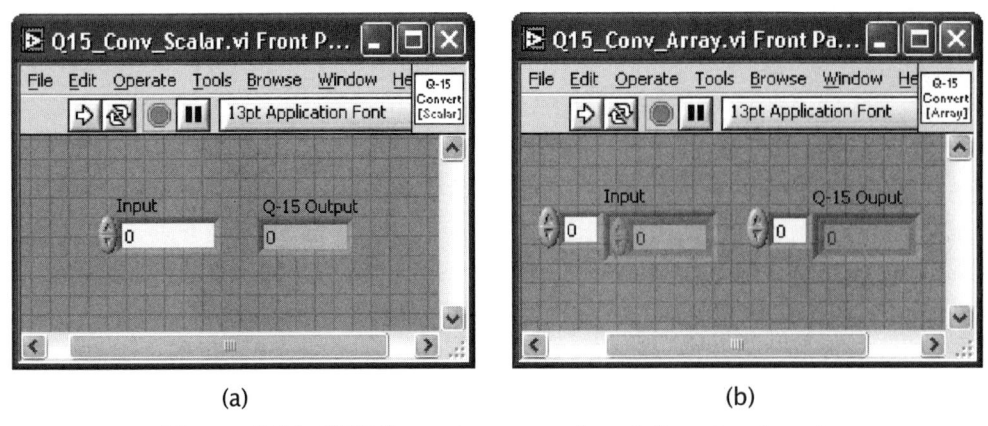

Figure 5-14: Q15 format conversion: (a) scalar input and output, and (b) array input and output.

L5.2.2 Creating a Polymorphic VI

The two VIs just created are integrated into a so called polymorphic VI so that one VI can handle both scalar and array inputs/outputs. A polymorphic VI is a collection of multiple VIs for different instances having the same input and output connector pane [3]. The multiplication function is a good example of polymorphism since it can be applied to two scalars, an array and a scalar, or two arrays.

To create a polymorphic VI, select **File → New → Other Document Types → Polymorphic VI**. This brings up a Polymorphic VI window, see Figure 5-15. Add the `Q15_Conv_Scalar` and `Q15_Conv_Array` VIs to include both the scalar and array cases. Edit the icon of the polymorphic VI as shown in Figure 5-15, then save the VI as *Q15_Conv.vi*. This polymorphic VI is used for the remaining part of this lab.

Next, a VI is presented to show how the overflow is checked. In the BD shown in Figure 5-16(a), two input values in the range between −1 and 1 are converted to Q15 format by using the polymorphic `Q15_conv` VI. These inputs are converted into a higher precision data type, long integer (I32), to avoid getting any saturation during their addition. Consider that LabVIEW automatically limits the output of numerical operations to the input range. The sum of the two input values is wired into the `In Range and Coerce` function to check whether they are in the allowable range. If the output does not fall in the range of I16, this indicates that an overflow has occurred. The FP in Figure 5-16(b) illustrates such an overflow.

Data Type and Scaling

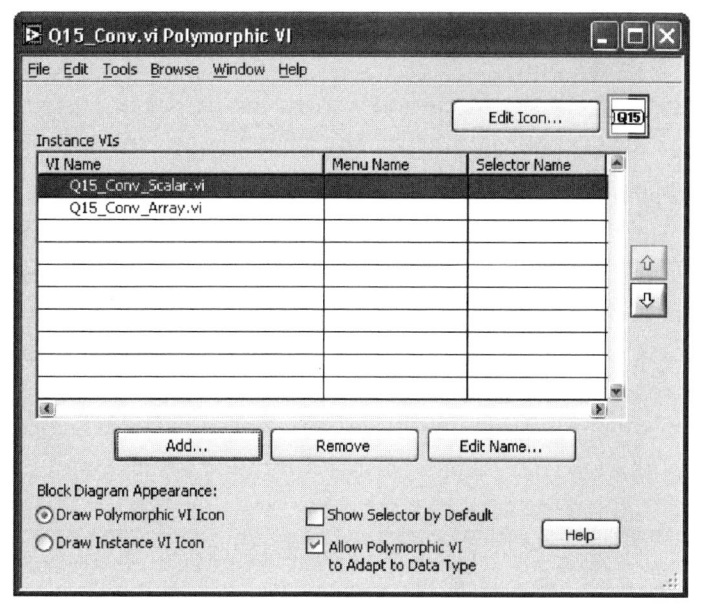

Figure 5-15: Creating a polymorphic VI.

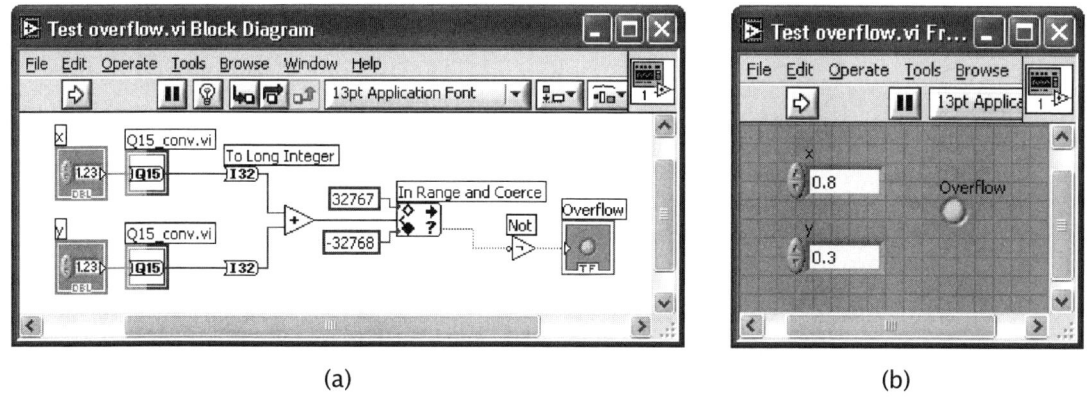

Figure 5-16: Test for overflow: (a) BD, and (b) FP.

L5.3 Scaling Approach

Scaling is the most widely used approach to overcome the overflow problem. In order to see how scaling works, let us consider a simple multiply/accumulate operation. Suppose there are four constants or coefficients that need to be multiplied with samples of an input signal. The worst possible overflow case would be the one where all the multiplicands C_k's and $x[n]$'s are 1's. For this case, the result $y[n]$ will be 4, given that $y[n] = \sum_{k=1}^{4} C_k x[n-k]$. Assuming that we have control over only the input

Lab 5

signal, the input samples should be scaled for the result or sum $y[n]$ to fall in the allowable range. A single half-scaling, or division by 2, reduces the input samples by one-half, and a double half-scaling reduces them further by one-quarter. Of course, this leads to less precision, but it is better than getting an erroneous outcome.

A simple method to implement the scaling approach is to create a VI that returns the necessary amount of scaling on the input. For multiply/accumulate type of operations, such as filtering or transforms, the worst case is multiplications and additions of all 1's. This means that the required number of scaling is dependent on the number of additions in the summation. To examine the worst case, one needs to obtain the required number of scaling so that all overflows disappear. This can be achieved by building a VI to compute the required number of scaling. For the example covered in this lab, such a VI is shown in Figure 5-17 and is named *Number of Scaling.vi*.

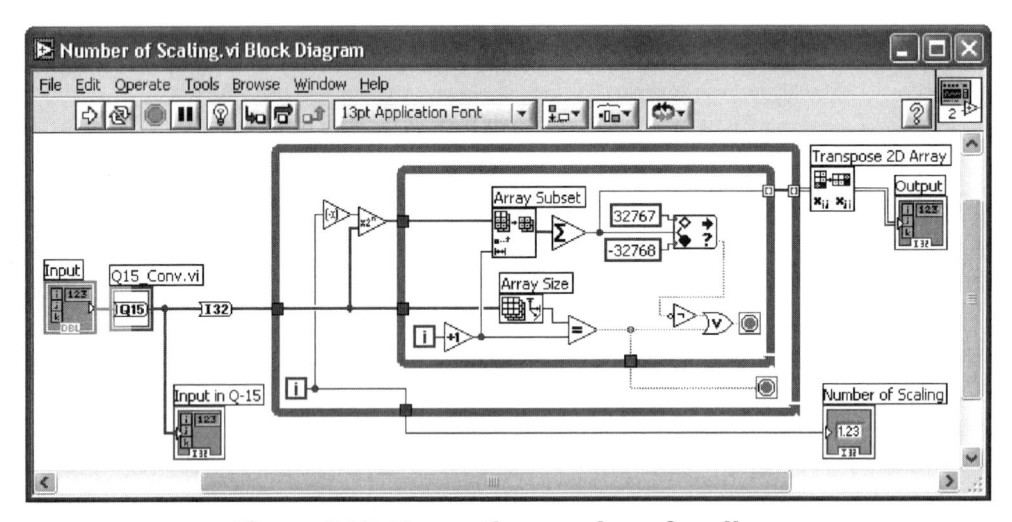

Figure 5-17: Computing number of scaling.

Here, the input is first converted into Q15 format via the Q15_Conv VI. Inside the outer loop, the input is scaled. In each iteration of the inner loop, a new input sample is taken into consideration and a summation is obtained. Then, the summation value is compared with the minimum and maximum values in the allowable range. If the summation value does not fall into this range, the inner loop is stopped and the input samples are scaled down for a next iteration. The number of scaling is also counted. After scaling the input, the summation is repeated. If another overflow occurs, the input sample is scaled down further. This process is continued until no overflow occurs. The final number of scaling is then displayed. Care must be taken not to scale

Data Type and Scaling

the input too many times; otherwise, the input signal gets buried in the quantization noise. Here, the auto-indexing is enabled to collect the output samples into an array corresponding to each `While Loop`. As a result, a 2D array is generated.

Figure 5-18 shows the FP for the computation of the number of scaling. The input signal consists of the samples of one period of a sinusoid with an amplitude of 0.8 sampled at 0.125 in terms of normalized frequency. The elements of the column shown in the output indicator represent the accumulated sums of the input samples. Notice that an overflow occurs at the third summation since the value is greater than the maximum value of a 16-bit signed integer, that is, 32767. The overflow disappears if the input is scaled down once by one-half. Thus, in this example, the required number of scaling to avoid any overflow is one.

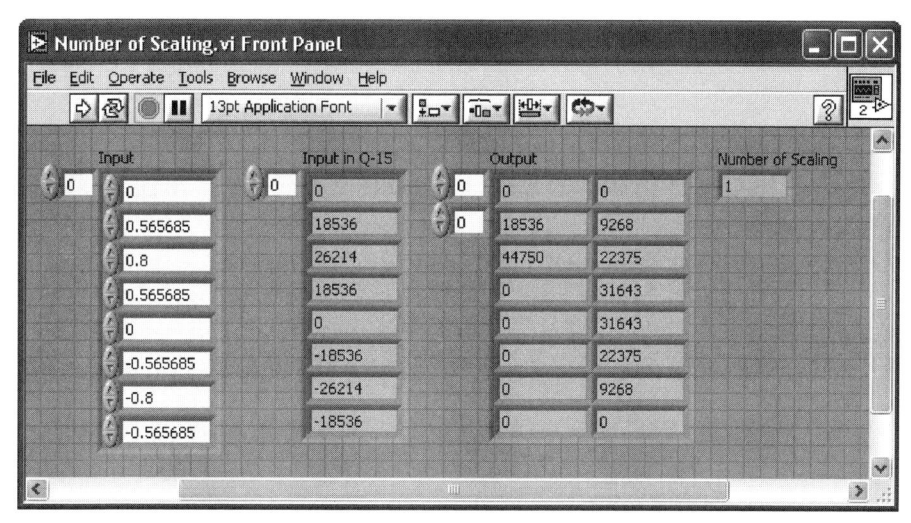

Figure 5-18: Number of scaling for one period of sinusoidal signal.

It is worth mentioning that, in addition to scaling the input, it is also possible to scale the filter coefficients or constants in convolution type of operations so that the outcome is forced to stay within the allowable range. In this case, the worst case for the input samples is assumed to be one. Note that scaling down the coefficients by one-half is equivalent to scaling down the input samples by one-half. An example of fixed-point digital filtering as well as coefficients scaling is examined in the following section.

L5.4 Digital Filtering in Fixed-Point Format

The analyses of overflow and scaling discussed above are repeated here by using the DFD Fixed-Point Tools VIs. These VIs allow the quantization of filter coefficients

Lab 5

and the fixed-point simulation of a digital filter. As an example of fixed-point digital filtering, the FIR lowpass filter designed in Lab 4 is used here.

L5.4.1 Design and Analysis of Fixed-Point Digital Filtering System

To design an FIR filter, place the `DFD Classical Filter Design` Express VI on the BD and enter the filter specifications on the configuration dialog box of this Express VI. To display the filter response, a total of four VIs as part of the DFD toolkit are used as shown in Figure 5-19.

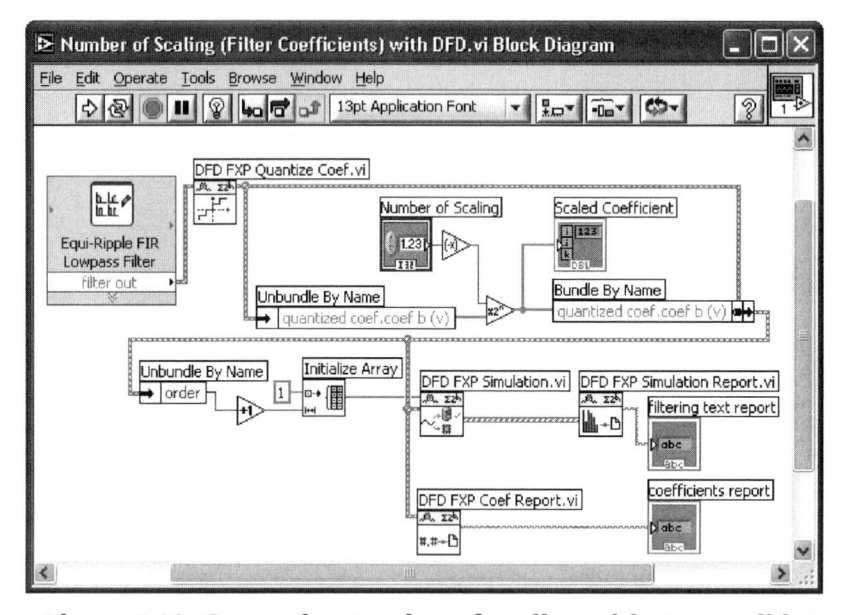

Figure 5-19: Computing number of scaling with DFD toolkit.

Let us explain each object of this BD. The `DFD FXP Quantize Coef` VI (**Functions → All Functions → Digital Filter Design → Fixed-Point Tools → DFD FXP Quantize Coef**) quantizes the filter coefficients according to the specified options. By default, a 16-bit word length is used for quantization. The b_k coefficients of the filter are unbundled by using the `Unbundle By Name` function. Right-click on the Unbundle By Name function and choose **Select Item → quantized coef → coef b(v)** from the shortcut menu. Next, the quantized coefficients are scaled down to match the number of scaling specified in the `Numeric Control`. The use of an array of indicators labeled as `Scaled Coefficient` is optional. This can be used to export the filter coefficients to other VIs easily. The original filter coefficients are replaced with the scaled coefficients by using the `Bundle by Name` function.

Data Type and Scaling

The DFD FXP Simulation VI (**Functions** → **All Functions** → **Digital Filter Design** → **Fixed-Point Tools** → **DFD FXP Simulation**) simulates the filter operation and generates its statistics using the fixed-point filter coefficient set. The filter statistics include min and max value, number of overflow/underflow, and number of operation. A text report of the filter statistics is generated via the DFD FXP Simulation Report VI (**Functions** → **All Functions** → **Digital Filter Design** → **Fixed-Point Tools** → **DFD FXP Simulation Report**) and displayed in the String Indicator. The DFD FXP Coef Report VI (**Functions** → **All Functions** → **Digital Filter Design** → **Fixed-Point Tools** → **DFD FXP Coef Report**) generates a text report on the quantized filter coefficients.

In order to consider the worst case scenario, an array of all ones are created and wired to act as the input of the filter. The length of this array is determined by the number of filter coefficients. The simulated result is shown in Figure 5-20. As seen in this figure, five overflows are reported at the output for the no scaling case. The sums of the filter coefficients C_k's are listed in Table 5-3 with the overflows highlighted.

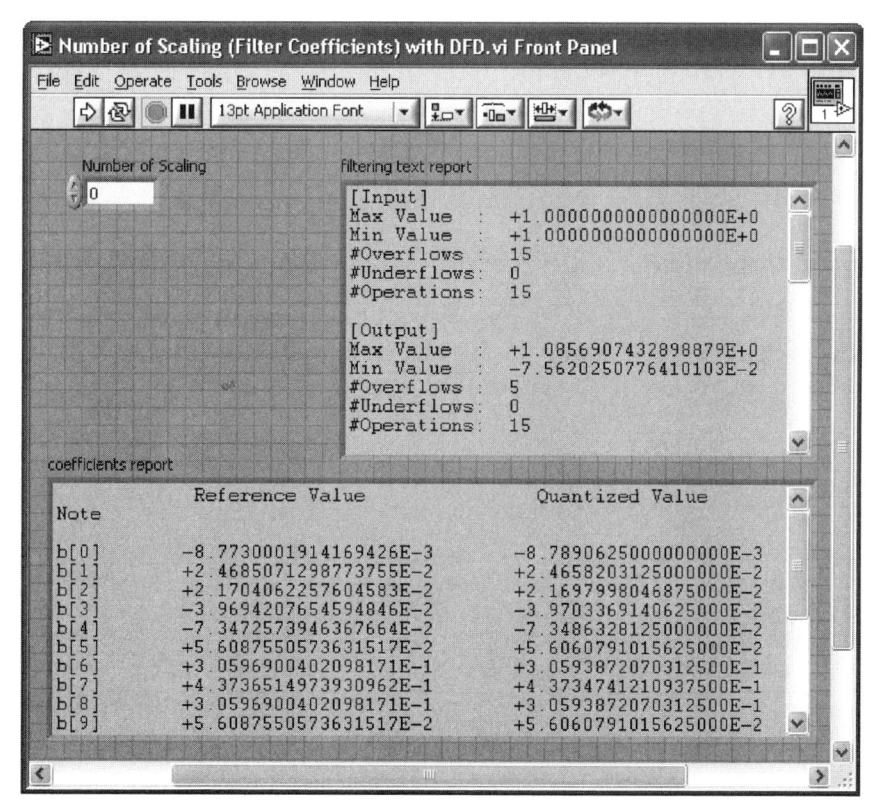

Figure 5-20: Fixed point analysis using DFD toolkit (no scaling).

Table 5-3: Scaling example.

C_k	$\sum C_k$	$\dfrac{C_k}{2}$	$\sum \dfrac{C_k}{2}$
−0.00878906	−0.00878906	−0.00439453	−0.00439453
0.0246582	0.01586914	0.0123291	0.00793457
0.021698	0.03756714	0.010849	0.01878357
−0.03970337	−0.00213623	−0.01985168	−0.00106811
−0.07348633	−0.07562256	−0.03674316	−0.03781127
0.05606079	−0.01956177	0.0280304	−0.00978087
0.30593872	0.28637695	0.15296936	0.14318849
0.43734741	0.72372436	0.21867371	0.3618622
0.30593872	1.02966308	0.15296936	0.51483156
0.05606079	1.08572387	0.0280304	0.54286196
−0.07348633	1.01223754	−0.03674316	0.5061188
−0.03970337	0.97253417	−0.01985168	0.48626712
0.021698	0.99423217	0.010849	0.49711612
0.0246582	1.01889037	0.0123291	0.50944522
−0.00878906	1.01010131	−0.00439453	0.50505069

Now, enter 1 as the number of scaling and run the VI. The outcome of this simulation after scaling by one-half is shown in Figure 5-21. No overflow is observed after this scaling. The scaled coefficient set is also shown in this figure. In addition, the sums of the scaled coefficients are listed in Table 5-3 indicating no overflow.

L5.4.2 Filtering System

As stated previously, the coefficients of the FIR filter need to be scaled by one-half to avoid overflows. For simplicity, two arrays of constants containing the scaled filter coefficients are used on the BD as shown in Figure 5-22. One way to create an array of constants corresponding to the filter coefficients is to change an array of indicators to an array of constants. To do this, an icon of indicator, labeled as `Scaled Coefficient` in Figure 5-19, is copied to the BD of a new VI. It should be verified that the array of indicators displays the coefficients before it is copied to a new VI. Right-click on the icon of the indicator on the BD and choose **Change to Constant**. This way an array of constants containing the filter coefficients is created.

A filter object is created from the coefficient or transfer function of the filter. This is done by placing the `DFD Build Filter from TF` VI (**Functions → All Functions → Digital Filter Design → Utilities → DFD Build Filter from TF**) and wiring the copied array constants to the numerator of the transfer function. Note that the denominator of the FIR transfer function is a single element array of size 1. Once the filter cluster is created, it is wired to the `DFD Filtering` VI to carry out the filtering operation.

Data Type and Scaling

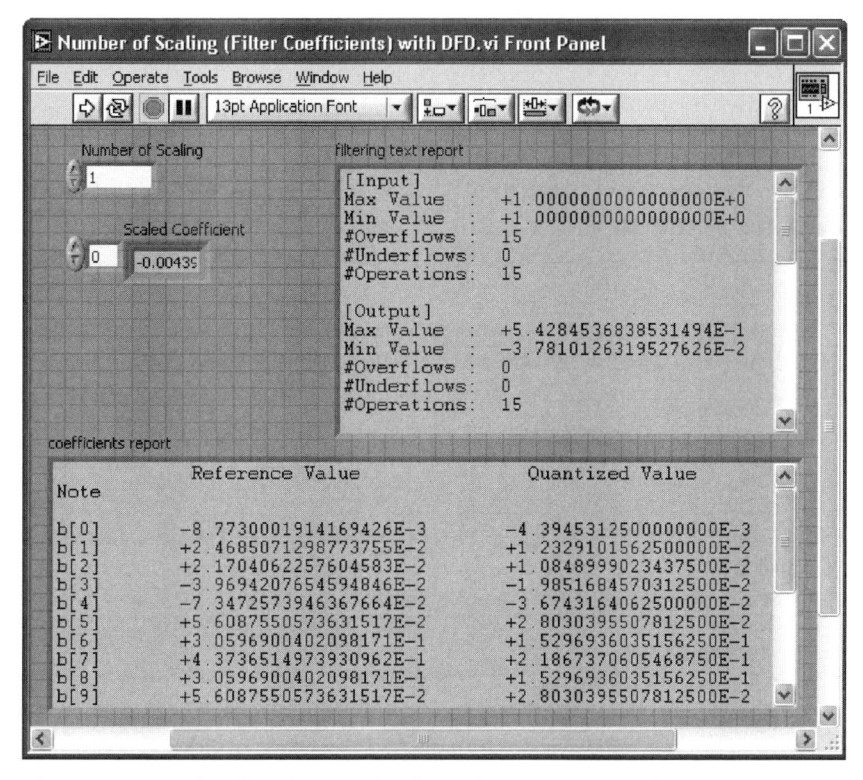

Figure 5-21: Fixed-point analysis using DFD toolkit (one scaling).

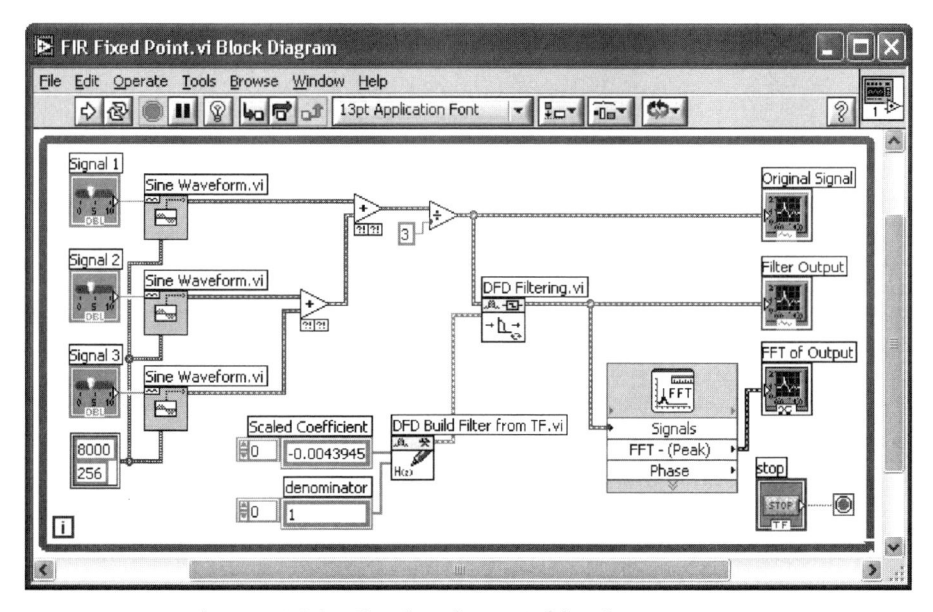

Figure 5-22: Fixed-point FIR filtering system.

Lab 5

The input to the `DFD Filtering` VI consists of three sinusoidal signals. The summed input is divided by 3 to make the range of the input values between −1 and 1, and then is connected to the FIR filter. The input and output signals are shown in Figure 5-23. It can be observed that the fixed-point version of the filter operates exactly the same way as the floating-point version covered in Lab 4. The difference in the scales is due to the use of the one-half scaled filter coefficients and one-third scaled input values.

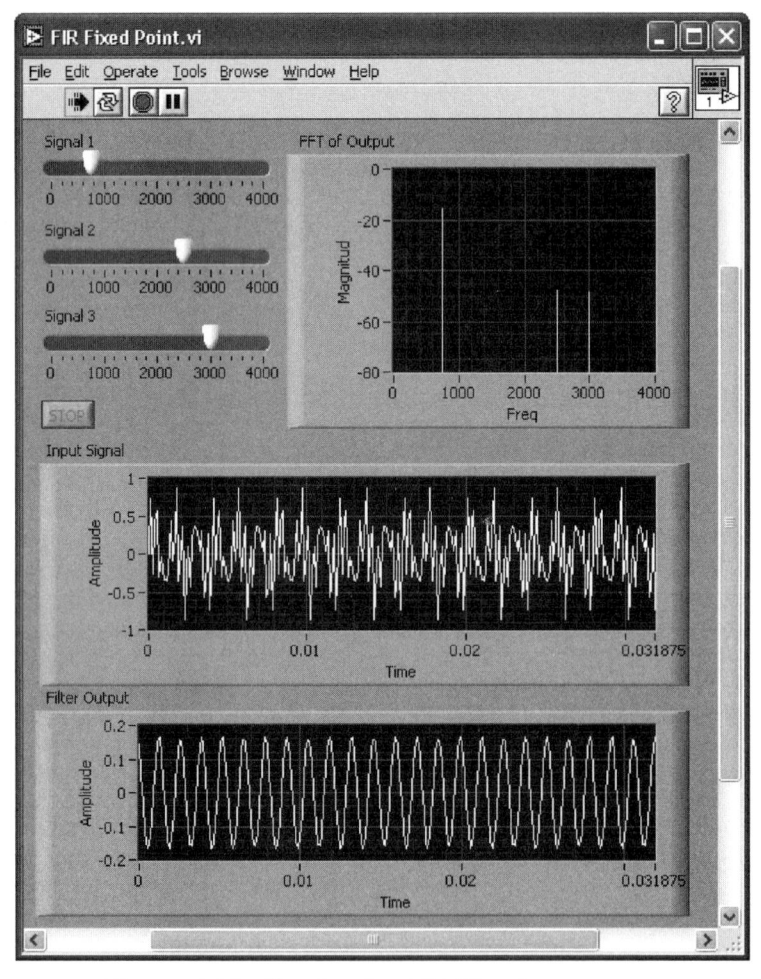

Figure 5-23: Fixed-point FIR filtering output.

L5.4.3 Fixed-Point IIR Filter Example

Considering that the stability of an IIR filter is sensitive to the precision used, an example is provided here to demonstrate this point. This example involves the fixed-point versions of an IIR filter corresponding to different filter forms.

Let us consider an IIR lowpass filter with the following specifications: passband response = 0.1 dB, passband frequency = 1200 Hz, stopband attenuation = 30 dB, stopband frequency = 2200 Hz, and sampling rate = 8000 Hz. The default form of the IIR filter designed by the DFD Classical Filter Design Express VI is the second-order cascade form. The filter can be converted to the direct form by using the DFD Convert Structure VI (**Functions** → **All Functions** → **Digital Filter Design** → **Conversion** → **DFD Convert Structure**). The DFD Convert Structure VI provides a total of 23 forms as the target structure. A Ring Constant is created by right-clicking on the target structure terminal of the VI, followed by choosing **Create** → **Constant**. Click on the created Ring Constant and select IIR Direct Form II as the target structure.

Next, the filter coefficients in the direct form are quantized by using the DFD FXP Quantize Coef VI. Notice that 16 bits is the default word length of the fixed-point representation in the absence of a configuration cluster constant. Different configurations of quantization can be set by creating and wiring a cluster constant at the Coefficient quantizer terminal of the DFD FXP Quantize Coef VI. The floating-point (unquantized) and fixed-point (quantized) filter clusters are wired to the DFD Filter Analysis Express VIs, which are configured to create the magnitude responses. These magnitude responses are placed into one Waveform Graph. This is done by creating a graphical indicator at the magnitude terminal of one of the Express VIs and then by wiring the output of the other Express VI to the same Waveform Graph. A Merge Signal function gets automatically located on the BD. This normally occurs when two or more dynamic data type wires are merged. The BD of the fixed-point IIR filter is shown in Figure 5-24.

Lab 5

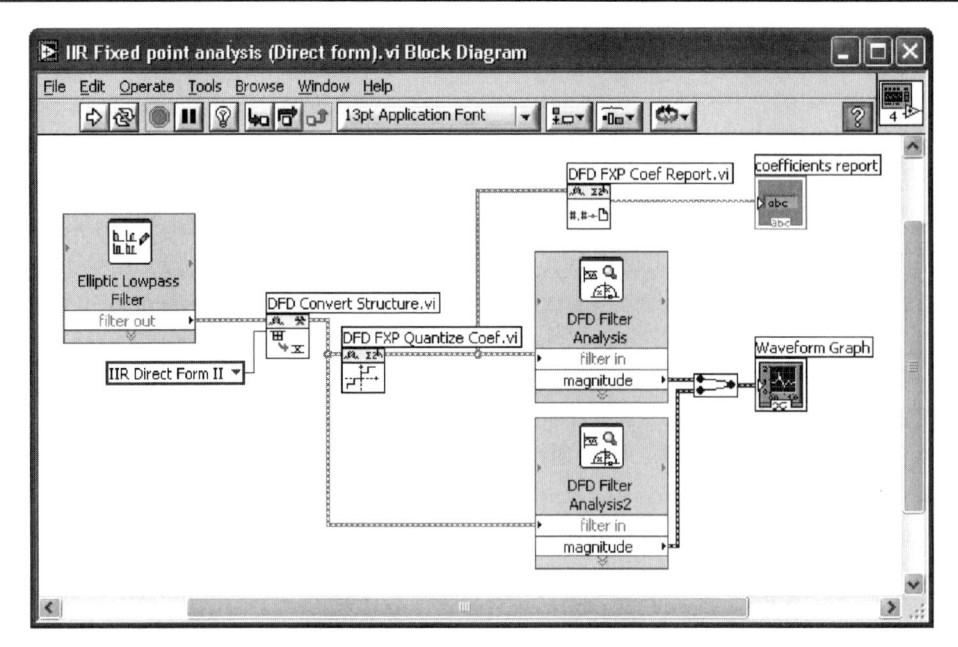

Figure 5-24: BD of fixed-point IIR filter in direct form.

The quantized filter object is also wired to a DFD FXP Coef Report VI (**Functions → All Functions → Digital Filter Design → Fixed-Point Tools → DFD FXP Coef Report**) to generate a text report. This report provides reference coefficients, quantized coefficients, and note sections such as overflow/underflow.

The FP of the VI after running the fixed-point filter is shown in Figure 5-25. Notice that the line style of the fixed-point plot is chosen as dotted for comparison purposes. This is done by right-clicking on the label or plotting in the plot legend and choosing **Line Style** from the shortcut menu.

From Figure 5-25, the magnitude response of the fixed-point version of the IIR filter is seen to be quite different than its floating-point version. This is due to the fact that one underflow and one overflow occur in the filter coefficients, causing the discrepancies in the responses.

Data Type and Scaling

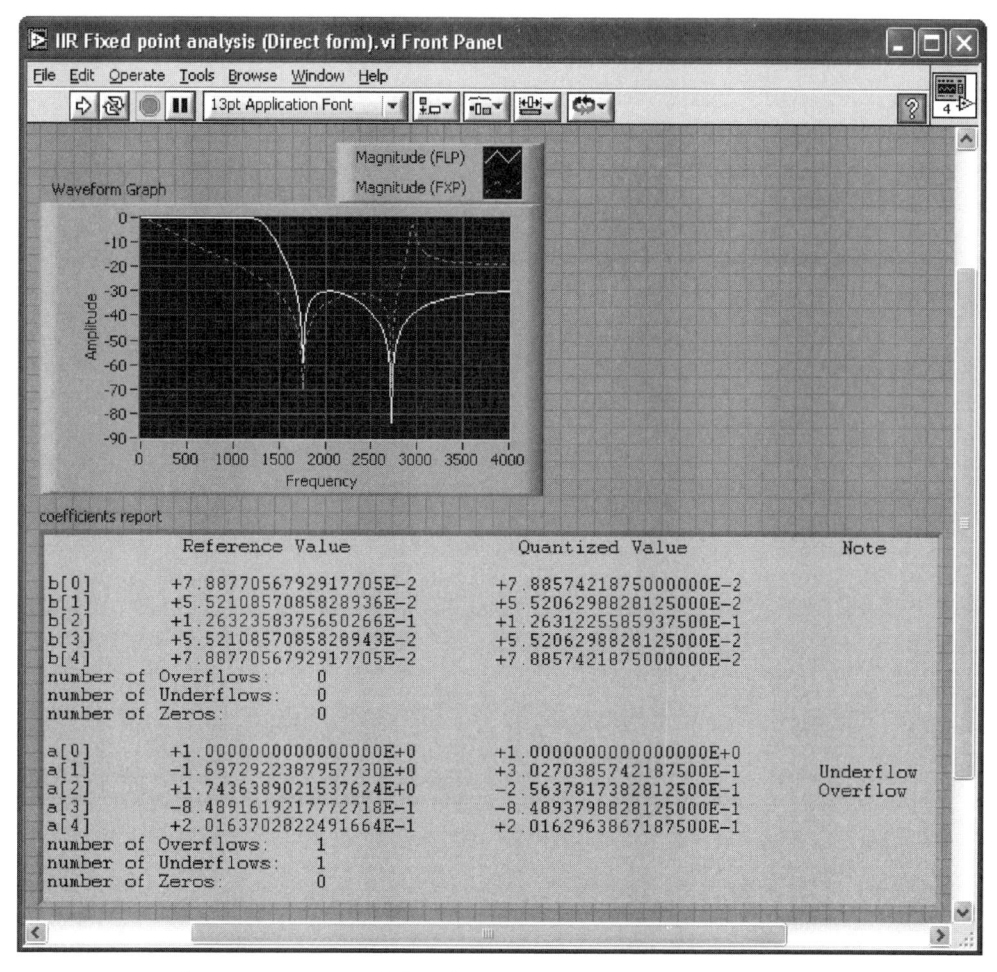

[Screenshot: IIR Fixed point analysis (Direct form).vi Front Panel showing Waveform Graph of Magnitude (FLP) and Magnitude (FXP) vs Frequency, and coefficients report]

```
coefficients report
              Reference Value            Quantized Value            Note
b[0]     +7.8877056792917705E-2     +7.8857421875000000E-2
b[1]     +5.5210857085828936E-2     +5.5206298828125000E-2
b[2]     +1.2632358375650266E-1     +1.2631225585937500E-1
b[3]     +5.5210857085828943E-2     +5.5206298828125000E-2
b[4]     +7.8877056792917705E-2     +7.8857421875000000E-2
number of Overflows:   0
number of Underflows:  0
number of Zeros:       0

a[0]     +1.0000000000000000E+0     +1.0000000000000000E+0
a[1]     -1.6972922387957730E+0     +3.0270385742187500E-1     Underflow
a[2]     +1.7436389021537624E+0     -2.5637817382812500E+0     Overflow
a[3]     -8.4891619217772718E-1     -8.4893798828125000E-1
a[4]     +2.0163702822491664E-1     +2.0162963867187500E-1
number of Overflows:   1
number of Underflows:  1
number of Zeros:       0
```

Figure 5-25: FP of fixed-point IIR filter in direct form.

Next, let us examine the fixed-point version of the IIR filter in the second-order cascade form. This can be achieved simply by removing the DFD Convert Structure VI from the BD shown in Figure 5-24. The magnitude responses of the floating-point and fixed-point versions are illustrated in Figure 5-26. These magnitude responses appear to be identical. Also, no overflow or underflow are observed. This indicates that the effect of quantization can be minimized by using the second-order cascade form.

Lab 5

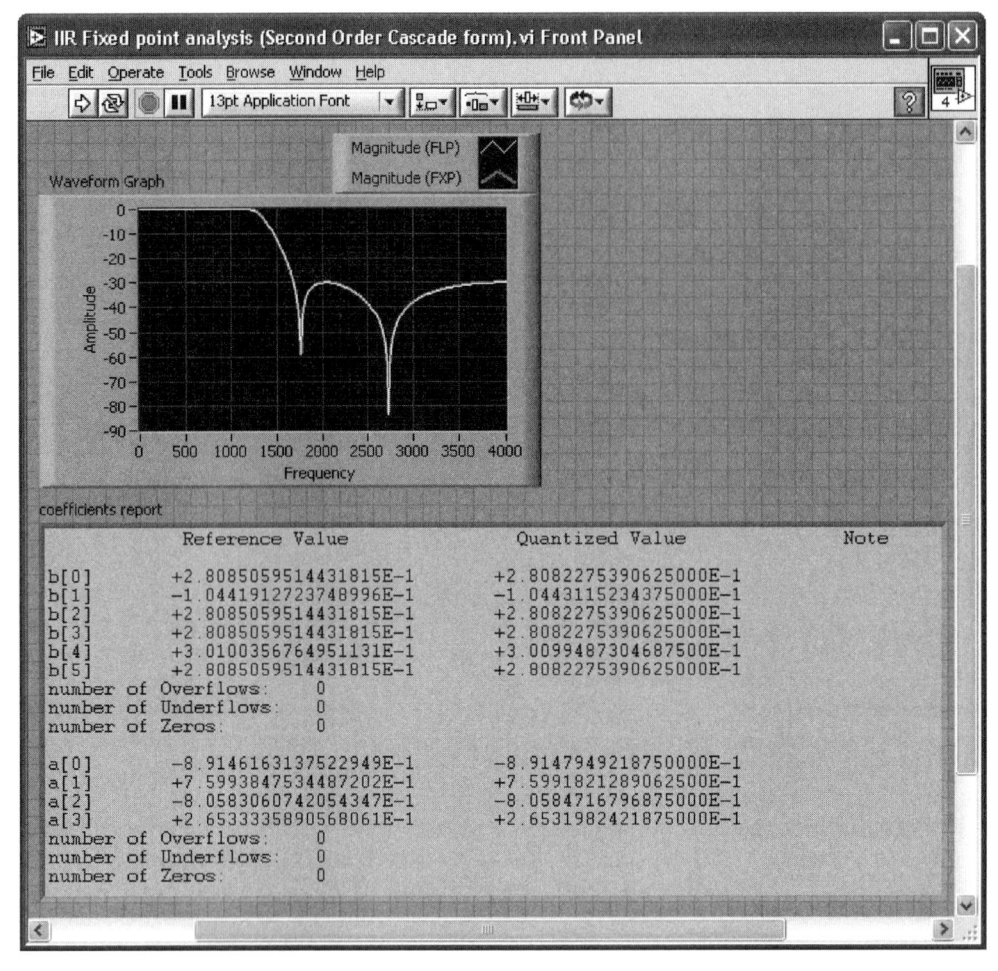

Figure 5-26: Fixed-point IIR filtering response.

L5.5 Bibliography

[1] National Instruments, *LabVIEW Data Storage*, Application Note 154, Part Number 342012C-01, 2004.

[2] Texas Instruments, *TMS320C6000 CPU and Instruction Set Reference Guide*, Literature Number: SPRU189F, 2000.

[3] National Instruments, *LabVIEW User Manual*, Part Number 320999E-01, 2003.

CHAPTER 6

Adaptive Filtering

Adaptive filtering is used in many applications, including noise cancellation and system identification. In most cases, the coefficients of an FIR filter are modified according to an error signal in order to adapt to a desired signal. In this chapter, a system identification and a noise cancellation system are presented wherein an adaptive FIR filter is used.

6.1 System Identification

In system identification, the behavior of an unknown system is modeled by accessing its input and output. An adaptive FIR filter can be used to adapt to the output of the unknown system based on the same input. As indicated in Figure 6-1, the difference in the output of the system, $d[n]$, and the output of the adaptive FIR filter, $y[n]$, constitutes the error term, $e[n]$, which is used to update the coefficients of the filter.

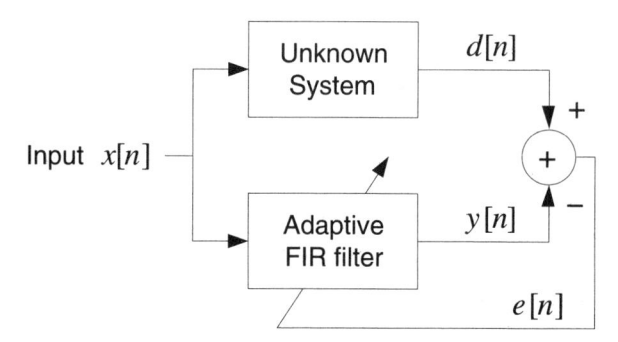

Figure 6-1: System identification system.

Chapter 6

The error term, or the difference between the outputs of the two systems, is used to update each coefficient of the FIR filter according to the equation below (known as the least mean square (LMS) algorithm [1]):

$$h_n[k] = h_{n-1}[k] + \delta e[n]x[n-k] \tag{6.1}$$

where h's denote the unit sample response or FIR filter coefficients, and δ a step size. This adaptation causes the output $y[n]$ to approach $d[n]$. A small step size will ensure convergence, but results in a slow adaptation rate. A large step size, though faster, may lead to skipping over the solution.

6.2 Noise Cancellation

A system for adaptive noise cancellation has two inputs, consisting of a noise-corrupted signal and a noise source. Figure 6-2 illustrates an adaptive noise cancellation system. A desired signal $s[n]$ is corrupted by a noise signal $v_1[n]$ which originates from a noise source signal $v_0[n]$. Bear in mind that the original noise source signal is altered as it passes through an environment or channel whose characteristics are unknown. For example, this alteration can be in the form of a low-pass filtering process. Consequently, the original noise signal $v_0[n]$ cannot be simply subtracted from the noise corrupted signal as there exists an unknown dependency between the two noise signals, $v_1[n]$ and $v_0[n]$. The adaptive filter is thus used to provide an estimate for the noise signal $v_1[n]$.

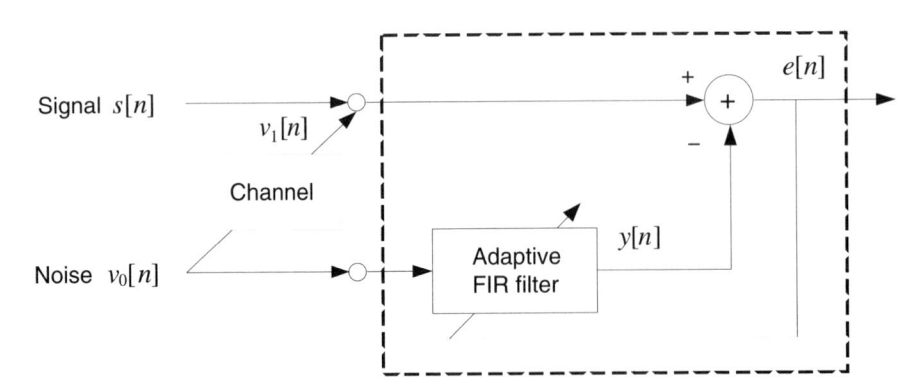

Figure 6-2: Noise cancellation system.

The weights of the filter are adjusted in the same manner stated previously. The error term of this system is given by

$$e[n] = s[n] + v_1[n] - y[n] \tag{6.2}$$

The error $e[n]$ approaches the signal $s[n]$ as the filter output adapts to the noise component of the input $v_1[n]$. To obtain an effective noise cancellation system, the sensor for capturing the noise source should be placed adequately far from the signal source.

6.3 Bibliography

[1] S. Haykin, *Adaptive Filter Theory*, Prentice-Hall, 1996.

Lab 6: Adaptive Filtering Systems

This lab covers adaptive filtering by presenting two adaptive systems using the LMS algorithm, one involving system identification and the other noise cancellation.

L6.1 System Identification

A seventh-order IIR bandpass filter having a passband from $\pi/3$ to $2\pi/3$ (radians) is used here to act as the unknown system. An adaptive FIR filter is designed to adapt to the response of this system.

L6.1.1 Point-By-Point Processing

Before building a system identification VI, it is useful to become familiar with the point-by-point processing feature of LabVIEW. As implied by its name, point-by-point processing is a scalar type of data processing. Point-by-point processing is suitable for real-time data processing tasks, such as signal filtering, since it allows inputs and outputs to be synchronized. On the other hand, in array-based processing, there exists a delay between data acquisition and processing [1].

Figure 6-3 shows the BD of an IIR filtering system utilizing point-by-point processing. A single sample of the input signal gets generated at each iteration of the While

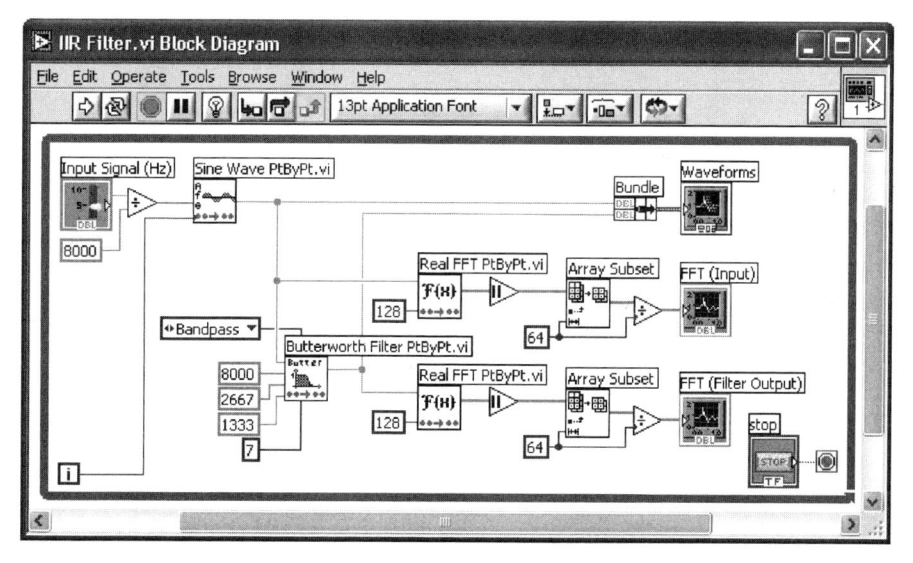

Figure 6-3: BD of IIR filtering system.

Lab 6

Loop by using the `Sine Wave PtByPt` VI (**Functions → All Functions → Analyze → Point By Point → Signal Generation PtByPt → Sine Wave PtByPt**). This VI requires a normalized frequency input. Thus, the signal frequency is divided by the sampling frequency, 8000 Hz, and is wired to the `f` terminal of the VI. Also, the iteration counter of the `While Loop` is wired to the `time` terminal of the VI.

The `Butterworth Filter PtByPt` VI (**Functions → All Functions → Analyze → Point By Point → Filters PtByPt → Butterworth Filter PtByPt**) is used here to serve as the IIR filter. The filter specifications need to be entered and wired to the VI. For example, right click on the `filter type` terminal and choose **Create → Constant** from the shortcut menu to create an `Enum Constant` (enumerated constant) which enumerates the filter types: lowpass, highpass, bandpass, and bandstop, in a pull-down menu. An enumerated constant is used to establish a list of string labels and corresponding numeric values.

The filtered output signal is then examined in both the time and frequency domains. The `Real FFT PtByPt` VI (**Functions → All Functions → Analyze → Point By Point → Frequency Domain PtByPt → Real FFT PtByPt**) is used to see the frequency response. Note that this VI collects a frame of the incoming samples to compute the FFT. A total of 128 input samples are used to generate the magnitudes of 128 complex FFT values. Only the first half of the values is displayed in normalized magnitude, considering that the second half is a mirror image of the first half. This is done by using the `Array Subset` function. Notice that the index is set to its default value, 0, in the absence of an input. The FFT outcome is then displayed in a `Waveform Graph`.

For the time domain observation of the signals, the `Bundle` function is used to combine the input and output signals and display them in the same `Waveform Chart` (**Controls → Graph Indicators → Waveform Chart**). Waveform Chart is discussed in a later subsection.

The FP of the IIR filtering system is shown in Figure 6-4. In order to display the two signals together in the same Waveform Chart, right-click on the chart and choose **Stack Plots**. To modify the display length, right-click on the plot area of the `Waveform Chart` and choose **Chart History Length…**. This brings up a dialog window to adjust the number of samples for the display. Enter 64 for the buffer length.

Let us change the properties of the FP objects. Rename the axes of the `Waveform Graphs` as shown in Figure 6-4. The scale factor needs to be modified in order to have a proper scaling of the frequency axis on the `Waveform Graphs`. The value of 4000/64 = 62.5 is used as the multiplier of the X axis to scale it in the range 0

Adaptive Filtering Systems

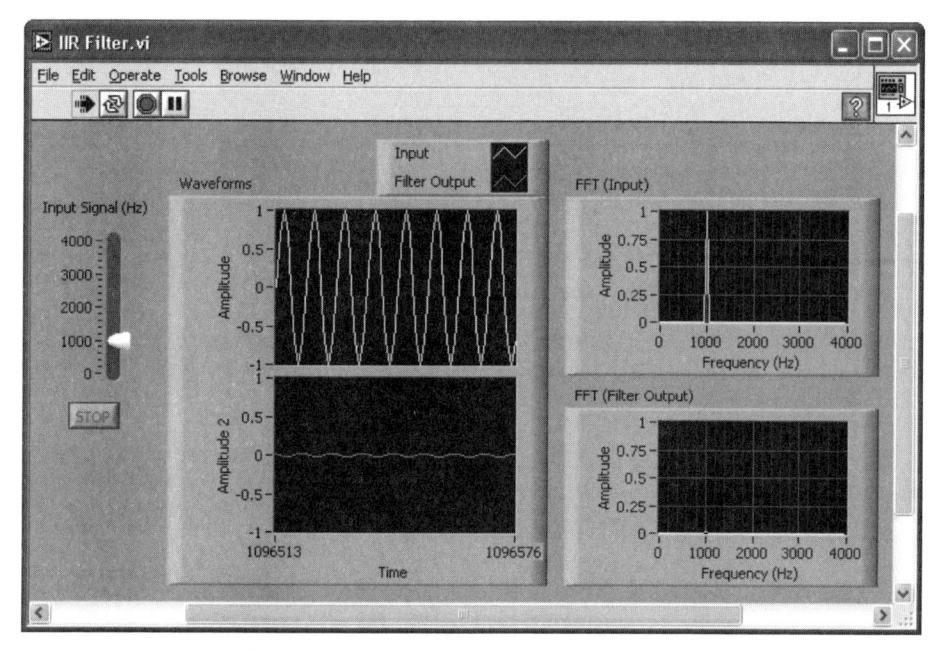

Figure 6-4: FP of IIR filtering system.

to π (radians), that is 4000 Hz. This is done by right-clicking on the `Waveform Graphs`, and choosing **Properties**. This brings up the Waveform Graph Property window. Click the **Scales** tab and choose the **Frequency (Hz)** axis to edit its property. Enter 62.5 for the multiplier under the `Scaling factors` field. The **Autoscale** option for the Y axis of the `Waveform Graphs` also needs to be disabled to observe the magnitude attenuation in the filter output.

The functionality of the filter can be verified by adjusting the frequency of the input signal and running the VI. Observe the frequency response to make sure that the output signal corresponds to the input signal in the passband (1333 to 2667 Hz), while the input signal in the stopband (0 to 1333 and 2667 to 4000 Hz) should get filtered out.

L6.1.2 Least Mean Square (LMS) algorithm

Figure 6-5 shows the BD of the `LMS` VI, which is built by using the point-by-point processing feature. The inputs of this VI include: desired output (`Input 1`), array of samples in a previous iteration (`x[n]`), input to the unknown system (`Input 2`), step size, and filter coefficient set of the previous iteration (`h[n]`) ordered from top to bottom. The outputs of this VI include: updated array (`x[n+1]`), error term, updated filter coefficient set (`h[n+1]`), and FIR filter output ordered from top to bottom.

Lab 6

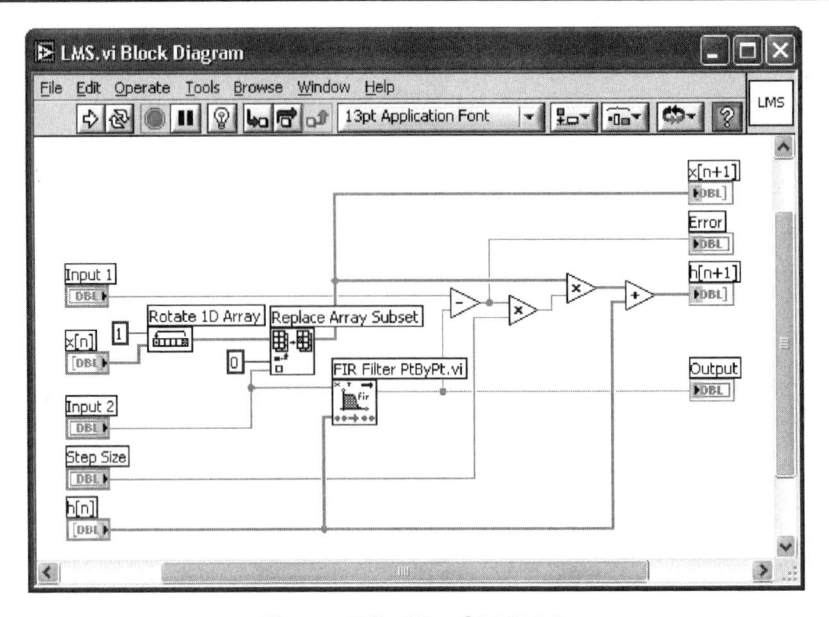

Figure 6-5: BD of LMS VI.

The two array functions, `Replace Array Subset` (**Functions → All Functions → Array → Replace Array Subset**) and `Rotate 1D Array` (**Functions → All Functions → Array → Rotate 1D Array**), act as a circular buffer where the input sample at index 0 gets replaced by a new incoming sample. To perform point-by-point processing, the `FIR Filter PtByPt` VI (**Functions → All Functions → Analyze → Point By Point → Filters PtByPt → FIR Filter PtByPt**) is used. This VI requires a single-element input and a coefficient array.

The `Subtract` function on the BD calculates the error or the difference between the desired signal and the output of the adaptive FIR filter. This error is multiplied by the step size δ and then by the elements in the input buffer to obtain the coefficient updates. Next, these updates are added to the previous coefficients $h[n]$'s to compute the updated coefficients $h[n+1]$'s as stated in Equation (6.1).

The icon of the LMS VI is edited as shown in Figure 6-5. The connector pane of the VI is also modified as shown in Figure 6-6 for it to be used as a subVI.

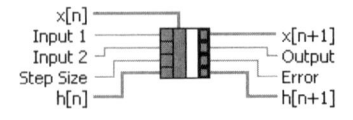

Figure 6-6: Connector pane of LMS VI.

Adaptive Filtering Systems

L6.1.3 Waveform Chart

A Waveform Graph plots an array of samples while a Waveform Chart takes one or more samples as its input and maintains a history so that a trajectory can be displayed, similar to an oscilloscope.

There are three different updating modes in a Waveform Chart. They include: Strip chart, Scope chart, and Sweep chart. The Strip chart mode provides a continuous data display. Once the plot reaches the right border, the old data are scrolled to the left and new data are plotted on the right edge of the plotting area. The Scope chart mode provides a data display from left to right. Then, it clears the plot and resumes it from the left border of the plotting area. This is similar to data display on an oscilloscope. The Sweep chart mode functions similar to the Scope chart mode except the old data are not cleared. The old and new data are separated by a vertical line. All three modes are illustrated in Figure 6-7. These modes are configured by right-clicking on the plot area and then by selecting **Advanced → Update Mode**.

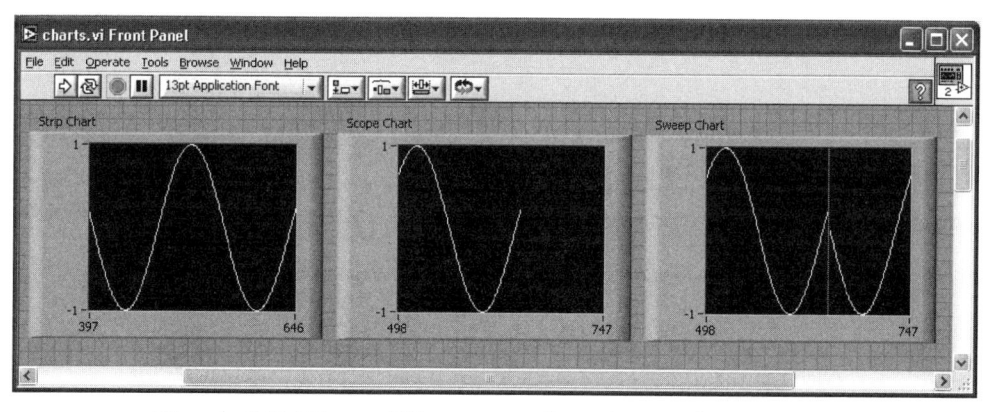

Figure 6-7: Three different modes of Waveform Chart.

The length of data displayed on the chart is changeable. To do so, right-click on the plot area and select **Chart History Length**. This brings up a dialog box to enter data length.

L6.1.4 Shift-Register and Feedback Node

The BD of the overall adaptive system is shown in Figure 6-8. Note that two Feedback Nodes, denoted by ⬅, are used. A Feedback Node is used to transfer data from one iteration to a next iteration inside a For Loop or a While Loop. A Feedback Node is automatically generated when the output of a VI is wired to its input inside a loop structure. By default, an initializer terminal is added onto the left

Lab 6

border of the loop for each Feedback Node. An initializer terminal is used to initialize the value to be transferred by the Feedback Node. If no initialization is needed, the terminal can be removed by right-clicking on it and unchecking **Initializer terminal**.

A Feedback Node can be replaced by a Shift Register. To achieve this, right-click on the Feedback Node. Then choose **Replace with Shift Register**. This adds a Shift Register at both sides of the While Loop. Also, the Shift Registers are wired to the terminals of the LMS subVI.

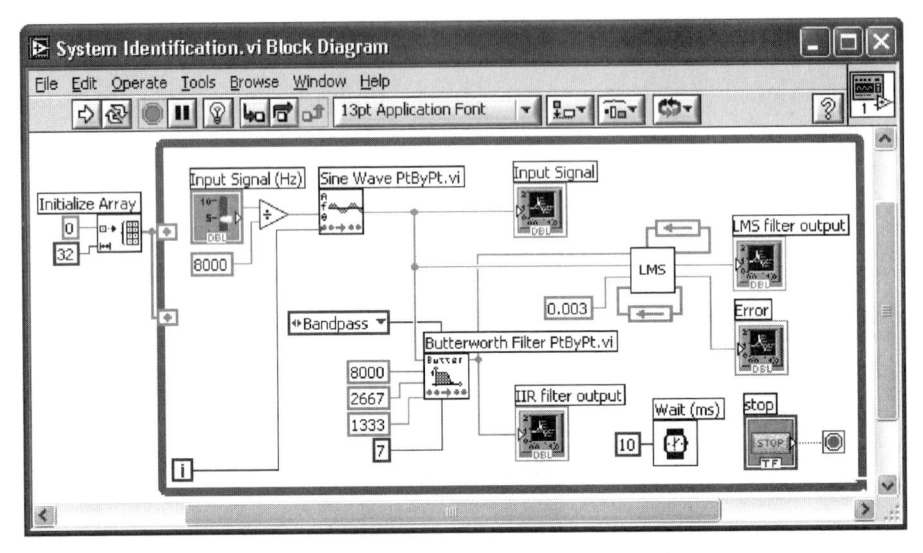

Figure 6-8: BD of system identification.

In the BD shown in Figure 6-8, the IIR filtering system built earlier is located to act as the unknown system. The filter coefficient array and the input data array are passed from one iteration to a next iteration by the Feedback Nodes to update the filter coefficients via the LMS algorithm. Both of these arrays are initialized with 32 zero values, considering that the number of filter taps is 32. This is done by wiring an Initialize Array function (**Functions → All Functions → Array → Initialize Array**) to the initializer terminal. The initialization array is configured to be of length 32 containing zero values.

Adaptive Filtering Systems

For the step size of the LMS algorithm, a `Numeric Constant` is created and wired. This value can be adjusted to control the speed of adaptation. In this example, 0.003 is used. Also, a `Wait(ms)` function is placed in the `While Loop` to delay the execution of the loop.

As shown in Figure 6-9, the output of the adaptive LMS filter adapts to the output of the IIR filter and thus the error between the outputs diminishes.

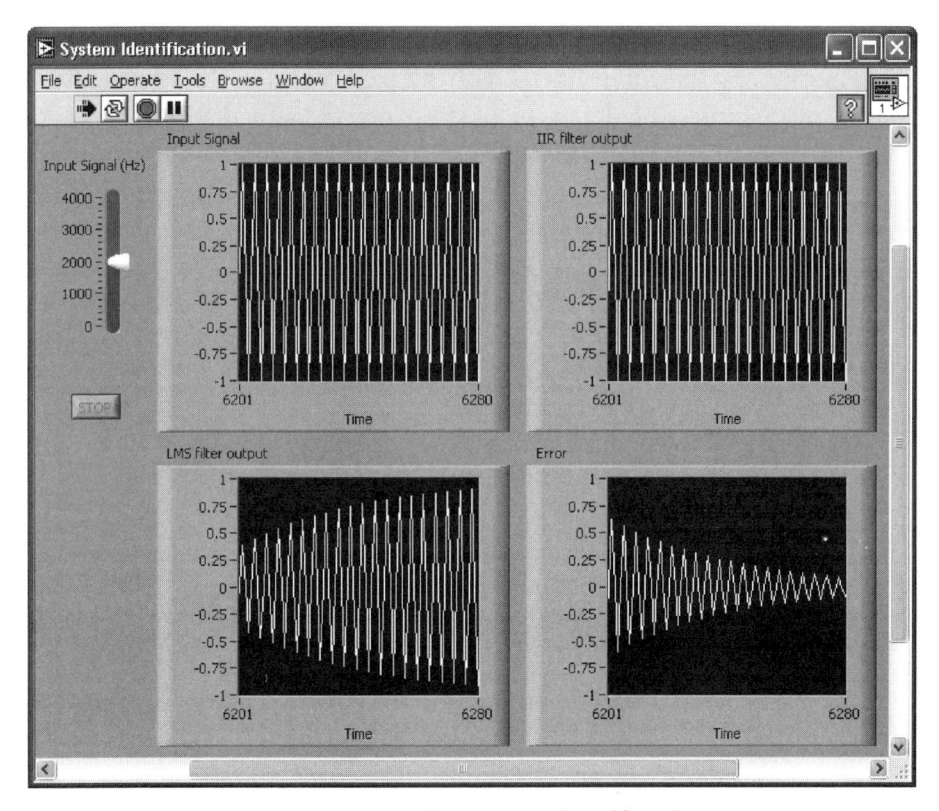

Figure 6-9: FP of system identification.

L6.2 Noise Cancellation

The design of a noise cancellation system is achieved similar to the system identification system mentioned above. A noise cancellation system takes two inputs: a noise-corrupted input signal and a reference noise signal. The BD of the adaptive noise cancellation system is shown in Figure 6-10. Again, as in the system identification example, the point-by-point processing feature is employed here. This requires using a `Get Waveform Components` function together with an `Index Array` function at the output of the noise and signal sources. The number of samples of the waveforms generated by the three `Sine Waveform` VIs is set to 1 for performing point-by-point processing. The data type of the Y component is still of array type with size 1. The `Index Array` function is used to extract a scalar element from the array. This ensures that the numerical operations are done in a point-by-point fashion.

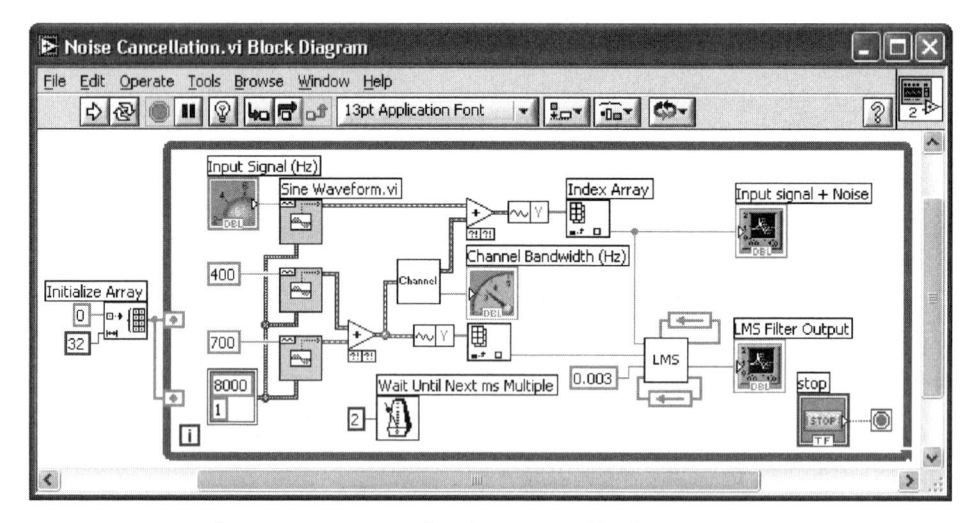

Figure 6-10: BD of noise cancellation system.

To be able to observe the adaptability of the system, a time-varying channel is added. The noise source, which consists of the sum of two sinusoidal waveforms (400 and 700 Hz), is passed through the channel before it is added to the input signal. The BD of the time-varying channel is shown in Figure 6-11.

Adaptive Filtering Systems

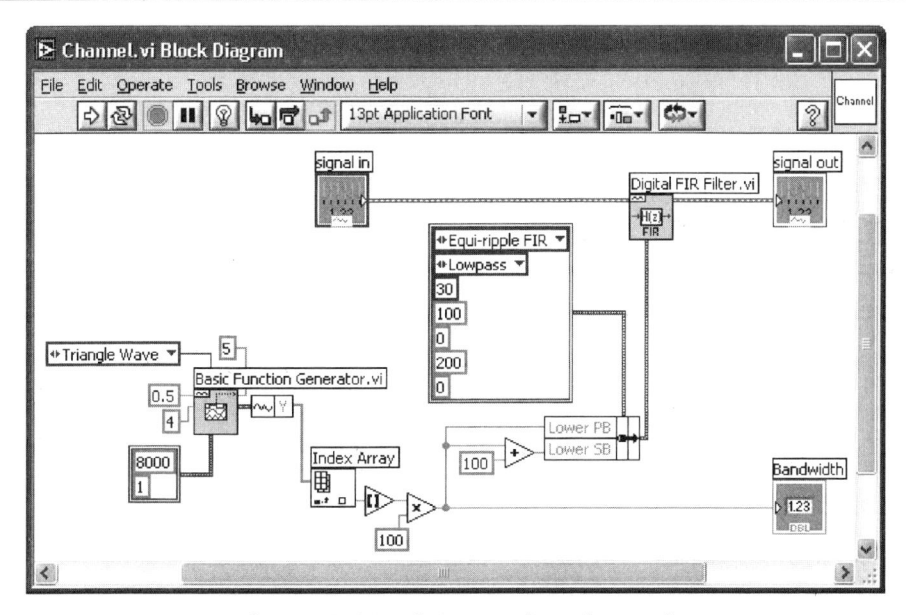

Figure 6-11: Time-varying channel.

The channel consists of an FIR lowpass filter with its passband and stopband varying according to a discretized triangular waveform. The reason for the discretization is to give the LMS algorithm enough time to adapt to the noise signal. The characteristic of the channel is varied with time by swinging the filter passband from 100 to 900 Hz. The bandwidth of the time-varying channel is shown in Figure 6-12.

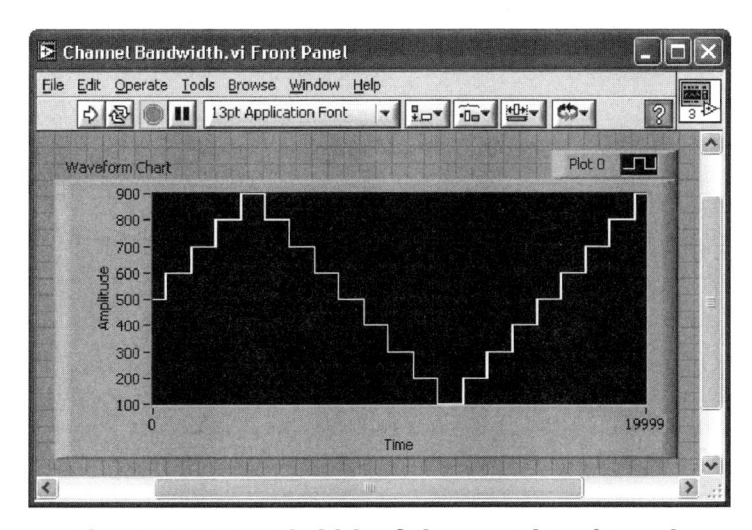

Figure 6-12: Bandwidth of time-varying channel.

Lab 6

The `Waveform Graph` shown in Figure 6-13 indicates that the adaptive filter adapts to its input by cancelling out the noise component as the characteristic of the channel is changed. As illustrated in Figure 6-13, the input signal to the system is a 50 Hz sinusoid and the noise varies in the range of 100–900 Hz. The step size δ may need to be modified depending on how fast the system is converging. It is necessary to make sure that the characteristic of the channel is not changing too fast to give the adaptive filter adequate time to adapt to the noise signal passed through it. As seen in Figure 6-13, the system adapts to the noise signal before the channel changes.

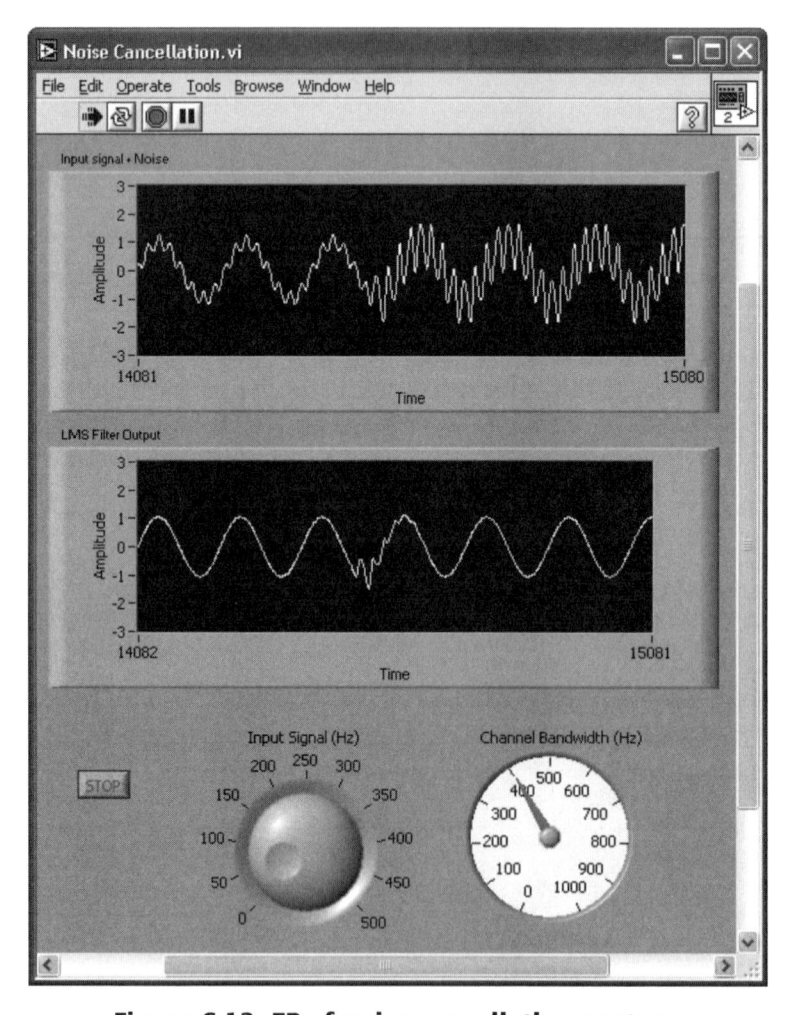

Figure 6-13: FP of noise cancellation system.

Adaptive Filtering Systems

Note that the noise-cancelled signal is available from the `Error` terminal of the LMS VI. If a DC input signal, in other words a 0 Hz signal, is applied to the system, the output of the adaptive filter becomes the error between the noise signal passed through the channel and the reference noise signal. This is illustrated in Figure 6-14.

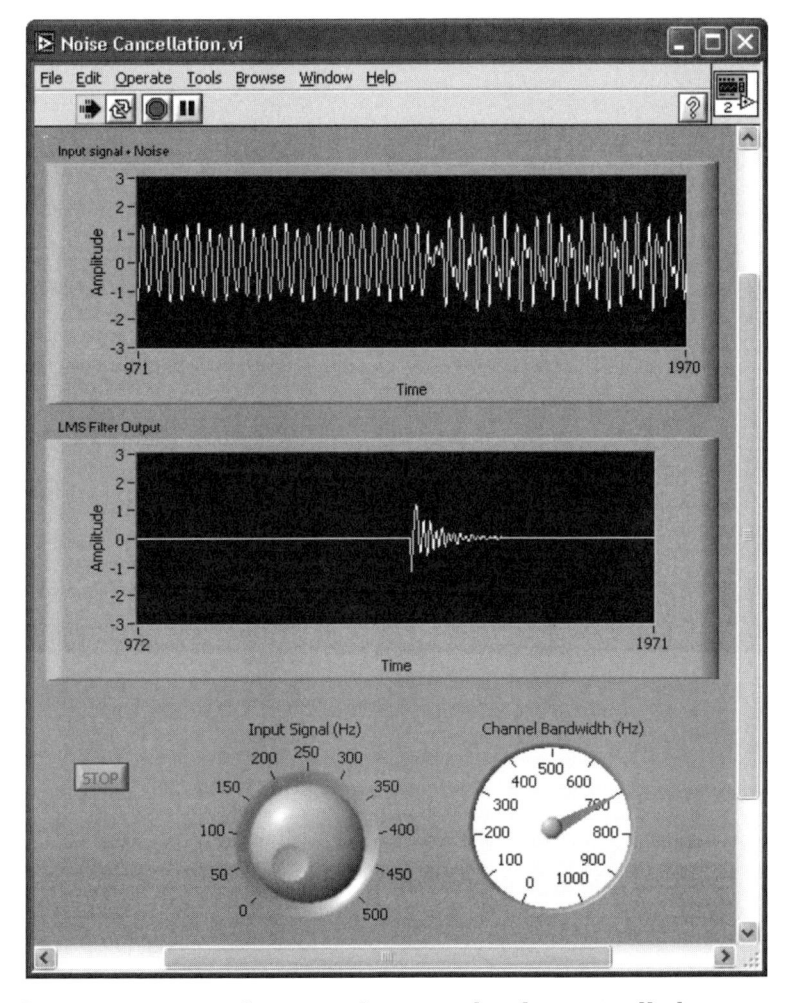

Figure 6-14: Error between input and noise cancelled output.

L6.3 Bibliography

[1] National Instruments, *LabVIEW Analysis Concepts*, Part Number 370192C-01, National Instruments, 2004.

CHAPTER 7

Frequency Domain Processing

Transformation of signals to the frequency domain is widely used in signal processing. In many cases, such transformations provide a more effective representation and a more computationally efficient processing of signals as compared to time domain processing. For example, due to the equivalency of convolution operation in the time domain to multiplication in the frequency domain, one can find the output of a linear system by simply multiplying the Fourier transform of the input signal by the system transfer function.

This chapter presents an overview of three widely used frequency domain transformations, namely fast Fourier transform (FFT), short-time Fourier transform (STFT), and discrete wavelet transform (DWT). More theoretical details regarding these transformations can be found in many signal processing textbooks, e.g. [1].

7.1 Discrete Fourier Transform (DFT) and Fast Fourier Transform (FFT)

Discrete Fourier transform (DFT) $x[k]$ of an N-point signal $x[n]$ is given by

$$\begin{cases} X[k] = \sum_{n=0}^{N-1} x[n] W_N^{nk}, & k = 0, 1, ..., N-1 \\ x[n] = \frac{1}{N} \sum_{n=0}^{N-1} X[k] W_N^{-nk}, & n = 0, 1, ..., N-1 \end{cases} \tag{7.1}$$

where $W_N = e^{-j\frac{2\pi}{N}}$. The above transform equations require N complex multiplications and $N-1$ complex additions for each term. For all N terms, N^2 complex multiplications and $N^2 - N$ complex additions are needed. As it is well known, the direct computation of (7.1) is not efficient.

Chapter 7

To obtain a fast or real-time implementation of (7.1), a fast Fourier transform (FFT) algorithm is often used which makes use of the symmetry properties of DFT. There are many approaches to finding a fast implementation of DFT; that is, there are many FFT algorithms. Here, we mention the approach presented in the *TI Application Report SPRA291* for computing a 2N-point FFT [2]. This approach involves forming two new N-point signals $x_1[n]$ and $x_2[n]$ from a 2N-point signal $g[n]$ by splitting it into an even and an odd part as follows:

$$x_1[n] = g[2n] \qquad 0 \leq n \leq N-1$$
$$x_2[n] = g[2n+1] \qquad (7.2)$$

From the two sequences $x_1[n]$ and $x_2[n]$, a new complex sequence $x[n]$ is defined to be

$$x[n] = x_1[n] + jx_2[n] \qquad 0 \leq n \leq N-1 \qquad (7.3)$$

To get $G[k]$, the DFT of $g[n]$, the equation

$$G[k] = X[k]A[k] + X^*[N-k]B[k] \qquad (7.4)$$
$$k = 0, 1, \ldots, N-1, \text{ with } X[N] = X[0]$$

is used, where

$$A[k] = \frac{1}{2}\left(1 - jW_{2N}^k\right) \qquad (7.5)$$

and

$$B[k] = \frac{1}{2}\left(1 + jW_{2N}^k\right) \qquad (7.6)$$

Only N points of $G[k]$ are computed from (7.4). The remaining points are found by using the complex conjugate property of $G[k]$, that is $G[2N - k] = G^*[k]$. As a result, a 2N-point transform is calculated based on an N-point transform, leading to a reduction in the number of operations.

7.2 Short-Time Fourier Transform (STFT)

Short-time Fourier transform (STFT) is a sequence of Fourier transforms of a windowed signal. STFT provides the time-localized frequency information for situations when frequency components of a signal vary over time, whereas the standard Fourier transform provides the frequency information averaged over the entire signal time interval.

Frequency Domain Processing

The STFT pair is given by

$$\begin{cases} X_{STFT}[m,n] = \sum_{k=0}^{L-1} x[k]g[k-m]e^{-j2\pi nk/L} \\ x[k] = \sum_{m}\sum_{n} X_{STFT}[m,n]g[k-m]e^{j2\pi nk/L} \end{cases} \quad (7.7)$$

where $x[k]$ denotes a signal, $g[k]$ an L-point window function. From (7.7), the STFT of $x[k]$ can be interpreted as the Fourier transform of the product $x[k]g[k-m]$. Figure 7-1 illustrates computing STFT by taking Fourier transforms of a windowed signal.

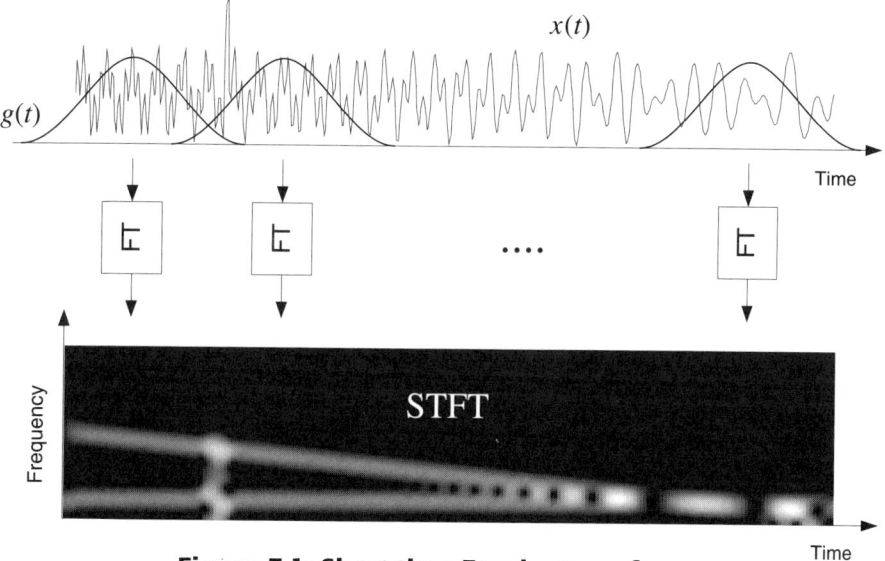

Figure 7-1: Short-time Fourier transform.

There exists a trade-off between time and frequency resolution in STFT. In other words, although a narrow-width window results in a better resolution in the time domain, it generates a poor resolution in the frequency domain, and vice versa. Visualization of STFT is often realized via its spectrogram, which is an intensity plot of STFT magnitude over time. Three spectrograms illustrating different time-frequency resolutions are shown in Figure 7-2. The implementation details of STFT are described in Lab 7.

Chapter 7

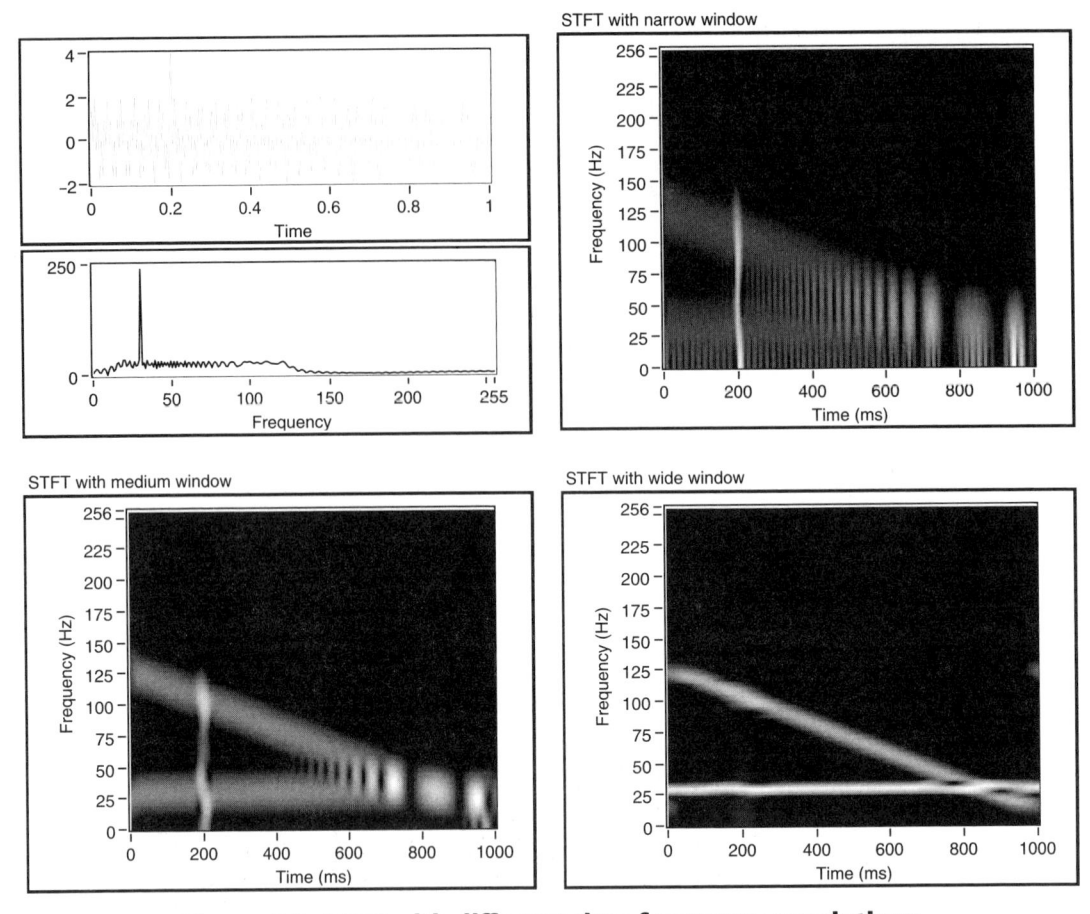

Figure 7-2: STFT with different time frequency resolutions.

7.3 Discrete Wavelet Transform (DWT)

Wavelet transform offers a generalization of STFT. From a signal theory point of view, similar to DFT and STFT, wavelet transform can be viewed as the projection of a signal into a set of basis functions named wavelets. Such basis functions offer localization in the frequency domain. In contrast to STFT having equally-spaced time-frequency localization, wavelet transform provides high frequency resolution at low frequencies and high time resolution at high frequencies. Figure 7-3 provides a tiling depiction of the time-frequency resolution of wavelet transform as compared to STFT and DFT.

Frequency Domain Processing

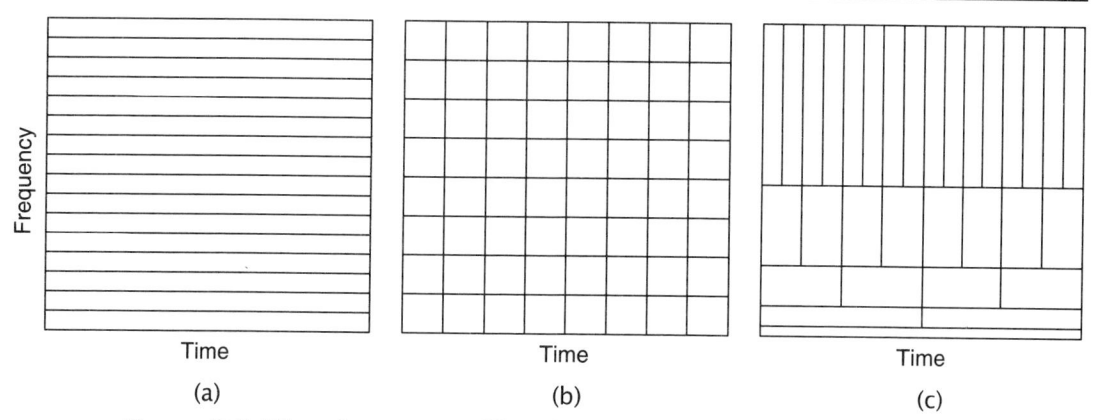

Figure 7-3: Time-frequency tiling for (a) DFT, (b) STFT, and (c) DWT.

The discrete wavelet transform (DWT) of a signal $x[n]$ is defined based on so-called approximation coefficients, $W_\varphi[j_0,k]$, and detail coefficients, $W_\psi[j,k]$, as follows:

$$W_\varphi[j_0,k] = \frac{1}{\sqrt{M}} \sum_n x[n] \varphi_{j_0,k}[n]$$
$$W_\psi[j,k] = \frac{1}{\sqrt{M}} \sum_n x[n] \psi_{j,k}[n] \quad \text{for } j \geq j_0 \quad (7.8)$$

and the inverse DWT is given by

$$x[n] = \frac{1}{\sqrt{M}} \sum_k W_\varphi[j_0,k] \varphi_{j_0,k}[n] + \frac{1}{\sqrt{M}} \sum_{j=j_0}^{J} \sum_k W_\psi[j,k] \psi_{j,k}[n] \quad (7.9)$$

where $n = 0,1,2,\ldots, M-1$, $j = 0,1,2,\ldots, J-1$, $k = 0,2,\ldots, 2^j - 1$, and M denotes the number of samples to be transformed. This number is selected to be $M = 2^J$, where J indicates the number of transform levels. The basis functions $\{\varphi_{j,k}[n]\}$ and $\{\psi_{j,k}[n]\}$ are defined as

$$\varphi_{j,k}[n] = 2^{j/2} \varphi[2^j n - k]$$
$$\psi_{j,k}[n] = 2^{j/2} \psi[2^j n - k] \quad (7.10)$$

where $\varphi[n]$ is called scaling function and $\psi[n]$ wavelet function.

For the implementation of DWT, the filter bank structure is often used. Figure 7-4 shows the decomposition or analysis filter bank for obtaining the forward DWT coefficients. The approximation coefficients at a higher level are passed through a highpass and a lowpass filter followed by a downsampling by two to compute both the

Chapter 7

detail and approximation coefficients at a lower level. This tree structure is repeated for a multi-level decomposition.

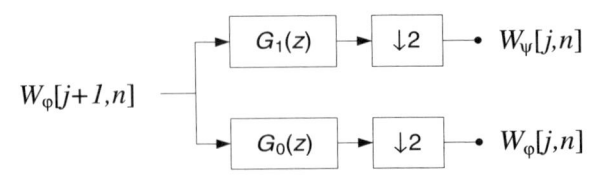

Figure 7-4: Discrete wavelet transform decomposition filter bank, G_0 lowpass and G_1 highpass decomposition filters.

Inverse DWT (IDWT) is obtained by using the reconstruction or synthesis filter bank shown in Figure 7-5. The coefficients at a lower level are upsampled by two and passed through a highpass and a lowpass filter. The results are added together to obtain the approximation coefficients at a higher level.

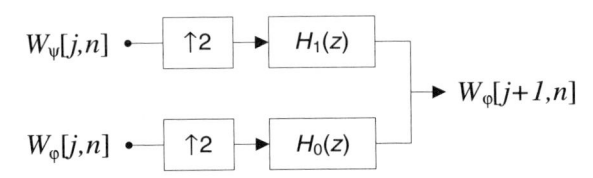

Figure 7-5: Discrete wavelet transform reconstruction filter bank, H_0 lowpass and H_1 highpass reconstruction filters.

7.4 Signal Processing Toolset

Signal Processing Toolset (SPT) is an add-on toolkit of LabVIEW that provides useful tools for performing time-frequency analysis [3]. SPT has three components: Joint Time-Frequency Analysis (JTFA), Super-resolution Spectral Analysis (SRSA), and Wavelet Analysis.

The VIs associated with STFT are included as part of the JTFA component. The SRSA component is based on the model-based frequency analysis normally used for situations when a limited number of samples are available. The VIs associated with the SRSA component include high-resolution spectral analysis and parameter estimation, such as amplitude, phase, damping factor, and damped sinusoidal estimation. The VIs associated with the Wavelet Analysis component include 1D and 2D wavelet transform as well as their filter bank implementations.

7.5 Bibliography

[1] C. Burrus, R. Gopinath, and H. Gao, *Wavelets and Wavelet Transforms A Primer*, Prentice-Hall, 1998.

[2] Texas Instruments, *TI Application Report SPRA291*.

[3] National Instruments, *Signal Processing Toolset User Manual*, Part Number 322142C-01, 2002.

Lab 7: FFT, STFT and DWT

This lab shows how to use the LabVIEW tools to perform FFT, STFT and DWT as part of a frequency domain transformation system.

L7.1 FFT versus STFT

To illustrate the difference between FFT and STFT transformations, a 512-point input signal is formed by combining three signals: a 75 Hz sinusoidal signal sampled at 512 Hz, a chirp signal with linearly decreasing frequency from 200 to 120 Hz, and an impulse signal having an amplitude of 2 for 500 ms located at the 256th sample. This composite signal is shown in Figure 7-6. The FFT and STFT graphs are also shown in

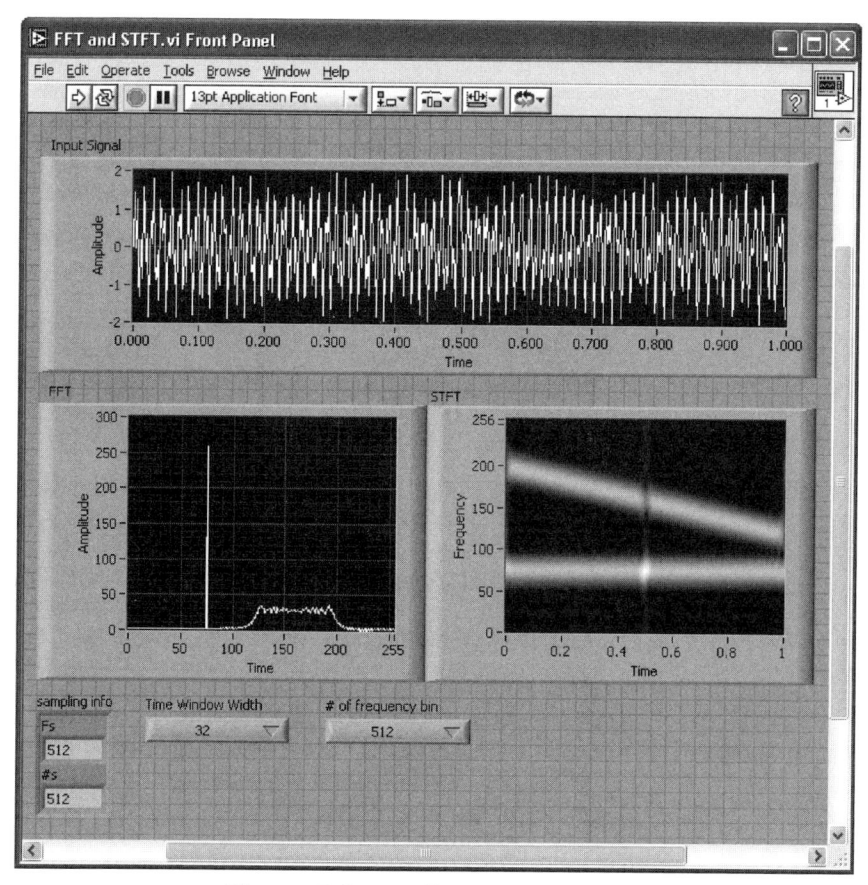

Figure 7-6: FP of FFT versus STFT.

this figure. The FFT graph shows the time averaged spectrum reflecting the presence of a signal from 120 to 200 Hz with one major peak at 75 Hz. As can be seen from this graph, the impulse having the short time duration does not appear in the spectrum. The STFT graph shows the spectrogram for a time increment of 1 and a rectangular window of width 32 by which the presence of the impulse can be detected.

As far as the FP is concerned, two Menu Ring controls (**Controls** → **All Controls** → **Ring & Enum** → **Menu Ring**) are used to input values via their labels. The labels and corresponding values of the ring controls can be modified by right clicking and choosing **Edit Items...** from the shortcut menu. This brings up the dialog window shown in Figure 7-7.

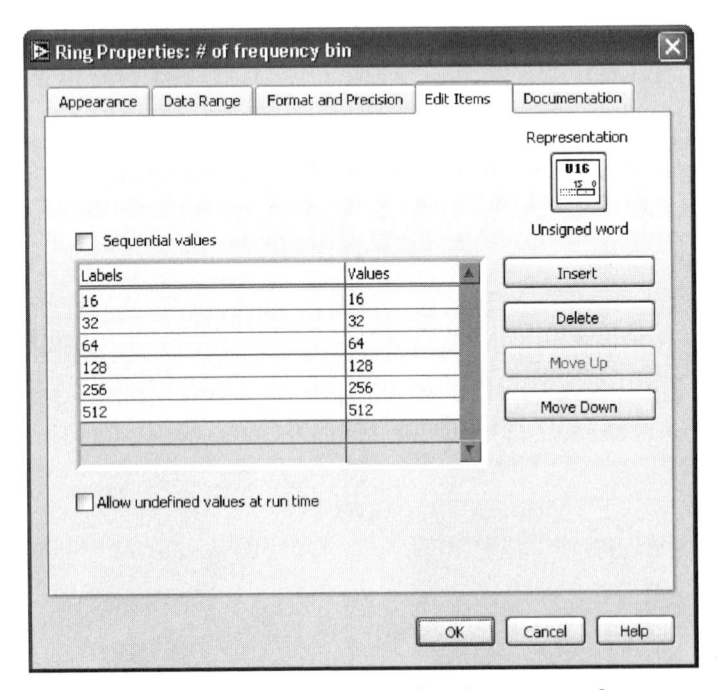

Figure 7-7: Properties of a ring control.

An Enum (enumerate) control acts the same as a Menu Ring control, except that values of an Enum control cannot be modified and are assigned sequentially. A Menu Ring or Enum can be changed to a Ring Constant or Enum Constant when used on a BD.

Several spectrograms with different time window widths are shown in Figure 7-8. Figure 7-8(a) shows an impulse (vertical line) at time 500 ms because of the relatively

FFT, STFT and DWT

time-localized characteristic of the window used. Even though a high resolution in the time domain is achieved with this window, the resolution in the frequency domain is so poor that the frequency contents of the sinusoidal and chirp signals cannot easily be distinguished. This is due to the Heisenberg's uncertainty principle [1], which states that if the time resolution is increased, the frequency resolution is decreased.

Now, let us increase the width of the time-frequency window. This causes the frequency resolution to become better while the time resolution becomes poorer. As a result, as shown in Figure 7-8(d), the frequency contents of the sinusoidal and chirp signals become better distinguished. It is also seen that as the time resolution becomes poorer, the moment of occurrence of the impulse becomes more difficult to identify.

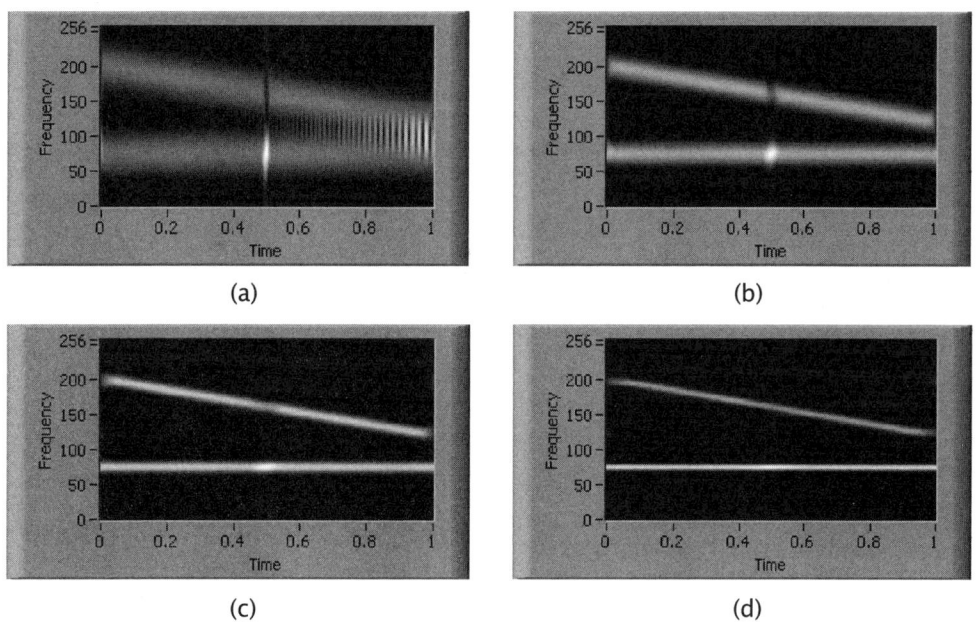

Figure 7-8: STFT with time window of width (a) 16, (b) 32, (c) 64, and (d) 128.

The BD of this example is illustrated in Figure 7-9. To build this VI, let us first generate the input signal with the specifications stated above. The sinusoidal waveform is generated by using the Sine Waveform VI (**Functions → All Functions → Analyze → Waveform Generation → Sine Waveform**) and the chirp signal is generated by using the Chirp Pattern VI (**Functions → All Functions → Analyze → Signal Processing → Signal Generation → Chirp Pattern**). Also, the impulse is generated by using the Impulse Pattern VI (**Functions → All Functions → Analyze → Signal Processing → Signal Generation → Impulse Pattern**). These signals are added together to form a

Lab 7

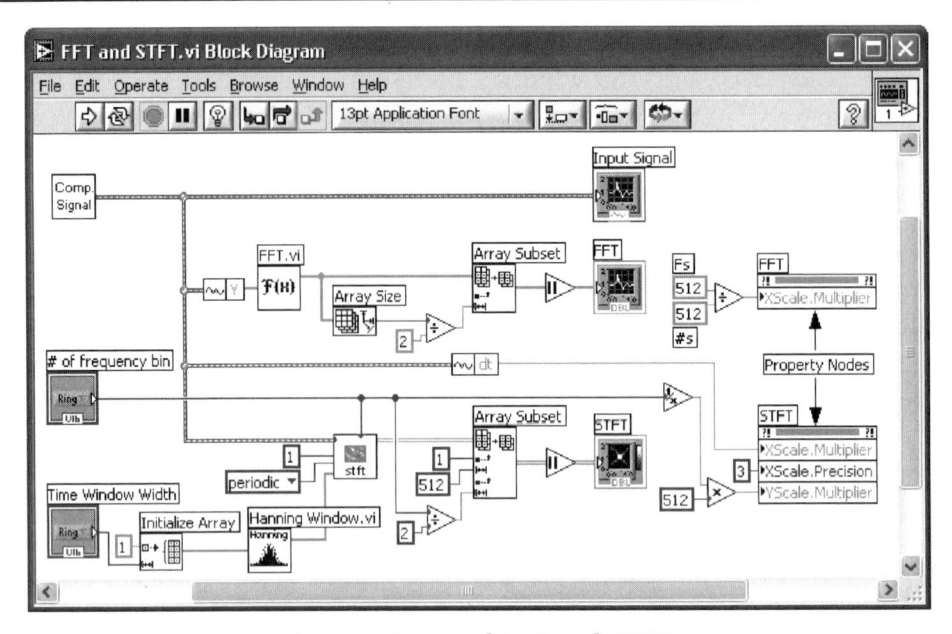

Figure 7-9: BD of FFT and STFT.

composite signal, see Figure 7-10. In order to use this VI as the signal source of the system, an output terminal in the connector pane is wired to the waveform indicator. Then, the VI is saved as *Composite Signal.vi*.

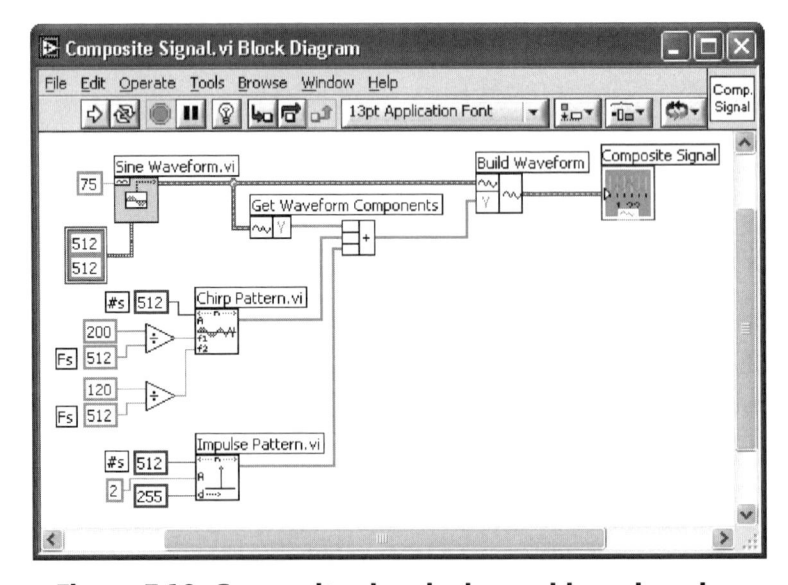

Figure 7-10: Composite signal: sine + chirp + impulse.

Now, let us create the entire transformation system using the `Composite Signal` VI just made. Create a blank VI, then select **Functions → All Functions → Select a VI...**. This brings up a window for choosing and locating a VI. Click *Composite Signal.vi* to insert it into the BD. The composite signal output is connected to three blocks consisting of a `Waveform Graph`, a `FFT`, and a `STFT` VI. The waveform data (Y component) is connected to the input of the `FFT` VI (**Functions → All Functions → Analyze → Signal Processing → Frequency Domain → FFT**). Only the first half of the output data from the `FFT` VI is taken since the other half is a mirror image of the first half. This is done by placing an `Array Subset` function and wiring to it one half of the signal length. The magnitude of the FFT output is then displayed in the `Waveform Graph`. Properties of a FP object, such as scale multiplier of a graph, can be changed programmatically by using a property node. Property nodes are discussed in the next subsection.

Getting the STFT output is more involved than FFT. The `STFT` VI (**Functions → All Functions → Advanced Signal Processing → Advanced JTFA → STFT**), which is part of the Signal Processing Toolkit (SPT), is used here for this purpose. To utilize the `STFT` VI, several inputs as well as the input signal need to be connected. These inputs are: time increment, extension, window, num of freq bin, initial condition, and final condition. The time increment input defines a step size for sliding a window along the time axis. A constant of 1 is used in the example shown in Figure 7-9. The extension input specifies the method to pad data at both ends of a signal to avoid abrupt changes in the transformed outcome. There exist five different extension options: zero padding, symmetric, periodic, spline, and user-defined. For the user-defined extension mode, an initial condition and a final condition input should be specified as well. The periodic mode is used in the example shown in Figure 7-9. The window input specifies which window to apply. In our example, a Hanning window is considered by passing an array of all 1s whose width is adjustable by the user through the `Hanning window` VI (**Functions → All Functions → Analyze → Signal Processing → Window → Hanning Window**). Similar to FFT, only one half of the frequency values are taken while the time values retain the original length. Additional details on using the `STFT` VI can be found in [2].

The output of the STFT is displayed in the `Intensity Graph` (**Controls → All Controls → Graph → Intensity Graph**). Right-click on the `Intensity Graph` and then uncheck the **Loose Fit** option under both **X Scale** and **Y Scale** from the shortcut menu. By doing this, the STFT output graph is fitted into the entire plotting area. Enable auto-scaling of intensity by right-clicking on the `Intensity Graph` and choosing **Z Scale → AutoScale Z**.

Lab 7

L7.1.1 Property Node

The number of FFT values varies based upon the number of samples. Similarly, the number of frequency rows of STFT varies based upon the number of frequency bins specified by the user. However, the scale of the frequency axis in FFT or STFT graphs should always remain between 0 and $f_s/2$, which is 256 Hz in the example, regardless of the number of frequency bins, as illustrated in Figure 7-6 and Figure 7-8. For this reason, the multiplier for the spectrogram scale needs to be changed depending on the width of the time window during run time.

A property node can be used to modify the appearance of an FP object. A property node can be created by right-clicking either on a terminal icon in a BD or an object in a FP, and then by choosing **Create → Property Node**. This way default elements, named **visible**, are created in a BD, which are linked to a corresponding FP object. Properties of FP objects can be set to either read or write. Note that by default a property node is set to read. Changing to the write mode can be done by right clicking on a property element and choosing **Change to Write**. The read/write mode of all elements can be changed together by choosing **Change all to Read/Write**.

To change the scale of the spectrogram graph, the value of the element **YScale.Multiplier** needs to be modified. Replace the element **visible** with **YScale.Multiplier** by clicking it and choosing **Y Scale → Offset and Multiplier → Multiplier**. The sampling frequency of the signal divided by the number of frequency bins, which defines the scale multiplier, is wired to the element **YScale.Multiplier** of the property node. Two more elements, **XScale.Multiplier** and **XScale.Precision**, are added to the property node for modifying the time axis multiplier and precision, respectively.

A property node of the FFT graph is also created and modified in a similar way, considering that the resolution of FFT is altered depending on the sampling frequency and number of input signal samples. The property nodes of the STFT and FFT graphs are shown in Figure 7-9. More details on using property nodes can be found in [3].

L7.2 DWT

In this transformation, the time-frequency window has high frequency resolution for higher frequencies and high time resolution for lower frequencies. This is a great advantage over STFT where the window size is fixed for all frequencies.

The BD of a 1D decomposition and reconstruction wavelet transform is shown in Figure 7-11. Three VIs including `Wavelet Filter VI`, `Discrete Wavelet Transform Ex VI`, and `Inverse Discrete Wavelet Transform Ex VI`, are used here from the advanced wavelet palette (**Functions → All Functions → Advanced Signal Processing → Advanced Wavelet**).

FFT, STFT and DWT

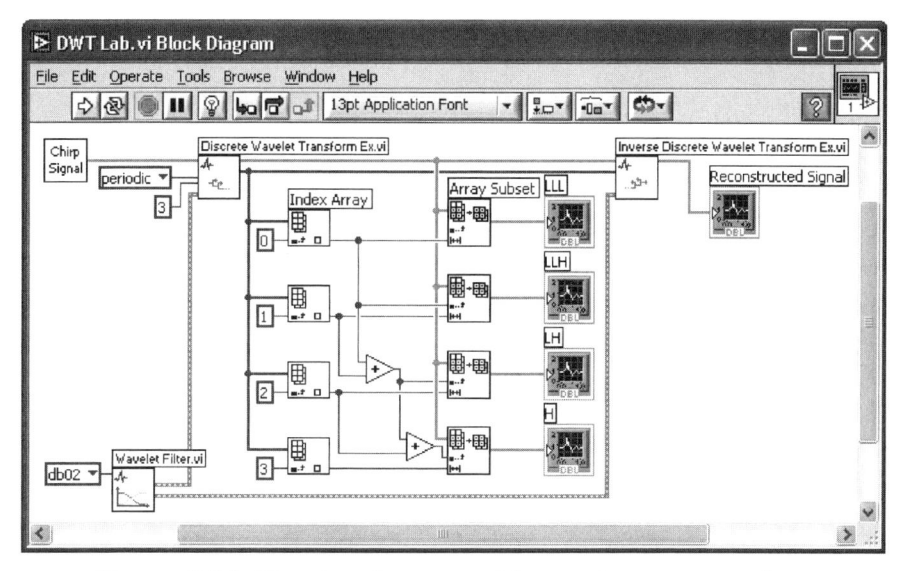

Figure 7-11: Wavelet decomposition and reconstruction.

A chirp type signal, shown in Figure 7-12, is considered to be the input signal source. This signal is designed to consist of four sinusoidal signals, each consisting of 128 samples with increasing frequencies in this order: 250, 500, 1000, 2000 Hz. This makes the entire chirp signal 512 samples. The Fourier transform of this signal is also shown in Figure 7-12.

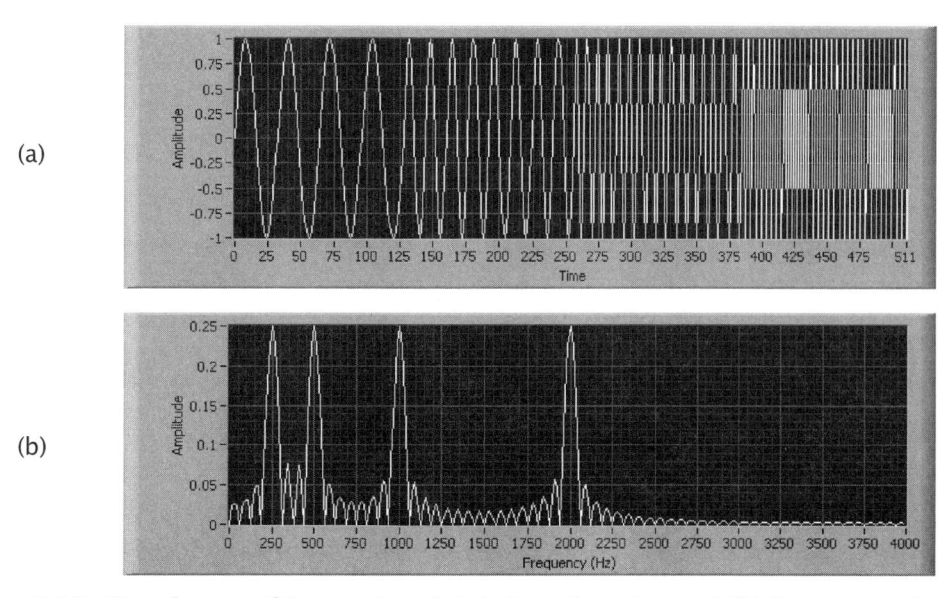

(a)

(b)

Figure 7-12: Waveforms of input signal: (a) time domain, and (b) frequency domain.

Lab 7

In Figure 7-13, the BD of this signal generation process is illustrated. Save this VI as *Chirp Signal.vi* to be used as a signal source subVI within the `DWT` VI. Note that the **Concatenate Inputs** option of the `Build Array` function should be chosen to build the 1D chirp signal. This VI has only one output terminal.

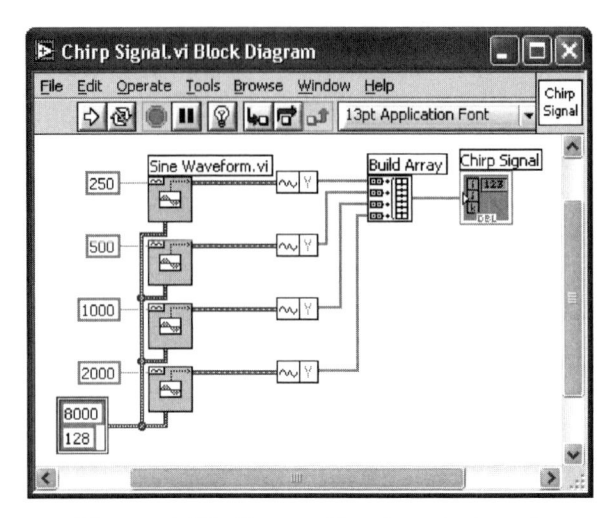

Figure 7-13: Generating input signal.

The `Discrete Wavelet Transform Ex` VI requires four inputs including input signal, extension, scales, and analysis filter. The input signal is provided by the `Chirp Signal` VI. For the extension input, the same options are available as mentioned earlier for STFT. The scales input specifies the level of decomposition. In the BD shown in Figure 7-11, a three-level decomposition is used via specifying a constant 3. The filter bank implementation for a three-level wavelet decomposition is illustrated in Figure 7-14. In this example, the Daubechies-2 wavelet is used. The coefficients of the filters are generated by the `Wavelet Filter` VI. This VI provides the coefficient sets for both the decomposition and reconstruction parts.

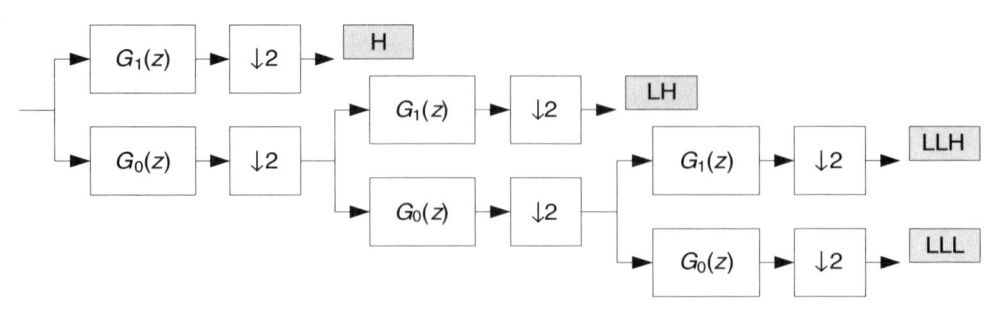

Figure 7-14: Waveform decomposition tree.

FFT, STFT and DWT

The result of the `Discrete Wavelet Transform Ex` VI is structured into a 1D array corresponding to the components of the transformed signal in this order LLL, LLH, LH, H, where L stands for low and H for high. The length of each component is also available from this VI. The wavelet decomposed outcome for each stage of the filter bank is shown in Figure 7-15. From the outcome, it can be observed that lower frequencies occur earlier and higher frequencies occur later in time. This demonstrates the fact that wavelet transform provides both frequency and time resolution, a clear advantage over Fourier transform.

The decomposed signal can be reconstructed by the `Inverse Discrete Wavelet Transform Ex` VI. From the reconstructed signal, shown in Figure 7-15, it is seen that the wavelet decomposed signal is reconstructed perfectly by using the synthesis or reconstruction filter bank.

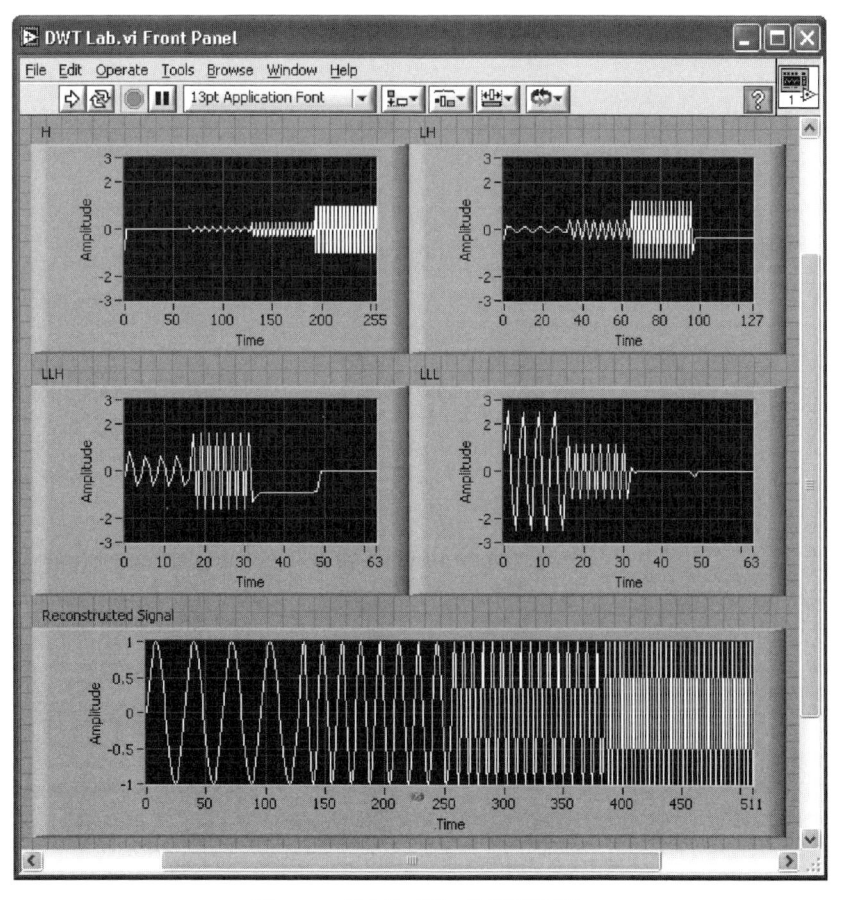

Figure 7-15: FP of DWT VI.

L7.3 Bibliography

[1] C. Burrus, R. Gopinath, and H. Gao, *Wavelets and Wavelet Transforms A Primer*, Prentice-Hall, 1998.

[2] National Instruments, *Signal Processing Toolset User Manual*, Part Number 322142C-01, 2002.

[3] National Instruments, *LabVIEW User Manual*, Part Number 320999E-01, 2003.

CHAPTER 8

DSP Implementation Platform: TMS320C6x Architecture and Software Tools

It is often computationally more efficient to implement some or most components of a signal processing system on a DSP processor. The choice of a DSP processor to use in a signal processing system is application dependent. There are many factors that influence this choice, including cost, performance, power consumption, ease-of-use, time-to-market, and integration/interfacing capabilities.

8.1 TMS320C6X DSP

The TMS320C6x family of processors, manufactured by Texas Instruments, is built to deliver speed. They are designed for million instructions per second (MIPS) intensive applications such as third generation (3G) wireless and digital imaging. There are many processor versions belonging to this family, differing in instruction cycle time, speed, power consumption, memory, peripherals, packaging, and cost. For example, the fixed-point C6416-600 version operates at 600 MHz (1.67 ns cycle time), delivering a peak performance of 4800 MIPS. The floating-point C6713-225 version operates at 225 MHz (4.4 ns cycle time), delivering a peak performance of 1350 MIPS.

Figure 8-1 shows a block diagram of the generic C6x architecture. The C6x central processing unit (CPU) consists of eight functional units divided into two sides: A and B. Each side has a .M unit (used for multiplication operation), a .L unit (used for logical and arithmetic operations), a .S unit (used for branch, bit manipulation and arithmetic operations), and a .D unit (used for loading, storing and arithmetic operations). Some instructions, such as ADD, can be done by more than one unit. There are sixteen 32-bit registers associated with each side. Interaction with the CPU must be done through these registers.

Chapter 8

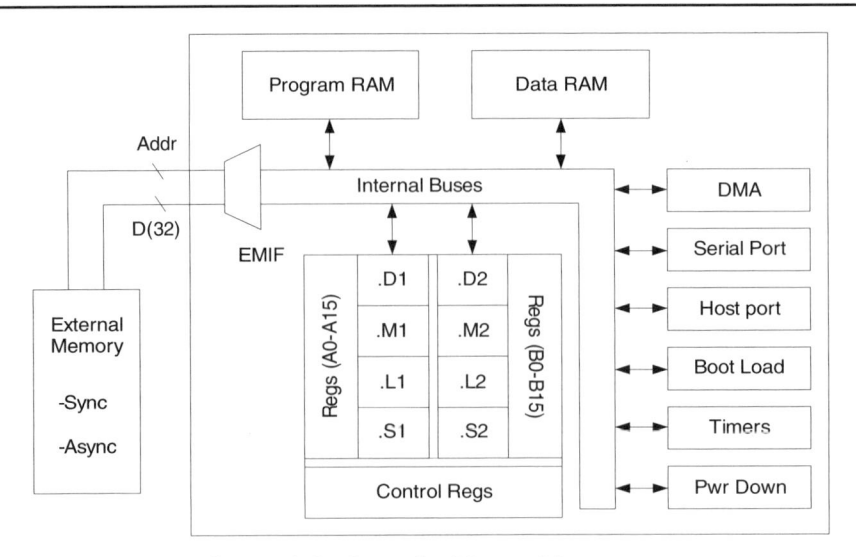

Figure 8-1: Generic C6x architecture.

As shown in Figure 8-2, the internal buses consist of a 32-bit program address bus, a 256-bit program data bus accommodating eight 32-bit instructions, two 32-bit data address buses (DA1 and DA2), two 32-bit (64-bit for C64 version) load data buses (LD1 and LD2), and two 32-bit (64-bit for the floating-point version) store data buses (ST1 and ST2). In addition, there are a 32-bit direct memory access (DMA) data and a 32-bit DMA address bus. The off-chip, or external, memory is accessed through a 20-bit address bus and a 32-bit data bus.

The peripherals on a typical C6x processor include external memory interface (EMIF), DMA, boot loader, multi-channel buffered serial port (McBSP), host port interface (HPI), timer, and power down unit. EMIF provides the necessary timing for accessing external memory. DMA allows the movement of data from one place in memory to another place without interfering with the CPU operation. Boot loader boots the code from off-chip memory or HPI to the internal memory. McBSP provides a high speed multi-channel serial communication link. HPI allows a host to access the internal memory. Timer provides two 32-bit counters. Power down unit is used to save power for durations when the CPU is inactive.

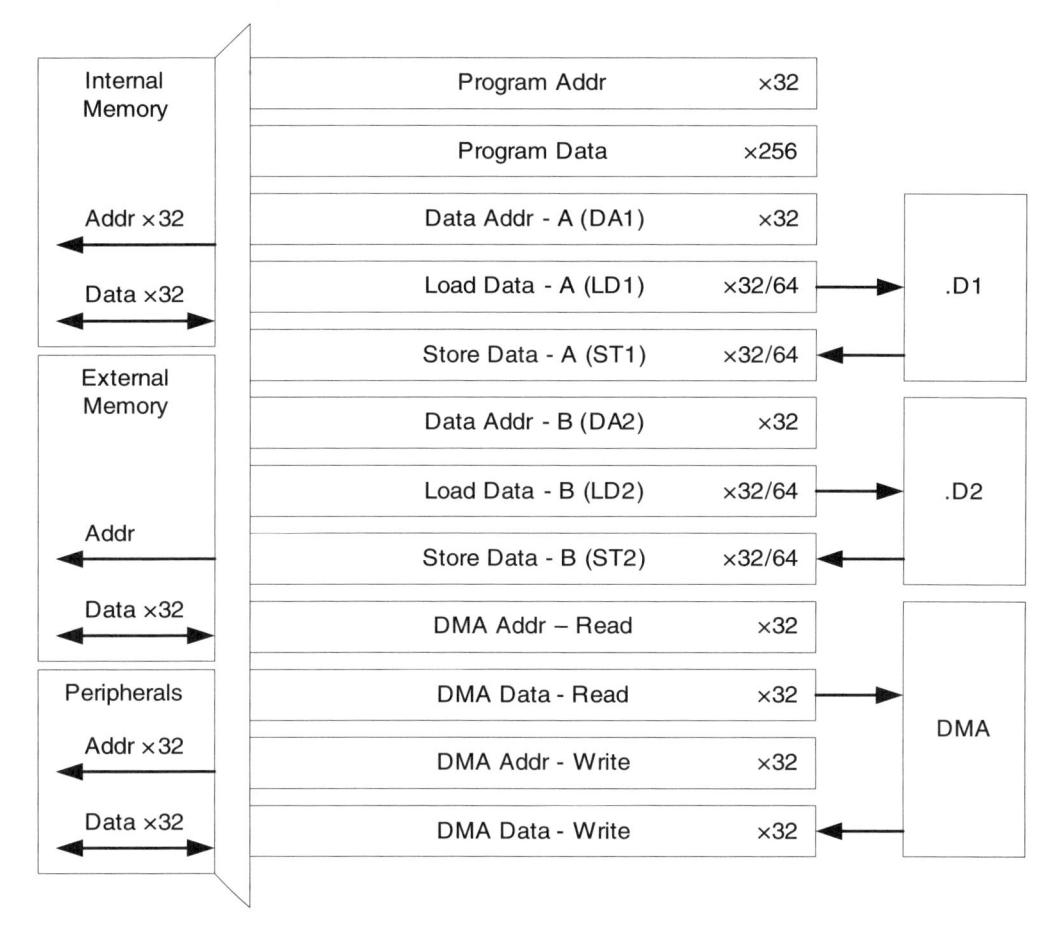

DSP Implementation Platform: TMS320C6x Architecture and Software Tools

Figure 8-2: C6x internal buses.

8.1.1 Pipelined CPU

In general, there are three basic steps to perform an instruction. They include: fetching, decoding, and execution. If these steps are done serially, not all of the resources on the processor, such as multiple buses or functional units, are fully utilized. In order to increase throughput, DSP CPUs are designed to be pipelined. This means that the foregoing steps are carried out simultaneously. Figure 8-3 illustrates the difference in the processing time for three instructions executed on a serial or nonpipelined and a pipelined CPU. As can be seen from this figure, a pipelined CPU requires fewer clock cycles to complete the same number of instructions.

The C6x architecture is based on the very long instruction word (VLIW) architecture. In such an architecture, several instructions are captured and processed

Chapter 8

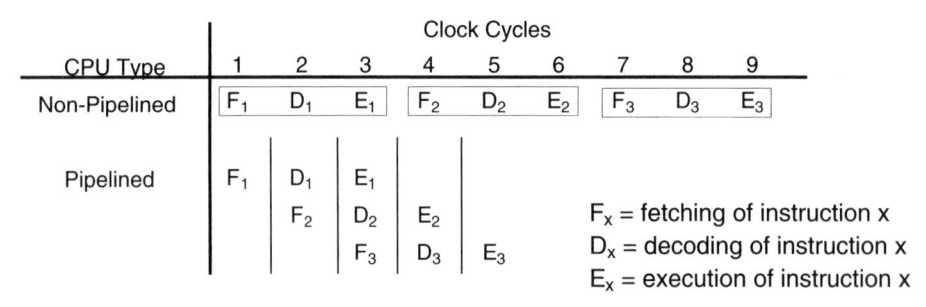

Figure 8-3: Pipelined versus nonpipelined CPU.

F_x = fetching of instruction x
D_x = decoding of instruction x
E_x = execution of instruction x

simultaneously. For more details on the TMS320C6000 architecture, the interested reader is referred to [1].

8.1.2 C64x DSP

The C64x is a more recently released DSP core, part of the C6x family, with higher MIPS power, operating at higher clock rates. This core can operate in the range of 300–1000 MHz clock rates, giving a processing power of 2400-8000 MIPS. The C64x speedups are achieved due to many enhancements, some of which are mentioned here.

Per CPU data path, the number of registers is increased from 16 to 32, A0–A31 and B0-B31. These registers support packed datatypes, allowing storage and manipulation of four 8-bit or two 16-bit values within a single 32-bit register.

Although the C64x core is code compatible with the earlier C6x cores, it can run additional instructions on packed datatypes, boosting parallelism. For example, the new instruction MPYU4 performs four, or quad, 8-bit multiplications, or the instruction MPY2 performs two, or dual, 16-bit multiplications in a single instruction cycle on a .M unit. This packed data processing capability is illustrated in Figure 8-4.

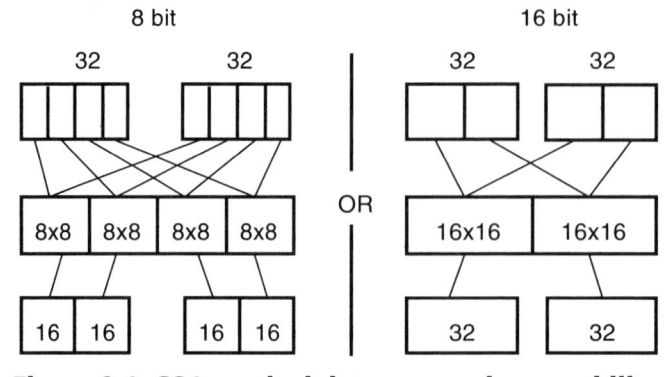

Figure 8-4: C64x packed data processing capability.

DSP Implementation Platform: TMS320C6x Architecture and Software Tools

Additional instructions have been added to each functional unit on the C64x for performing special purpose functions encountered in wireless and digital imaging applications. In addition, the functionality of each functional unit on the C64x has been improved leading to a greater orthogonality, or generality, of operations. For example, the .D unit can perform 32-bit logical operation just as the .S and .L units, or the .M unit can perform shift and rotate operations just as the .S unit.

8.2 C6x DSK Target Boards

Given the availability of a DSP starter kit (DSK) board, appropriate components of a DSP system can be run on an actual C6x processor.

8.2.1 Board configuration and peripherals

As shown in Figure 8-5, the C6713 DSK board is a DSP system which includes a C6713 floating-point DSP chip operating at 225 MHz with 8 Mbytes of on-board synchronous dynamic RAM (SDRAM), 512 Kbytes of flash memory, and a 16-bit stereo codec AIC23. The codec is used to convert an analog input signal to a digital signal for the DSP manipulation. The sampling frequency of the codec can be changed from 8 kHz to 96 kHz. The C6416 DSK board includes a C6416 fixed-point DSP chip operating at 600 MHz with 16 Mbytes of on-board SDRAM, 512 Kbytes of flash memory, and an AIC23 codec.

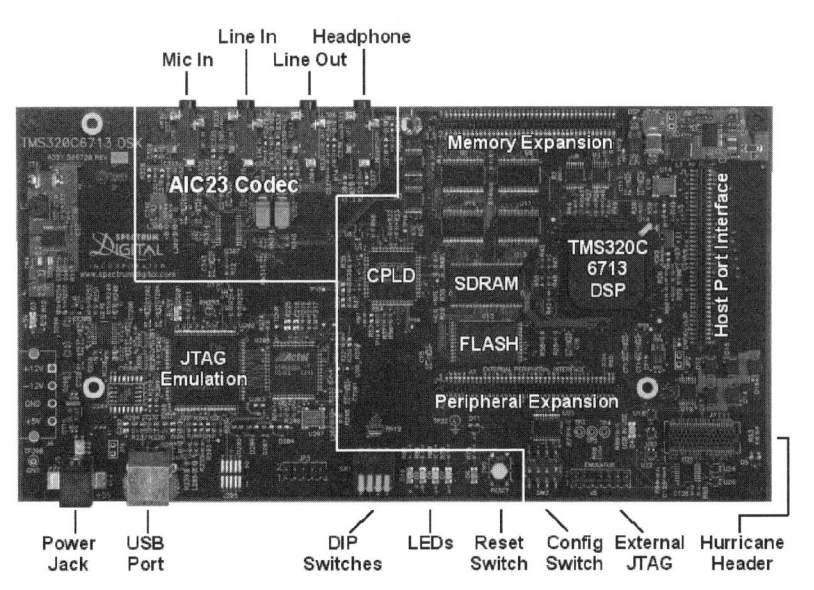

Figure 8-5: C6713 DSK board [2].

Chapter 8

8.2.2 Memory Organization

The external memory used by a DSP processor can be either static or dynamic. Static memory (SRAM) is faster than dynamic memory (DRAM), but it is more expensive, since it takes more space on silicon. SDRAM (synchronous DRAM) provides a compromise between cost and performance.

Given that the address bus is 32 bits wide, the total memory space of C6x consists of 2^{32} = 4 Gbytes. For the lab exercises in this book, the DSK board is configured based on the memory map 1 shown in Figure 8-6.

Address	Memory Map 1	Block Size (Bytes)
0000 0000	Internal RAM (L2)	64K
0001 0000	Reserved	24M
0180 0000	EMIF control regs	32
0184 0000	Cache Configuration regs	4
0184 4000	L2 base addr & count regs	32
0184 4020	L1 base addr & count regs	32
0184 5000	L2 flush & clean regs	32
0184 8200	CE0 mem attribute regs	16
0184 8240	CE1 mem attribute regs	16
0184 8280	CE2 mem attribute regs	16
0184 82C0	CE3 mem attribute regs	16
0188 0000	HPI control regs	4
018C 0000	McBSP0 regs	40
0190 0000	McBSP1 regs	40
0194 0000	Timer0 regs	12
0198 0000	Timer1 regs	12
019C 0000	Interrupt selector regs	12
01A0 0000	EDMA parameter RAM	2M
01A0 FFE0	EDMA control regs	32
0200 0000	QDMA regs	20
0200 0020	QDMA pseudo-regs	20
3000 0000	McBSP0 data	64M
3400 0000	McBSP1 data	64M
8000 0000	CE0, SDRAM	16M
9000 0000	CE1, 8-bit ROM	128K
9008 0000	CE1, 8-bit I/O port	4
A000 0000	CE2-Daughtercard	256M
B000 0000	CE3-Daughtercard	256M
1000 0000		

Figure 8-6: C6x DSK memory map [3].

If a program fits into the on-chip or internal memory, it should be run from there to avoid delays associated with accessing off-chip or external memory. If a program is too big to fit into the internal memory, most of its time-consuming portions should be placed into the internal memory for efficient execution. For repetitive codes, it is recommended that the internal memory is configured as cache memory. This allows accessing external memory as seldom as possible and hence avoiding delays associated with such accesses.

8.3 DSP Programming

Programming most DSP processors can be done either in C or assembly. Although writing programs in C would require less effort, the efficiency achieved is normally less than that of programs written in assembly. Efficiency means having as few instructions or as few instruction cycles as possible by making maximum use of the resources on the chip.

In practice, one starts with C coding to analyze the behavior and functionality of an algorithm. Then, if the required processing time is not met by using the C compiler optimizer, the time-consuming portions of the C code are identified and converted into assembly, or the entire code is rewritten in assembly. In addition to C and assembly, the C6x allows writing code in linear assembly. Figure 8-7 illustrates code efficiency versus coding effort for three types of source files on the C6x: C, linear assembly, and hand optimized assembly. As can be seen, linear assembly provides a good compromise between code efficiency and coding effort.

		Typical Efficiency	Coding Effort
C → Compiler Optimizer		50–80%	Low
Linear ASM → Assembly Optimizer		80–100%	Med
ASM → Hand Optimized		100%	High

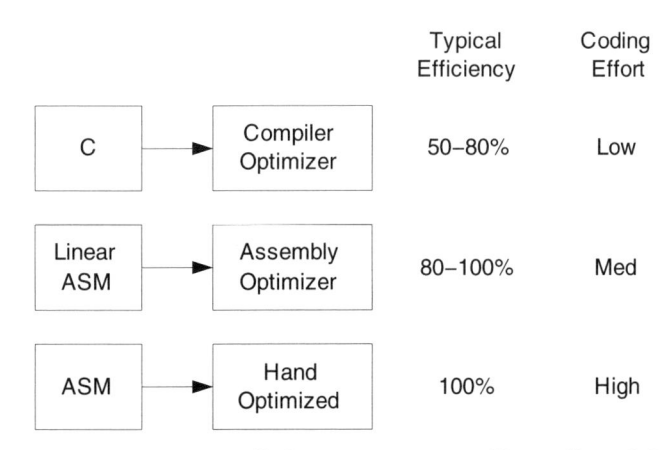

Figure 8-7: Code efficiency versus coding effort [1].

Chapter 8

More efficient code is obtained by performing assembly programming, fully utilizing the pipelined feature of the CPU. Details regarding programming in assembly/linear assembly and code optimization are discussed in [1].

8.3.1 Software Tools: Code Composer Studio

The assembler is used to convert an assembly file into an object file (.obj extension). The assembly optimizer and the compiler are used to convert, respectively, a linear assembly file and a C file into an object file. The linker is used to combine object files, as instructed by the linker command file (.cmd extension), into an executable file. All the assembling, linking, compiling, and debugging steps have been incorporated into an integrated development environment (IDE) called Code Composer Studio (CCS or CCStudio). CCS provides an easy-to-use graphical user environment for building and debugging C and assembly codes on various target DSPs. Figure 8-8 shows the steps involved for going from a source file (.c extension for C, .asm for assembly, and .sa for linear assembly) to an executable file (.out extension).

During its set-up, CCS can be configured for different target DSP boards (such as C6713 DSK, C6416 DSK, C6xxx Simulator). The version used in the book is CCS 2.2, the latest version at the time of this writing. CCS provides a file management environment for building application programs. It includes an integrated editor for editing C and assembly files. For debugging purposes, it provides breakpoints, data monitoring and graphing capabilities, profiler for benchmarking, and probe points to stream data to and from a target DSP.

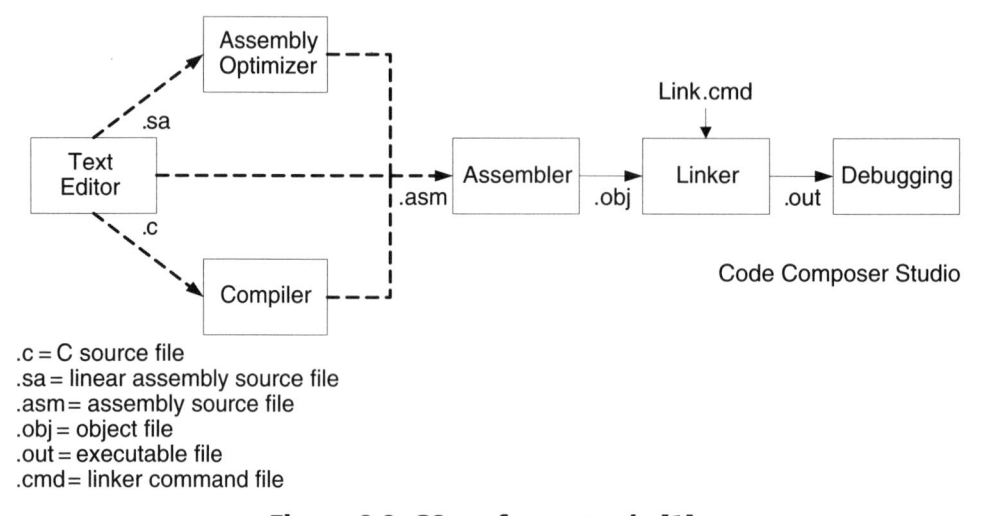

.c = C source file
.sa = linear assembly source file
.asm = assembly source file
.obj = object file
.out = executable file
.cmd = linker command file

Figure 8-8: C6x software tools [1].

8.3.2 Linking

Linking places code, constant, and variable sections into appropriate locations in memory as specified in a linker command file. Also, it combines several object files into a final executable file. A typical command file corresponding to the DSK memory map 1 is shown below in Figure 8-9.

```
MEMORY
{
    VECS:   o=00000000h   l=00000200h      /* interrupt vectors */
    PMEM:   o=00000200h   l=0000FE00h      /* Internal RAM (L2) mem */
    BMEM:   o=80000000h   l=01000000h      /* External Memory CE0, SDRAM, 16 Mbytes */
}

SECTIONS
{
    .intvecs    > 0h
    .text       > PMEM
    .far        > PMEM
    .stack      > PMEM
    .bss        > PMEM
    .cinit      > PMEM
    .pinit      > PMEM
    .cio        > PMEM
    .const      > PMEM
    .data       > PMEM
    .switch     > PMEM
    .sysmem     > PMEM
}
```

Figure 8-9: A typical command file.

The first part, MEMORY, provides a description of the type of physical memory, its origin and its length. The second part, SECTIONS, specifies the assignment of various code sections to the available physical memory. These sections are defined by directives such as .text, .data, etc.

8.3.3 Compiling

The build feature of CCS can be used to perform the entire process of compiling, assembling, and linking in one step. The compiler allows four levels of optimizations.

Debugging and full-scale optimization cannot be done together, since they are in conflict; that is, in debugging, information is added to enhance the debugging process, while in optimizing, information is minimized or removed to enhance code efficiency. In essence, the optimizer changes the flow of C code, making program debugging very difficult.

It is thus a good programming practice to first verify the proper functionality of code by using the compiler with no optimization. Then, use full optimization to make the code efficient. It is recommended that an intermediary step be taken in which some optimization is done without interfering with source level debugging. This intermediary step can re-verify code functionality before performing full optimization.

8.4 Bibliography

[1] N. Kehtarnavaz, *Real-Time Digital Signal Processing Based on the TMS320C6000*, Elsevier, 2005.

[2] Texas Instruments, "C6713 DSK Tutorial", *Code Composer Studio 2.2 C6713 DSK Help*, 2003.

[3] Texas Instruments, TMS320C6711, TMS320C6711B, TMS320C6711C, TMS320C6711D *Floating-Point Digital Signal Processors*, SPRS088L, 2004.

Lab 8: Getting Familiar with Code Composer Studio

L8.1 Code Composer Studio

This tutorial lab introduces the basic features of CCS one needs to know in order to build and debug a program running on a DSP processor. To become familiar with all of its features, refer to the *TI CCS Tutorial* [1] and *TI CCS User's Guide* manuals [2].

This lab demonstrates how a simple DSP program can be compiled and linked by using CCS. The algorithm consists of a sinewave generator using a difference equation. As part of this example, debugging and benchmarking issues are also covered. Knowledge of C programming is required for this and next lab. These labs may be skipped if the reader is only interested in the LabVIEW implementation.

Note that the accompanying CD provides the lab programs separately for the DSP target boards C6416 and C6713 DSK, as well as the simulator.

L8.2 Creating Projects

Let us consider all the files required to create an executable file; that is, .c (c), .asm (assembly), .sa (linear assembly) source files, a .cmd linker command file, a .h header file, and appropriate .lib library files. The CCS code development process begins with the creation of a so-called project to integrate and manage all these required files for generating and running an executable file. After opening CCS by double-clicking the CCS icon on the Windows® desktop, the **Project View** panel can be seen on the left-hand side of the CCS window. This panel provides an easy mechanism for building a project. In this panel, a project file (.pjt extension) can be created or opened to contain not only the source and library files but also the compiler, assembler, and linker options for generating an executable file.

To create a project, choose the menu item **Project → New** from the CCS menu bar. This brings up the dialog box **Project Creation**, as shown in Figure 8-10. In the dialog box, navigate to the working folder, here considered to be C:*ti**myprojects*, and type a project name in the field **Project Name**. Then, click the button **Finish** for CCS to create a project file named *Lab08.pjt*. All the files necessary to run a program should be added to the project.

Lab 8

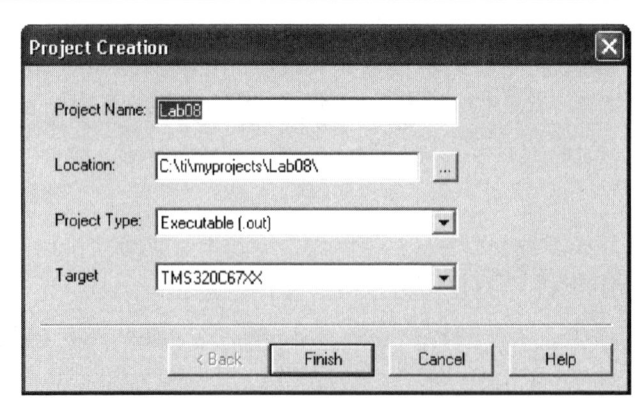

Figure 8-10: Creating a new project.

CCS provides an integrated editor which allows the creation of source files. Some of the features of the editor are color syntax highlighting, code block marking in parentheses and braces, parenthesis/brace matching, control indentions, and find/replace/search capabilities. It is also possible to add files to a project from Windows Explorer using the drag-and-drop approach. An editor window is brought up by choosing the menu item **File → New → Source File**. For this lab, let us type the following C code into the editor window:

```
#include <stdio.h>
#include <math.h>

#define PI          3.141592

void main()
{
    short i, gain;
    float fs, f;
    float y[3], a[2], b1, x;

    short *output;
    output = (short *) 0x0000FF00;

    // Coefficient Initialization

    fs = 8000;                          // Sampling frequency
    f = 500;                            // Signal frequency
    gain = 100;                         // Amplitude gain

    a[0] = -1;
    a[1] = 2 * cos(2*PI*f/fs);
    b1 = gain;
```

```
// Initial Conditions

y[1] = y[2] = 0;
x = 1;

printf("BEGIN\n");

for ( i = 0; i < 128; i++ )
{
        y[0] = b1*x + a[1]*y[1] + a[0]*y[2];
        y[2] = y[1];
        y[1] = y[0];

        x = 0;

        output[i] = (short) y[0];
}

printf("END\n");
}
```

This code generates a sinusoidal waveform $y[n]$ based on the following difference equation:

$$y[n] = B_1 x[n-1] + A_1 y[n-1] + A_0 y[n-2] \tag{8.1}$$

where $B_1 = 1$, $A_0 = -1$, $A_1 = 2\cos(\theta)$, and $x[n]$ is a delta function. The frequency of the waveform is given by [1]

$$f = \frac{f_s}{2\pi} \arccos(A_1/2) \tag{8.2}$$

By changing the coefficient A_1, the frequency can be altered. By changing the coefficient B_1, the gain can be altered.

Save the created source file by choosing the menu item **File → Save**. This brings up the dialog box **Save As**, as shown in Figure 8-11. In the dialog box, go to the field **Save as type** and select **C Source Files (*.c)** from the pull-down list. Then, go to the field **File name**, and type **sinewave.c**. Finally, click **Save** to save the code into a C source file named *sinewave.c*.

Lab 8

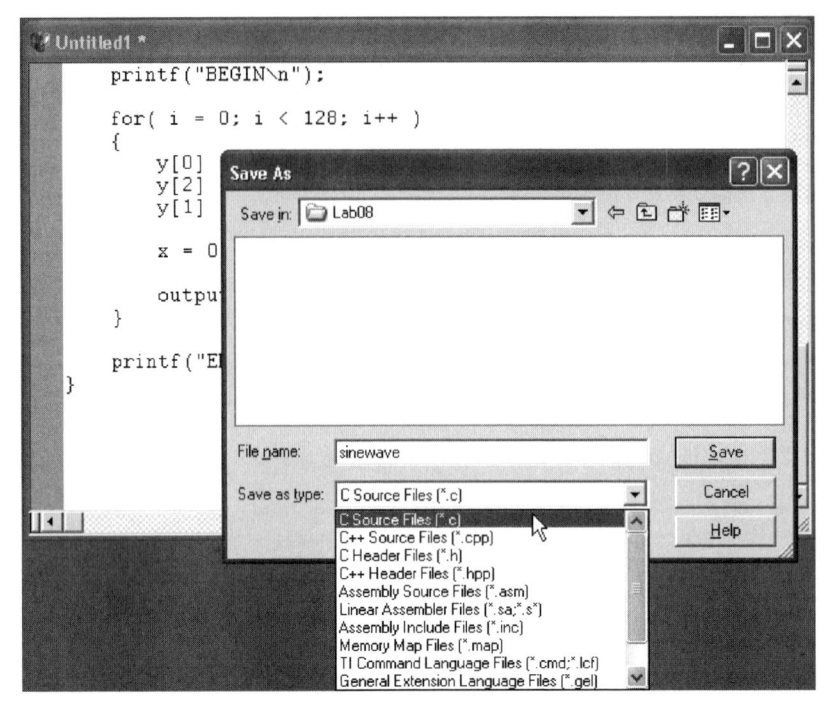

Figure 8-11: Creating a source file.

In addition to source files, a linker command file must be specified to create an executable file and to conform to the memory map of the target DSP. A linker command file can be created by choosing **File → New → Source File**. For this lab, let us type the command file shown in Figure 8-12. This file can also be downloaded from the accompanying CD, which is configured based on the DSK memory map. Save the editor window into a linker command file by choosing **File → Save** or by pressing <Ctrl + S>. This brings up the dialog box **Save As**. Go to the field **Save as type** and select **TI Command Language Files (*.cmd)** from the pull-down list. Then, type Lab08.cmd in the field **File name** and click **Save**.

Getting Familiar with Code Composer Studio

```
MEMORY
{
    PMEM:   o=00000000h   l=0000FF00h   /* Internal RAM (L2) mem  */
    PMEM2:  o=0000FF00h   l=00000100h   /* Defined for output value*/
    BMEM:   o=80000000h   l=01000000h   /* CE0, SDRAM, 16 MBytes  */
}

SECTIONS
{
    .text     >    PMEM
    .far      >    PMEM
    .stack    >    PMEM
    .bss      >    PMEM
    .cinit    >    PMEM
    .pinit    >    PMEM
    .cio      >    PMEM
    .const    >    PMEM
    .data     >    PMEM
    .switch   >    PMEM
    .sysmem   >    PMEM
}}
```

Figure 8-12: Linker command file for Lab 8.

Now that the source file *sinewave.c* and the linker command file *Lab08.cmd* are created, they should be added to the project for compiling and linking. To do this, choose the menu item **Project → Add Files to Project**. This brings up the dialog box **Add Files to Project**. In the dialog box, select **sinewave.c** and click the button **Open**. This adds *sinewave.c* to the project. In order to add *Lab08.cmd*, choose **Project → Add Files to Project**. Then, in the dialog box **Add Files to Project**, set **Files of type** to **Linker Command File (*.cmd)**, so that *Lab08.cmd* appears in the dialog box. Next, select **Lab08.cmd** and click the button **Open**. In addition to *sinewave.c* and *Lab08.cmd* files, the run-time support library should be added to the project. To do so, choose **Project → Add Files to Project**, go to the compiler library folder, here assumed to be the default option *C:\ti\c6000\cgtools\lib*, select **Object and Library Files (*.o*,*.l*)** in the box **Files of type**, then select *rts6700.lib* and click **Open**. If running on the TMS320C6416, select *rts6400.lib* instead.

After adding all the source files, the command file and the library file to the project, it is time either to build the project or to create an executable file for the target DSP. This is achieved by choosing the **Project → Build** menu item. CCS compiles, assembles, and links all of the files in the project. Messages about this process are shown in a panel at the bottom of the CCS window. When the building process is completed without any errors, the executable file *Lab08.out* is generated. It is also possible to do

Lab 8

incremental builds—that is, rebuilding only those files changed since last build, by choosing the menu item **Project → Rebuild**. The CCS window provides shortcut buttons for frequently used menu options, such as **build** 🗒 and **rebuild all** 🗓.

Although CCS provides default build options, all the compiler, assembler, and linker options can be changed via the menu item **Project → Build Options**. Among many compiler options shown in Figure 8-13, particular attention should be paid to the optimization level options. There are four levels of optimization (0, 1, 2, 3), which control the type and degree of optimization. Note that in some cases, debugging is not possible due to optimization. Thus, it is recommended to debug a program first to make sure that it is logically correct before performing any optimization.

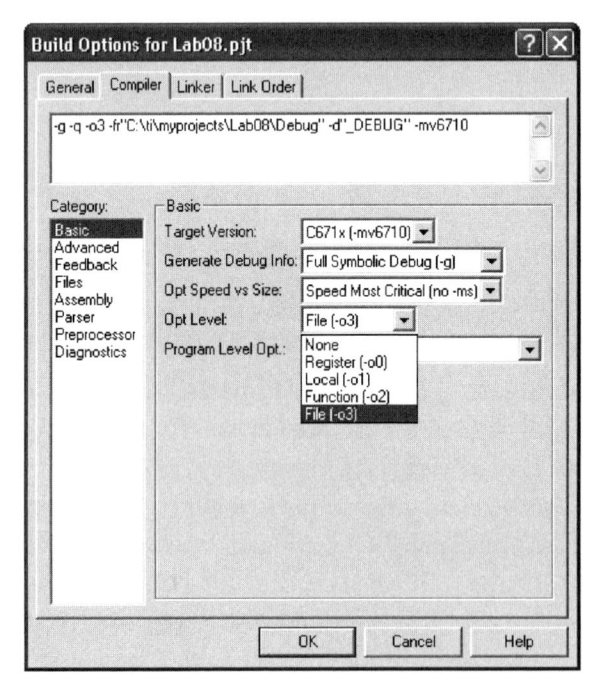

Figure 8-13: Build and compiler options.

Another important compiler option is the **Target Version** option. When implementing on the floating-point target DSP (TMS320C6713 DSK), go to the **Target Version** field and select **C671x (-mv6710)** from the pull-down list. For the fixed-point target DSP (TMS320C6416 DSK), select **C64xx (-mv6400)**.

When a message stating a compilation error appears, click **Stop Build** and scroll up in the build area to see the syntax error message. Double-click on the red text that describes the location of the syntax error. Notice that the *sinewave.c* file opens, and

Getting Familiar with Code Composer Studio

the cursor appears on the line that has caused the error, see Figure 8-14. After correcting the syntax error, the file should be saved and the project rebuilt.

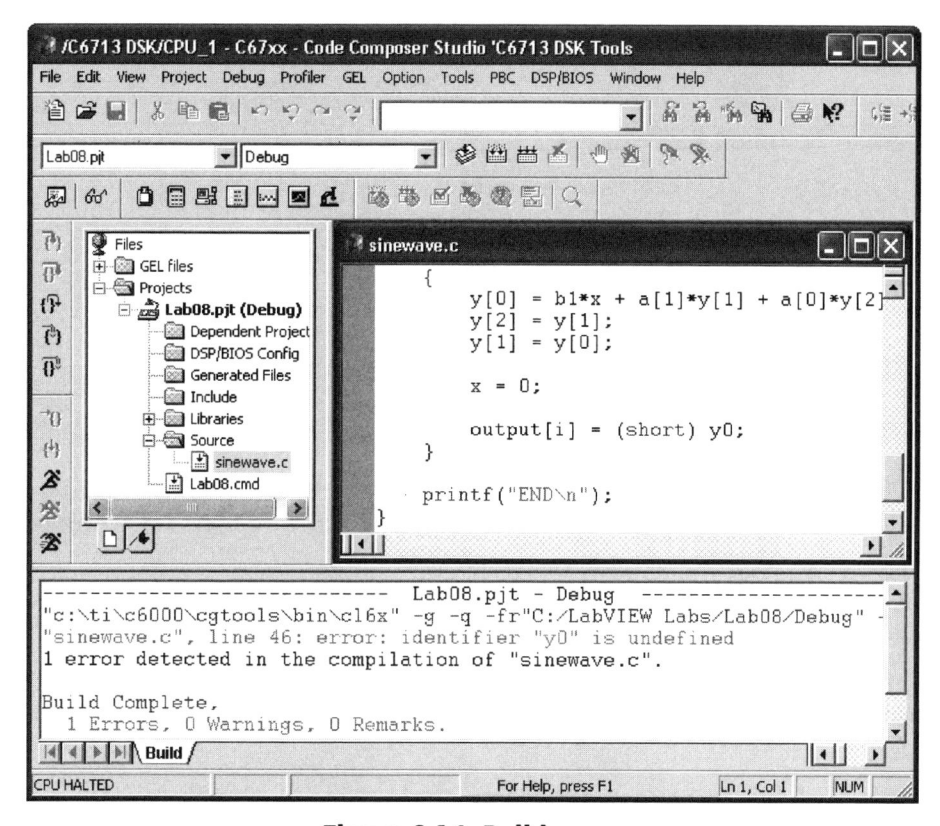

Figure 8-14: Build error.

L8.3 Debugging Tools

Once the build process is completed without any errors, the program can be loaded and executed on the target DSP. To load the program, choose **File → Load Program**, select the program *Lab08.out* just rebuilt, and click **Open**. To run the program, choose the menu item **Debug → Run**. You should be able to see BEGIN and END appearing in the **Stdout** window.

Now, let us view the content of the memory at a specific location. To do so, select **View → Memory** from the menu. The dialog box **Memory Window Options** should appear. This dialog box allows one to specify various attributes of the **Memory** window. Go to the **Address** field and enter 0x0000FF00. Then, select **16-bit Signed Int** from the pull-down list in the **Format** field and click **OK**. Note that a global variable

Lab 8

name should be used in the **Address** field. A **Memory** window appears as shown in Figure 8-15. The contents of the CPU, peripheral, DMA, and serial port registers can also be viewed by selecting **View → Registers → Core Registers**.

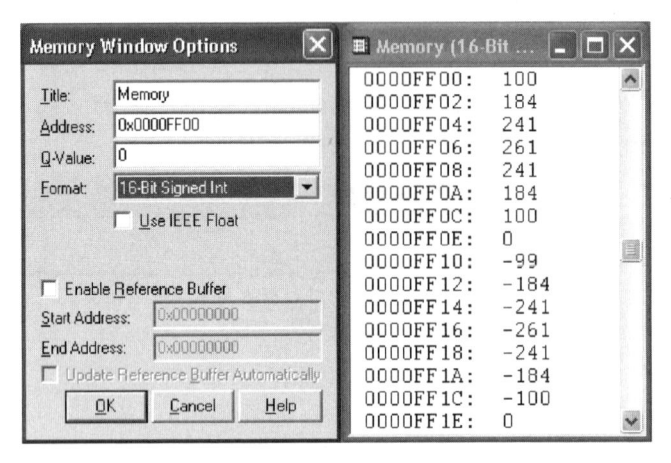

Figure 8-15: Memory Window Options dialog box and Memory window.

Data stored in the DSP memory can be saved to a PC file. CCS provides a probe point capability, so that a stream of data can be moved from the DSP to a PC host file or vice versa. In order to use this capability, a probe point should be set within the program by placing the mouse cursor at the line where a stream of data needs to be transferred and by clicking the button **Probe Point**. Choose **File → File I/O** to invoke the dialog box **File I/O**. Select the tab **File Output** for saving the data file, then click the button **Add File** and type the name of the data file. Next, the file should be connected to the probe point by clicking the button **Add Probe Point**. In the **Probe Point** field, select the probe point to make it active, then connect the probe point to the PC file through **File Out:...** in the field **Connect To**. Click the button **Replace** and then the button **OK**, see Figure 8-16. Finally, enter the memory location in the **Address** field and the data type in the **Length** field. For storing the data in short format, 64 words needs to be stated in the length field for 128 short data. A probe point connected to a PC file is shown in Figure 8-17.

Getting Familiar with Code Composer Studio

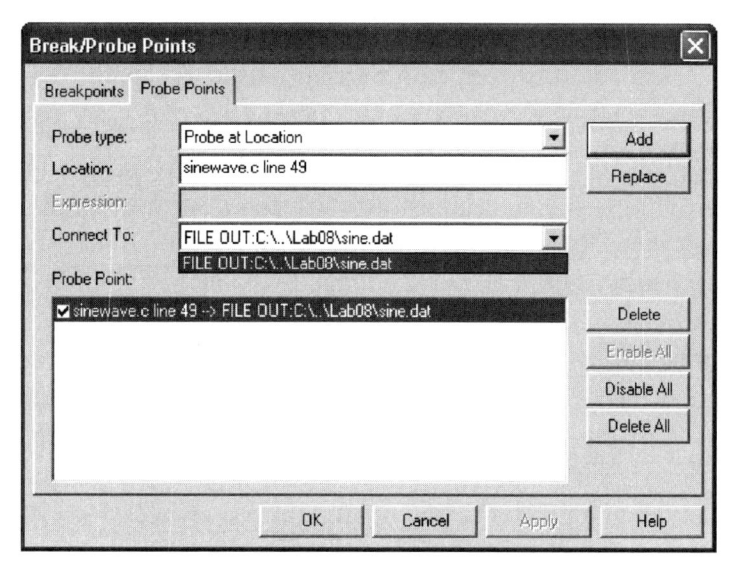

Figure 8-16: Probe Points window.

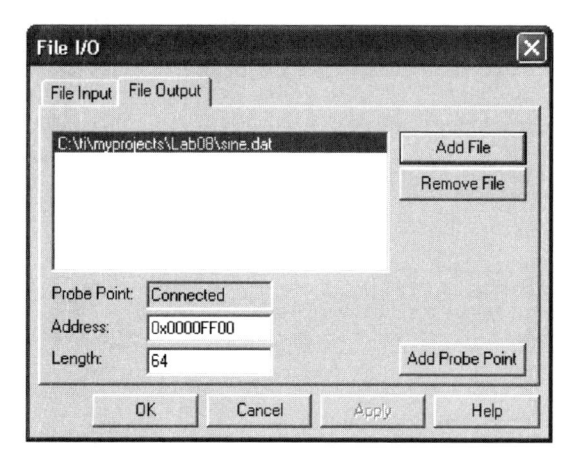

Figure 8-17: File I/O window.

The probe point capability can be used to simulate the execution of a program in the absence of a live signal. A valid PC file should have the correct file header and extension. The file header should conform to the following format:

```
MagicNumber Format StartingAddress PageNum Length
```

MagicNumber is fixed at 1651. Format indicates the format of samples in the file: 1 for hexadecimal, 2 for integer, 3 for long, and 4 for float. StartingAddress and PageNum are determined by CCS when a stream of data is saved into a PC file. Length indicates the number of samples in memory. A valid data file should have the

Lab 8

extension *.dat*. Data files having the same format can be transferred by choosing **File → Data → Load...**, instead. However, data transfer with this capability of CCS needs to be reinvoked manually while the probe point does data updates automatically.

A graphical display of data often provides better feedback about the behavior of a program. CCS provides a signal analysis interface to monitor a signal. Let us display the array of values at 0x0000FF00 as a signal or a time graph. To do so, select **View → Graph → Time/Frequency** to view the **Graph Property Dialog** box. Field names appear in the left column. Go to the **Start Address** field, click it and type 0x0000FFFF. Then, go to the **Acquisition Buffer Size** field, click it and enter 128. Also, enter 128 in the **Display Data Size** field. Finally, click on **DSP Data Type**, select **16-bit signed integer** from the pull-down list, and click **OK**, see Figure 8-18. A graph window appears based on the properties selected. This window is illustrated in Figure 8-19. Properties of the graph window can be changed by right-clicking on it and selecting **Properties** at any time during the debugging process.

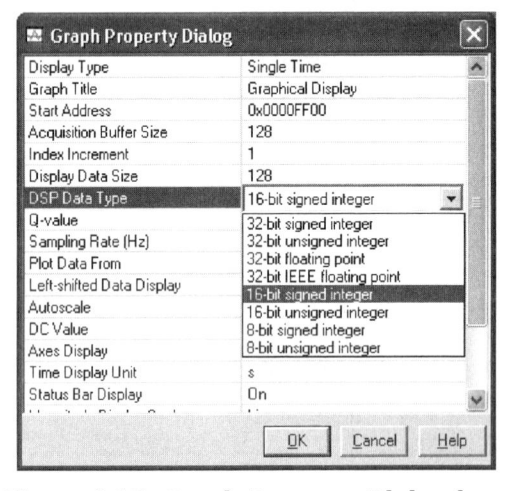

Figure 8-18: Graph Property Dialog box.

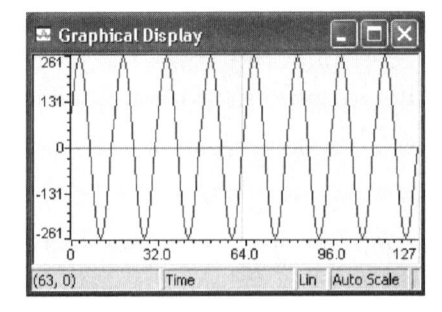

Figure 8-19: Graphical Display window.

176

Getting Familiar with Code Composer Studio

To access a specific location of the DSP memory, a memory address can be assigned directly to a pointer. It is necessary to typecast the pointer to short since the values are of that type. In the code shown, a pointer is assigned to 0x0000FF00.

When developing and testing programs, one often needs to check the value of a variable during program execution. This can be achieved by using breakpoints and watch windows. To view the values of the pointer in *sinewave.c* before and after the pointer assignment, choose **File → Reload Program** to reload the program. Then, double-click on *sinewave.c* in the **Project View** panel to bring up the source file, see Figure 8-20. You may wish to make the window larger to see more of the file in one place. Next, put the cursor on the line that reads `output = (short *) 0x0000FF00` and press <F9> to set a breakpoint. To open a watch window, choose **View → Watch Window** from the menu bar. This will bring up a **Watch Window** with the local variables listed in the Watch Locals tab. To add a new expression to the **Watch Window**, select the Watch 1 tab, then type `output` (or any expression you desire to examine) in the Name column. Then, choose **Debug → Run** or press <F5>. The program stops at the breakpoint and the Watch Window displays the value of the pointer. This is the value before the pointer is set to 0x0000FF00. By pressing <F10> to step over the line, or the shortcut button, one should be able to see the value 0x0000FF00 in the Watch Window.

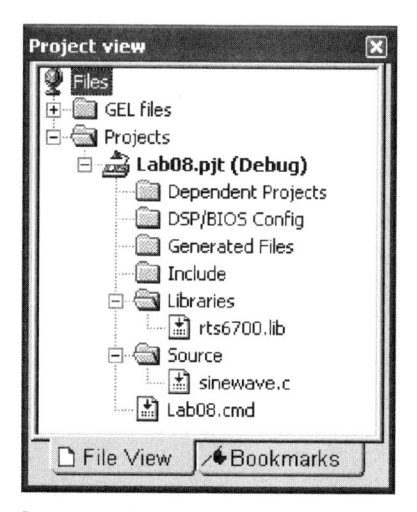

Figure 8-20: Project View panel.

Lab 8

To add a C function that sums the values, we can simply pass a pointer to an array and have a return type of integer. The following C function can be used to sum the values and return the result:

```c
#include <stdio.h>
#include <math.h>

#define PI          3.141592

void main()
{
    short i, gain;
    int ret;
    float fs, f;
    float y[3], a[2], b1, x;

    short *output;
    output = (short *) 0x0000FF00;

    // Coefficient Initialization

    fs = 8000;                          // Sampling frequency
    f = 500;                            // Signal frequency
    gain = 100;                         // Amplitude gain

    a[0] = -1;
    a[1] = 2 * cos(2*PI*f/fs);
    b1 = gain;

    // Initial Conditions

    y[1] = y[2] = 0;
    x = 1;

    printf("BEGIN\n");

    for ( i = 0; i < 128; i++ )
    {
        y[0] = b1*x + a[1]*y[1] + a[0]*y[2];
        y[2] = y[1];
        y[1] = y[0];

        x = 0;

        output[i] = (short) y[0];
    }

    ret = sum(output,128);
```

```c
        printf("Sum = %d\n", ret);
        printf("END\n");
}

int sum(const short* array,int N)
{
    int count,sum;
    sum = 0;

    for(count=0 ; count < N ; count++)
        sum += array[count];

    return(sum);
}
```

As part of the debugging process, it is normally required to benchmark or time a program. In this lab, let us determine how much time it takes for the function `sum()` to run. To achieve this benchmarking, reload the program and choose **Profiler → Start New Session**. This will bring up **Profile Session Name**. Type a session name, MySession by default, then click **OK**. The **Profile** window showing the code size and the cycle statistics will be docked at the bottom of CCS. Resize this window by dragging its edges or undock it so that all the columns can be seen. Now select the code area of the function to be benchmarked, then right-click and choose **Profile Function → in MySession Session** from the shortcut menu. The name of the function is added to the list in the **Profile** window. The same step can be achieved by clicking **Profile All Functions** on the MySession Profile window, and deleting unnecessary functions from the list. Finally, press <F5> to run the program. Examine the number of cycles shown in Figure 8-21 for `sum()`. It should be about 161 cycles (the exact number may vary slightly). This is the number of cycles it takes to execute the function `sum()`.

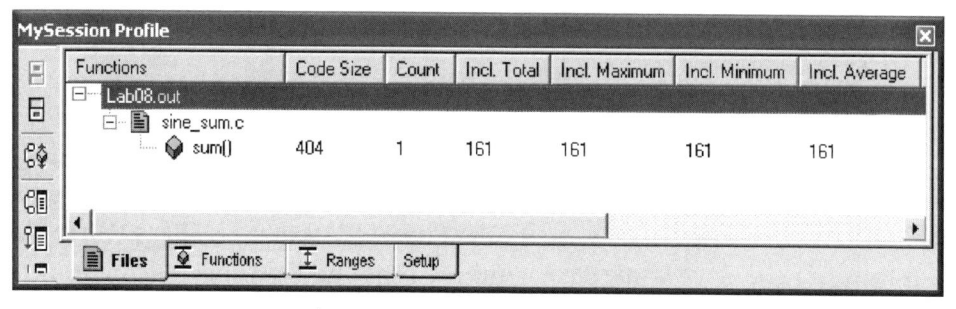

Figure 8-21: Profile window.

Lab 8

There is another way to benchmark codes using breakpoints. Double-click on the file *sinewave.c* in the **Project View** panel and choose **View** → **Mixed Source/ASM** to list the assembled instructions corresponding to the C code lines. Set a breakpoint at the calling line by placing the cursor on the line that reads `ret = sum(point,128)`, then press <F9> or double-click **Selection Margin** located on the left-hand side of the editor. Set another breakpoint at the next line as indicated in Figure 8-22. Once the breakpoints are set, choose **Profiler** → **Enable Clock** to enable a profiler clock. Then, choose **Profiler** → **View Clock** to bring up a window displaying **Profile Clock**. Now, press <F5> to run the program. When the program is stopped at the first breakpoint, reset the clock by double-clicking the inner area of the **Profile Clock** window. Finally, click **Step Out** or **Run** in the **Debug** menu to execute and stop at the second breakpoint. Examine the number of clocks in the **Profile Clock** window. It should read 4189. The difference in the number of cycles between the breakpoint and the profile approaches originates from the extra procedures for calling functions, for example passing arguments to functions, storing return addresses, branching back from functions, and so on.

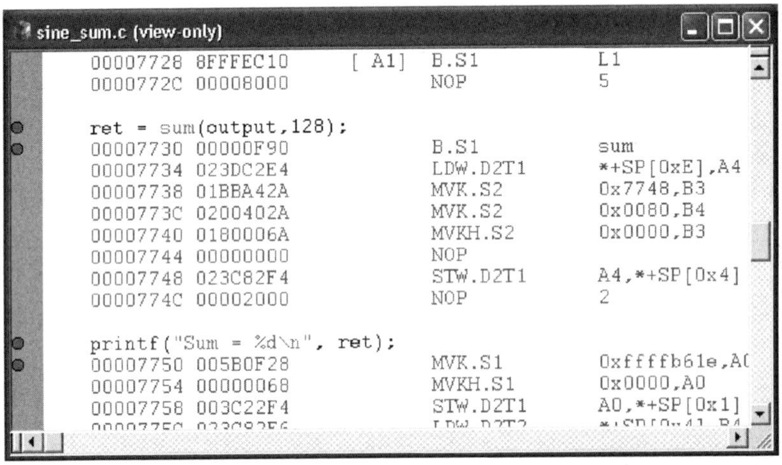

Figure 8-22: Profiling code execution time using breakpoints.

A workspace containing breakpoints, probe points, graphs and watch windows can be stored and recalled later. To do so, choose **File** → **Workspace** → **Save Workspace As**. This will bring up the **Save Workspace** window. Type the workspace name in the **File name** field, then click **Save**.

Table 8-1 provides the number of cycles it takes to run the `sum()` function using several different builds. When a program is too large to fit into the internal memory,

Getting Familiar with Code Composer Studio

it has to be placed into the external memory. Although the program in this lab is small enough to fit in the internal memory, it is also placed in the external memory to study the change in the number of cycles. To move the program into the external memory, open the command file *Lab08.cmd* and replace the line `.text > PMEM` with `.text > BMEM`. As seen in Table 8-1, this build slows down the execution to 37356 cycles. In the second build, the program resides in the internal memory and the number of cycles is hence reduced to 4180. By increasing the optimization level, the number of cycles can be further decreased to 161. At this point, it is worth pointing out that the stated numbers of cycles in this lab correspond to the C6713 DSK with CCS version 2.2. The numbers of cycles vary depending on the DSK target and CCS version used.

Table 8-1: Number of cycles for different builds.

Type of Build	Code size	Number of Cycles
C program in external memory	148	37356
C program in internal memory	148	4180
–o0	60	1085
–o1	60	1078
–o2	128	198
–o3	404	161

L8.4 Bibliography

[1] Texas Instruments, *TMS320C6000 Code Composer Studio Tutorial*, Literature ID# SPRU 301C, 2000.

[2] Texas Instruments, *Code Composer Studio User's Guide*, Literature ID# SPRU 328B, 2000.

CHAPTER 9

LabVIEW DSP Integration

A DSP system designed in LabVIEW can be placed entirely or partially on a hardware platform. This chapter discusses the implementation process on a DSP hardware platform consisting of a TMS320C6713 or TMS320C6416 DSK board. Such an implementation or integration is made possible by using the LabVIEW toolkit DSP Test Integration for TI DSP.

9.1 Communication with LabVIEW: Real-Time Data Exchange (RTDX)

Communication between LabVIEW and a C6x DSK board is achieved by using the real-time data exchange (RTDX™) feature of the TMS320C6x DSP. This feature allows one to exchange data between a DSK board and a PC host (running LabVIEW) without stopping program execution on the DSP side. This data exchange is done either via the joint test action group (JTAG) connection, or the universal serial bus (USB) port emulating the JTAG connection.

RTDX can be configured in two modes: noncontinuous and continuous. In noncontinuous mode, data is written to a log file on the host. This mode is normally used for recording purposes. In continuous mode, data is buffered by the RTDX host library. This mode is normally used for continuously displaying data. Here, to view the processed data on the PC/LabVIEW side, RTDX is configured to be in continuous mode.

9.2 LabVIEW DSP Test Integration Toolkit for TI DSP

The DSP Test Integration for TI DSP toolkit provides a set of VIs which enable interfacing between LabVIEW and Code Composer Studio [1]. The VIs provided in

Chapter 9

this toolkit are categorized into two groups: CCS Automation and CCS Communication. These VI groups are listed in Table 9-1.

Table 9-1: List of VIs in the LabVIEW DSP Test Integration toolkit.

CCS Automation VIs	CCS Communication VIs
`CCS Open Project` VI	`CCS RTDX Read` VI
`CCS Build` VI	`CCS RTDX Write` VI
`CCS Download Code` VI	`CCS RTDX Enable` VI
`CCS Run` VI	`CCS RTDX Enable Channel` VI
`CCS Halt` VI	`CCS RTDX Disable` VI
`CCS Close Project` VI	`CCS RTDX Disable Channel` VI
`CCS Window Visibility` VI	`CCS Memory Read` VI
`CCS Reset` VI	`CCS Memory Write` VI
	`CCS Symbol to Memory Address` VI

The VIs in the CCS Automation group allow automating the CCS code execution process through LabVIEW. They include: (a) open CCS, (b) build project, (c) reset CPU, (d) load program, (e) run code, (f) halt CPU, and (g) close CCS. The flow of these steps is the same as those in CCS.

The VIs in the CCS Communication group allow exchange of data through the RTDX channel. For example, the `CCS RTDX write` VI and `CCS RTDX read` VI are used for writing and reading data to and from the DSP side, respectively. Note that these VIs are polymorphic. Therefore, data types (such as single precision, double precision, or integer) and data formats (such as scalar or array) should be matched in LabVIEW and CCS in order to establish a proper LabVIEW DSP integration.

9.3 Combined Implementation: Gain Example

In this section, a LabVIEW DSP integration example is presented to show the basic steps that are required for a combined LabVIEW and DSP implementation. From the main dialog of LabVIEW, open the NI Example Finder shown in Figure 9-1 by choosing **Help → Find Examples**.

Open the `Gain_dsk6713` VI by clicking on **Directory Structure** from the category **Browse according to** of the **Browse** tab, and by choosing **DSPTest → dsk6713 → Gain → Gain_dsk6713.vi**. If using a C6416 DSK, open the dsk6416 folder.

In this example, an input signal along with a gain factor are sent from the LabVIEW side to the DSP side. On the DSP side, the input signal is multiplied by the gain

LabVIEW DSP Integration

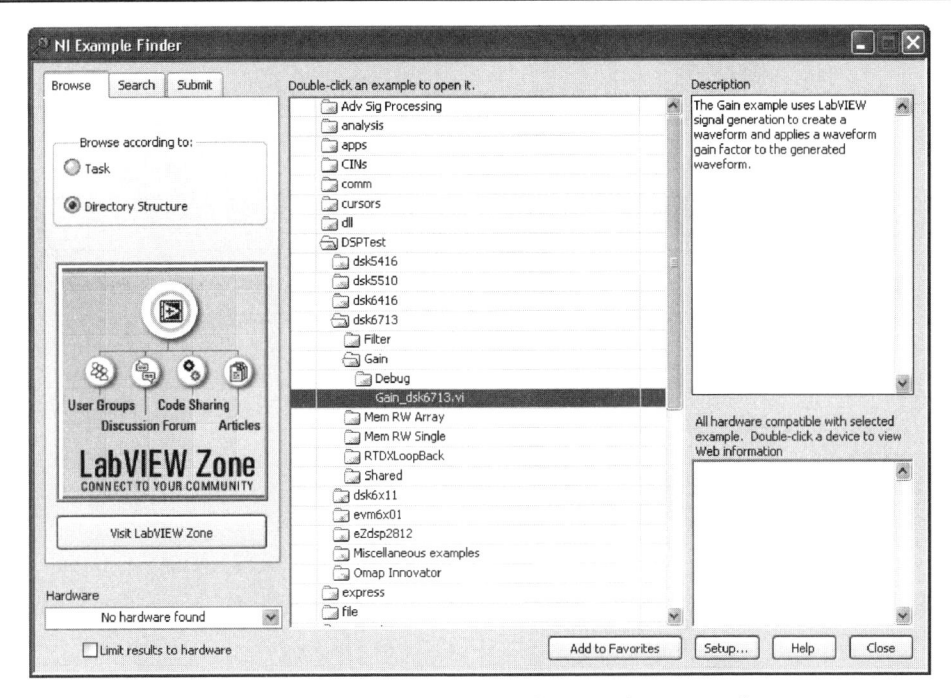

Figure 9-1: NI Example Finder—Gain example.

factor, then sent back to the LabVIEW side. The gain factor, the frequency of the input signal, and the signal type can be altered by the controls specified in the FP. Also, the original and scaled signals are displayed in the FP as shown in Figure 9-2.

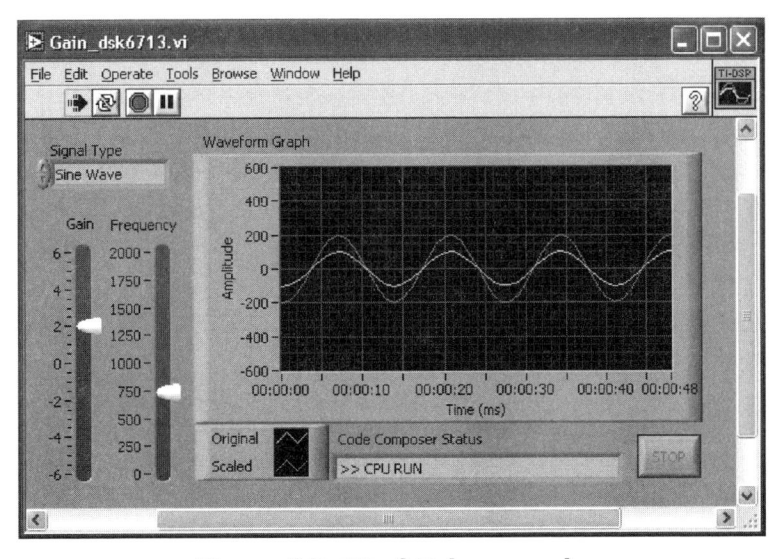

Figure 9-2: FP of Gain example.

9.3.1 LabVIEW Configuration

To better understand the LabVIEW DSP integration process, let us examine the BD of the `Gain_dsk6713` VI, which is shown in Figure 9-3.

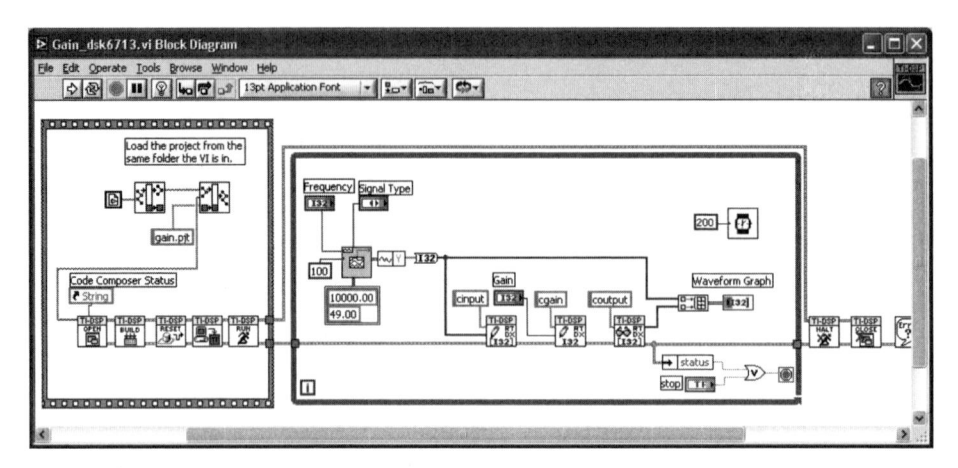

Figure 9-3: BD of Gain example.

There are two major sections associated with this BD. The first section consisting of a `Stacked Sequence` structure, shown to the left side of the `While Loop`, corresponds to the CCS automation process. This section includes a `CCS Open Project` VI, a `CCS Build` VI, a `CCS Reset` VI, a `CCS Download Code` VI, and a `CCS Run` VI. In addition, a `CCS Halt` VI and a `CCS Close Project` VI, shown to the right side of the `While Loop`, are a part of the CCS automation process. The three functions (`Strip Path`, `Build Path`, and `Current VI's Path`) of the File I/O palette are used in the `Stacked Sequence` structure to create a file path to a CCS project file that can be opened from the CCS side. With these VIs and functions in place, the process of opening CCS, building a project, loading a program to the DSP and running it on the DSP can be controlled from the LabVIEW side. The CCS Automation VIs are located in the DSP Test Integration Palette (**Functions → All Functions → DSP Test Integration**). Note that the CCS automation process just described can be used for all the LabVIEW DSP integration examples presented in Lab 9.

The second section of the BD shown in the `While Loop` involves signal generation and CCS RTDX communication. The `Basic Function Generator` VI is used to generate waveform samples. Two `CCS RTDX read` VIs and one `CCS RTDX write` VI are located in the `While Loop`. The channel name of each CCS RTDX VI is wired to this VI. This allows the generated samples to be continuously sent to

LabVIEW DSP Integration

the DSP side, and the DSP processed samples to be continuously read from the DSP side. The scaled signal is displayed in a `Waveform Graph`. Note that the original and scaled signals are of array type, while the gain factor is scalar. Thus, one of the `CCS RTDX read/write` VIs is set to 32-bit integer array, indicated by [I32] on its icon, and the other is set to 32-bit integer, indicated by I32 on its icon.

9.3.2 DSP Configuration

A CCS Project implemented on the DSP side should include four components: a linker command file, an interrupt service table which defines the interrupt vector for RTDX, the RTDX library along with the run-time support library, and the source code as shown in Figure 9-4.

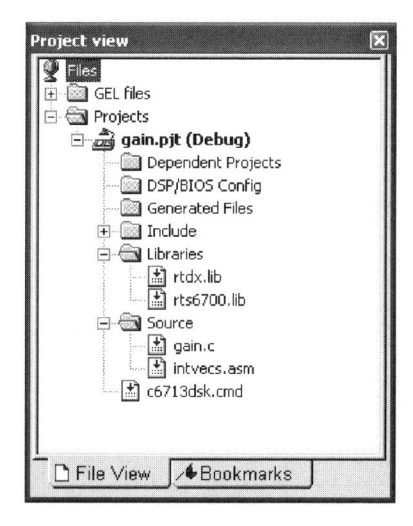

Figure 9-4: Project view of CCS.

The source code of the Gain example running on the DSP side is shown below.

```
#include <rtdx.h>                       /* RTDX                    */
#include "target.h"                     /* TARGET_INITIALIZE()     */

#define kBUFFER_SIZE 49

RTDX_CreateInputChannel(cinput);
RTDX_CreateInputChannel(cgain);
RTDX_CreateOutputChannel(coutput);

// Gain value scales the wavefrom
void Gain (int *output, int *input, int gain)
{
```

Chapter 9

```
    int i;
    for(i=0; i<kBUFFER_SIZE; i++)
        output[i]=input[i]*gain;
}

void main()
{
    int input[kBUFFER_SIZE];
    int output[kBUFFER_SIZE];
    int gain = 10;

    // Target initialization for RTDX
    TARGET_INITIALIZE();

    /*enable RTDX channels*/
    RTDX_enableInput(&cgain);
    RTDX_enableInput(&cinput);
    RTDX_enableOutput(&coutput);

    for (;;)     /* Infinite message loop. */
    {
        /* Read new gain value if one exists */
        if (!RTDX_channelBusy(&cgain))
            RTDX_readNB(&cgain, &gain, sizeof(gain));
        /* Wait for input waveform */
        while(!RTDX_read(&cinput, input, sizeof(input)));

        Gain (output, input, gain);

        /* Write scaled data back to host. */
        RTDX_write(&coutput, output, sizeof(output));
    }
}
```

In this code, several application program interfaces (APIs), which are part of the CCS RTDX library, are used to allow data exchange between the DSP and LabVIEW side. First, the `RTDX_CreateInputChannel()` and `RTDX_CreateOutputChannel()` APIs are used to declare the input and output channels. Second, the DSP board is initialized with the `TARGET_INITIALIZE()` API. Both of the RTDX channels are enabled by the `RTDX_enableInput()` and `RTDX_enableOutput()` APIs. To get scalar data from the LabVIEW to the DSP side, the `RTDX_readNB()` API is used with the arguments being channel, buffer pointer, and buffer size. In addition, the `RTDX_read()` API is used with the arguments being channel, array pointer, and array size. The `RTDX_write()` API is used to send data back to the LabVIEW side.

Bear in mind that the name assigned to the RTDX communication channel should be the same as the one used in LabVIEW. Also, the data types of polymorphic VIs, i.e. CCS RTDX Read VI and CCS RTDX Write VI, as well as the array lengths should be the same as the ones defined in LabVIEW. For example, the input array input[] in the above source code should be defined as follows:

```
int input[kBUFFER_SIZE];
```

That is, input[] must be declared as 32-bit integer, and the array size must be configured to be kBUFFER_SIZE, which is specified as 49 at the beginning of the Gain example code.

9.4 Bibliography

[1] National Instruments, *LabVIEW DSP Test Toolkit for TI DSP User's Manual*, Literature Number 323452A-01, 2002.

Lab 9: DSP Integration Examples

This lab includes four DSP integration examples. These examples correspond to the DSP systems built by LabVIEW in the previous labs, that is, digital filtering, integer arithmetic, adaptive filtering, and frequency processing.

L9.1 CCS Automation

Figure 9-5 illustrates the CCS automation process. In this lab, all the examples are assumed to have the sub-diagrams shown to the left and right of the `While Loop` and thus are not explicitly mentioned.

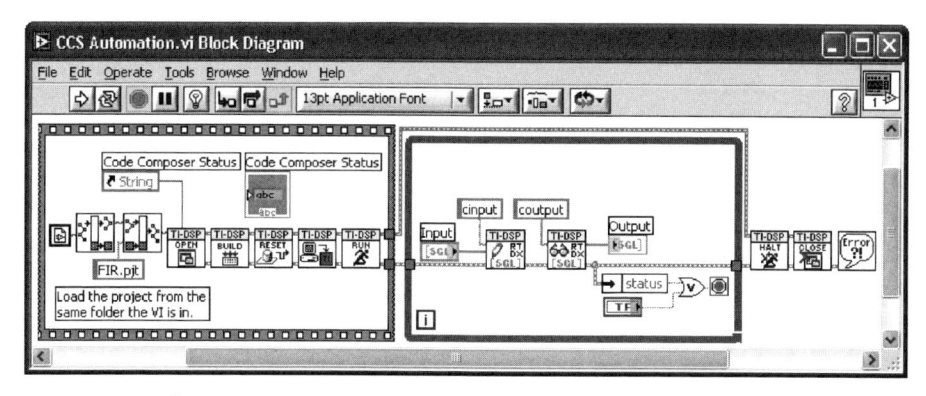

Figure 9-5: Generic structure of CCS Automation.

Let us explain the CCS automation process in more detail. In order to specify a project to be used by the CCS Automation VIs, a file path to the project file should be built. Two methods of creating a path are mentioned here. The first method involves using a relative path by assuming that the CCS project file is located in the folder where the VI resides. Place a `Current VI's Path` function (**Functions → All Functions → File I/O → File Constants → Current VI's Path**) in the BD to get the entire file path of the project file and wire the output of this VI to the `path` terminal of a `Strip Path` function (**Functions → All Functions → File I/O → Strip Path**), which returns a stripped path by removing the VI's name from the path. The stripped path is wired to the `base path` terminal of the `Build Path` function (**Functions → All Functions → File I/O → Build Path**). This function appends the name of the file, wired to the `name or relative path` terminal of the function as a string constant, to the stripped path. Now, the output of the function indicates the entire path of the

CCS project file. The created file path is then wired to the `Path to Project` terminal of the `CCS Open Project` VI to allow access by the CCS Automation VIs.

The second method of creating a project path involves using an absolute path. Wire a `Path Constant` (**Functions → All Functions → File I/O → File Constants → Path Contant**) to the `Path to Project` terminal of the `CCS Open Project` VI or create a file constant by right-clicking and choosing **Create → Constant** on the `Path to Project` terminal of the VI. Enter the absolute path of the CCS project file in the `Path Constant`. The absolute path can also be generated by browsing the project file path. To do this, right-click on the `Path Constant` and choose **Browse for Path...** from the shortcut menu. A file dialog box appears to select the path via file browsing.

Next, place the CCS Automation VIs (`CCS Open Project` VI, `CCS Build` VI, `CCS Reset` VI, `CCS Download Code` VI, and `CCS Run` VI) from the DSP Test Integration Palette (**Functions → All Functions → DSP Test Integration**) in the order shown in Figure 9-5. These VIs are wired to each other via the terminals `CCS references out` (or `dup CCS references`) and `error out` to the terminals `CCS references in` and `error in`. The VIs are used to open a CCS project, build a project, reset CPU, download a program to the DSP, and run the program. The CCS references cluster, wired to all the CCS Automation VIs, contains the CCS IDE references, while the error in/out cluster carries the error information of the CCS Automation VIs. Consequently, if an error occurs in one of the CCS Automation VIs, the error information is passed through the CCS Automation VIs to the `Simple Error Handler` VI (**Functions → All Functions → Time & Dialog → Simple Error Handler**) located at the end of the CCS Automation VIs. This VI displays a description of the error.

A `String Indicator` is placed in the FP in order to display the current status of the CCS automation process. A `Control Reference` for this indicator is created by right-clicking on it and choosing **Create → Reference** from the shortcut menu. This reference should be wired to the `Status String Refnum` terminal of the `CCS Open Project` VI in order to post a status string to this indicator.

Now, let us explain the exchange of data between LabVIEW and the DSP. Data is continuously exchanged using the CCS Communication VIs, `CCS RTDX Read` VI and `CCS RTDX Write` VI, when both the VI and CCS are running. As mentioned earlier, data types should be carefully configured on the LabVIEW and CCS sides since the `CCS RTDX Read` VI and the `CCS RTDX Write` VI are data type polymorphic. The read/write data type can be specified from the **Select Type** menu as part

DSP Integration Examples

of the shortcut menu. This menu can be brought up by right-clicking on the `CCS RTDX Read` VI or `CCS RTDX Write` VI. Another way to change the data type is by using a Polymorphic VI Selector. This selector can be displayed by right-clicking on it and choosing **Visible Items → Polymorphic VI Selector** from the shortcut menu. The string constants indicate the names of the RTDX channels that are wired to the `CCS RTDX Read` VI or `CCS RTDX Write` VI.

The execution of the `While Loop` structure is stopped by pressing a stop button in the FP or if an error is generated by the CCS Automation or CCS Communication VIs. In such cases, the CCS needs to be halted and closed. This is done by locating and wiring a `CCS Halt` VI to a `CCS Close Project` VI. These VIs appear to the right side of the `While Loop` structure.

L9.2 Digital Filtering

In this section, the filtering code written in C is used to run the filtering block or component of the Lab 4 filtering system on the DSP.

L9.2.1 FIR Filter

The BD of the FIR lowpass filtering system that was built in Lab 4 is illustrated in Figure 9-6.

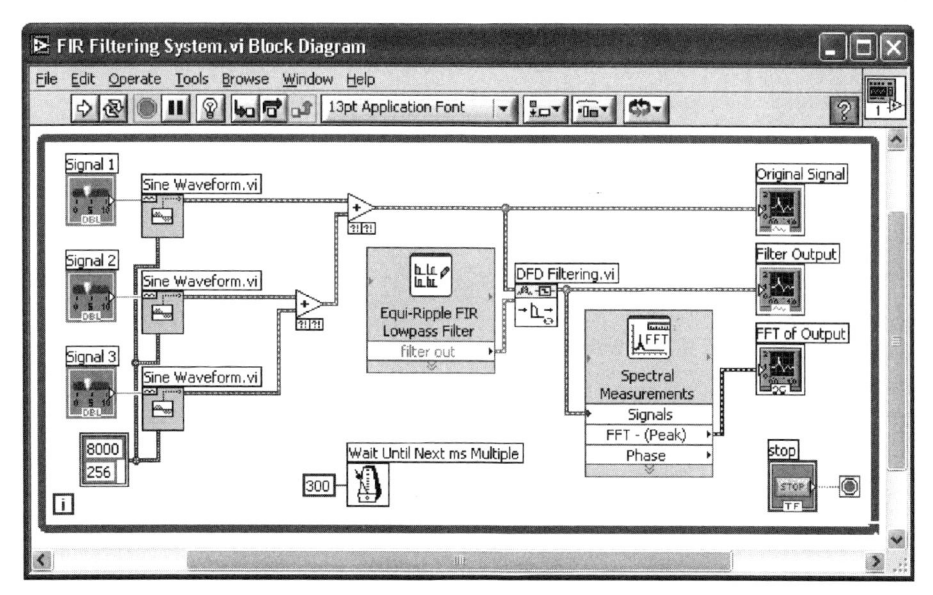

Figure 9-6: Filtering system in Lab 4.

Lab 9

Let us modify this BD to send the generated samples to the DSP and then to receive the filtered samples from the DSP. This is achieved by inserting the CCS automation process to the left and right sections of the While Loop structure. As indicated in Figure 9-7, a portion of the DFD Filter VI is replaced with the CCS RTDX Write VI and the CCS RTDX Read VI. Both of these VIs are configured to write and read single precision floating-point array data, which means configuring the polymorphic VIs as CCS RTDX Write Array SGL and CCS RTDX Read Array SGL, refer to Figure 9-7. Consider that the number of samples in the sampling info cluster is reduced to 128 in order to reduce the time associated with the RTDX communication.

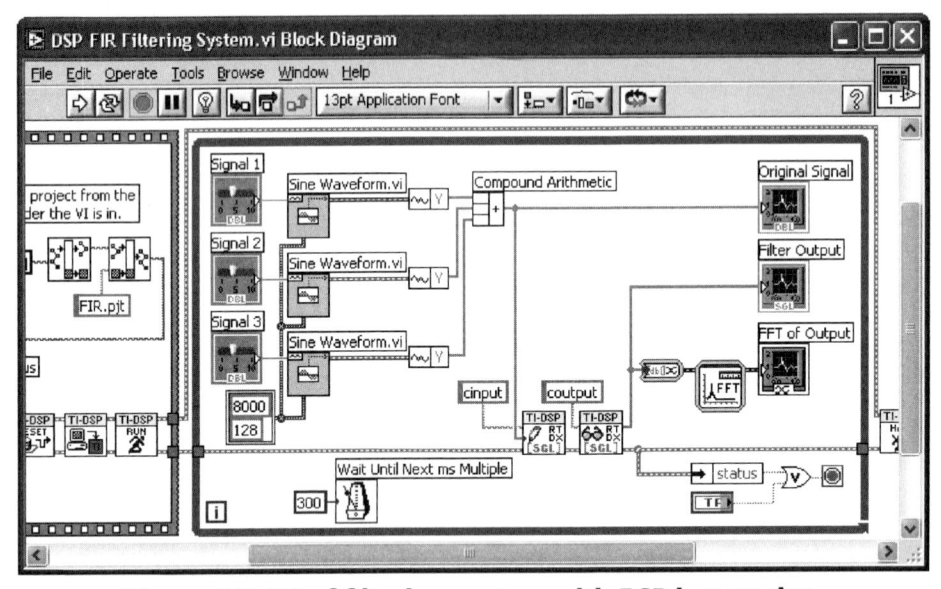

Figure 9-7: BD of filtering system with DSP integration.

An array of signal samples consisting of the sum of the three sinusoids is wired to the Data terminal of the CCS RTDX Write Array SGL VI. Also, a string constant containing the name of the input channel, cinput, is wired to the Channel terminal.

In the CCS RTDX Read Array SGL VI, the data transmitted via RTDX is read from the Data terminal of the VI. This terminal is wired to a Waveform Graph as well as to a Spectral Measurements Express VI for frequency analysis.

194

DSP Integration Examples

To check the status of errors generated by the CCS Automation or CCS Communication VIs, observe the `status` element of the error cluster. This is made possible by locating an `Unbundle By Name` function (**Functions** → **All Functions** → **Cluster** → **Unbundle By Name**). Wire the `error out` cluster from the `CCS RTDX Read Array SGL` VI to the `Unbundle By Name` function. This way the `status` element of the `error out` cluster is selected by default. The result of an OR operation of two Boolean values, corresponding to the `status` element of the cluster and a stop button, is wired to the conditional terminal of the `While Loop`. Whenever the stop button is pressed or an error occurs while accessing CCS or communicating via RTDX, the execution of the loop stops.

Notice the importance of the timeout value of the `CCS RTDX Read Array SGL` VI. If the RTDX communication speed is too slow or the number of data samples is large, the timeout value should be changed to avoid getting a RTDX error as shown in Figure 9-8. The default timeout value is 2000. To change the timeout value, wire a `Numeric Constant` to the `timeout` terminal of the `CCS RTDX Read Array SGL` VI, and enter a desired timeout value in milliseconds. Save the completed VI as *DSP FIR Filtering System.vi*.

Figure 9-8: RTDX error.

Lab 9

Next, let us state how to operate in CCS. Create a new project and name it *FIR.pjt*. Add the linker command file *c6713dsk.cmd*, the interrupt service vector *intvecs.asm*, the source code *FIR.c*, and the library files *rtdx.lib* (*ti\c6000\rtdx\lib*) and *rts6700.lib* (*ti\c6000\cgtools\lib*), into the project. The linker command file and interrupt service vector are located in the folder *ti\examples\dsk6713\rtdx\shared*. This folder should also include the header file *target.h*. The path to this folder needs to be added in **Include Search Path** of **Build Options**, as shown in Figure 9-9.

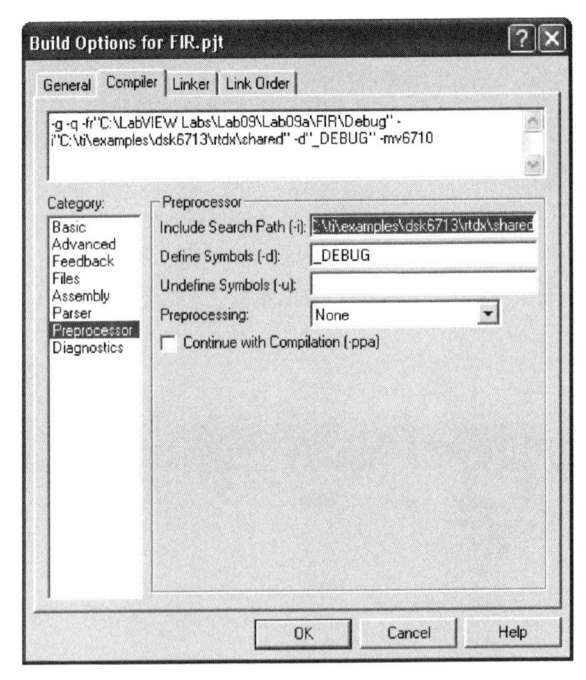

Figure 9-9: Build Options of CCS.

The FIR filtering C source code is shown below:

```
#include <rtdx.h>              /* RTDX                  */
#include "target.h"            /* TARGET_INITIALIZE()   */

#define kBUFFER_SIZE 128
#define N 15

float b[N] = {-0.008773, 0.0246851, 0.0217041, -0.0396942, -0.0734726,
    0.0560876, 0.305969, 0.437322, 0.305969, 0.0560876, -0.0734726,
    -0.0396942, 0.0217041, 0.0246851, -0.008773};
float samples[N];

RTDX_CreateInputChannel(cinput);
```

DSP Integration Examples

```c
RTDX_CreateOutputChannel(coutput);

void FIR(float *input, float *output)
{
    int i, j;
    float result;

    for( j = 0; j < kBUFFER_SIZE; j++ )
    {
        for(i = N-1; i > 0; i-- )
            samples[i] = samples[i-1];

        samples[0] = input[j];

        result = 0;
        for( i = 0 ; i < N ; i++ )
            result += samples[i] * b[i];

        output[j] = result;
    }
}

void main()
{
    float input[kBUFFER_SIZE];
    float output[kBUFFER_SIZE];
    int i;

    for( i = 0; i < N ; i++ )
        samples[i] =0;

    // Target initialization for RTDX
    TARGET_INITIALIZE();

    /*enable RTDX channels*/
    RTDX_enableInput(&cinput);
    RTDX_enableOutput(&coutput);

    for (;;)     /* Infinite message loop. */
    {
        while(!RTDX_read(&cinput, input, sizeof(input)));

        FIR(input, output);

        /* Write filtered data back to host. */
        RTDX_write(&coutput, output, sizeof(output));
    }
}
```

Lab 9

After creating the VI for the signal source and the CCS project for the FIR filtering block, run the VI from LabVIEW. One should see the outcome depicted in Figure 9-10. Notice that the amplitudes of the frequency components in the stopband (2200-4000 Hz) appear attenuated by 30 dB. This agrees with the filter specification.

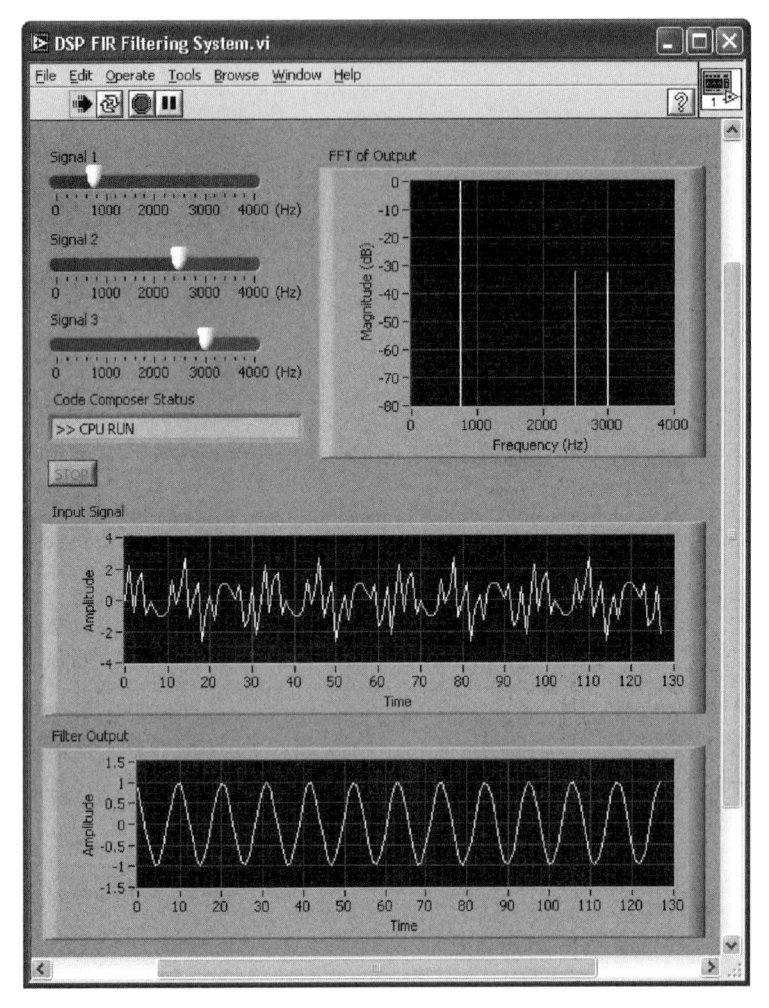

Figure 9-10: FP of FIR filtering system with DSP integration.

L9.2.2 IIR Filter

The bandpass IIR filter designed in Lab 4 is modified here. The previously used DSP FIR Filtering System VI is modified in order to run the IIR filtering project, *IIR.pjt*, on the DSP. The modified VI is then saved as *DSP IIR Filtering System.vi*.

DSP Integration Examples

As mentioned in Chapter 4, by default, the `DFD Classical Filter Design Express VI` of the DFD toolkit provides the IIR filter coefficients in the second-order cascade form. In the C source code shown below, the IIR filter comprises three second-order IIR filters in cascade. The advantage of the second-order cascade form lies in its lower sensitivity to coefficient quantization. In this implementation, the output from a second-order filter becomes the input to a next second-order filter.

```c
#include <rtdx.h>                         /* RTDX                    */
#include "target.h"                       /* TARGET_INITIALIZE()     */

#define kBUFFER_SIZE 128

float a1[2]={-0.955505, 0.834882};
float b1[3]={0.545337, -0.735242, 0.545337};

float a2[2]={0.954255, 0.834810};
float b2[3]={0.545337, 0.734702, 0.545337};

float a3[2]={-0.000622, 0.372609};
float b3[3]={0.545337, 0, -0.545337};

float IIRwindow1[3] = {0,0,0};
float y_prev1[2] = {0,0};

float IIRwindow2[3] = {0,0,0};
float y_prev2[2] = {0,0};

float IIRwindow3[3] = {0,0,0};
float y_prev3[2] = {0,0};

RTDX_CreateInputChannel(cinput);
RTDX_CreateOutputChannel(coutput);

void main()
{
    float input[kBUFFER_SIZE];
    float output[kBUFFER_SIZE];
    int i, n;
    float ASUM, BSUM;

    // Target initialization for RTDX
    TARGET_INITIALIZE();

    /*enable RTDX channels*/
    RTDX_enableInput(&cinput);
    RTDX_enableOutput(&coutput);

    for (;;)       /* Infinite message loop. */
```

Lab 9

```c
{
        while(!RTDX_read(&cinput, input, sizeof(input)));

        // IIR filtering

        for(i=0; i<kBUFFER_SIZE; i++)
        {
            // Stage #1

            for(n=2;n>0;n--)
               IIRwindow1[n] = IIRwindow1[n-1];

            IIRwindow1[0] = input[i];

            BSUM = 0;
            for(n = 0; n <= 2; n++)
            {
                    BSUM += b1[n] * IIRwindow1[n];
            }

            ASUM = 0;
            for(n = 0;n <= 1; n++)
            {
               ASUM += a1[n] * y_prev1[n];
            }

            y_prev1[1] = y_prev1[0];
            y_prev1[0] = BSUM - ASUM;

            // Stage #2

            for(n=2;n>0;n--)
               IIRwindow2[n] = IIRwindow2[n-1];

            IIRwindow2[0] = y_prev1[0];

            BSUM = 0;
            for(n = 0; n <= 2; n++)
            {
                    BSUM += b2[n] * IIRwindow2[n];
            }

            ASUM = 0;
            for(n = 0;n <= 1; n++)
            {
               ASUM += a2[n] * y_prev2[n];
            }
```

```
            y_prev2[1] = y_prev2[0];
            y_prev2[0] = BSUM - ASUM;

            // Stage #3

            for(n=2;n>0;n--)
               IIRwindow3[n] = IIRwindow3[n-1];

            IIRwindow3[0] = y_prev2[0];

            BSUM = 0;
            for(n = 0; n <= 2; n++)
            {
                    BSUM += b3[n] * IIRwindow3[n];
            }

            ASUM = 0;
            for(n = 0;n <= 1; n++)
            {
               ASUM += a3[n] * y_prev3[n];
            }

               output[i] = BSUM - ASUM;
            y_prev3[1] = y_prev3[0];
            y_prev3[0] = output[i];

        }

        /* Write data back to host. */
        RTDX_write(&coutput, output, sizeof(output));
    }
}
```

The output of the IIR bandpass filter is depicted in Figure 9-11. Considering that the passband of this filter is between 1333 and 2667 Hz, the amplitude of any signal in the stopband is attenuated by about 25 dB, which matches the outcome in Lab 4.

Lab 9

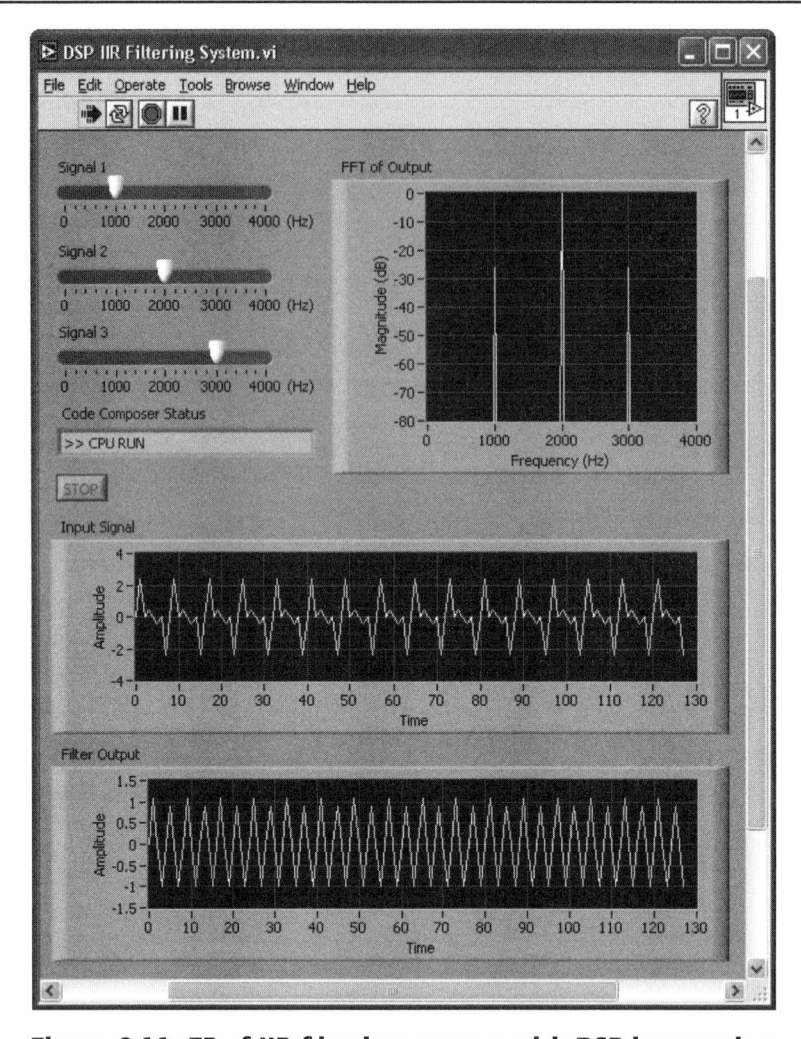

Figure 9-11: FP of IIR filtering system with DSP integration.

L9.3 Fixed-Point Implementation

In this section, an example is shown to demonstrate fixed-point arithmetic operations on the DSP. The FIR filtering system in the previous section is modified here to achieve fixed-point filtering on the DSP.

The BD of the fixed-point version of the FIR filtering system is shown in Figure 9-12. In this BD, the amplitude of the sum of the three sinusoids is multiplied by 10000 to represent it as a 16-bit integer while not exceeding the representable range of 16-bit integer numbers.

DSP Integration Examples

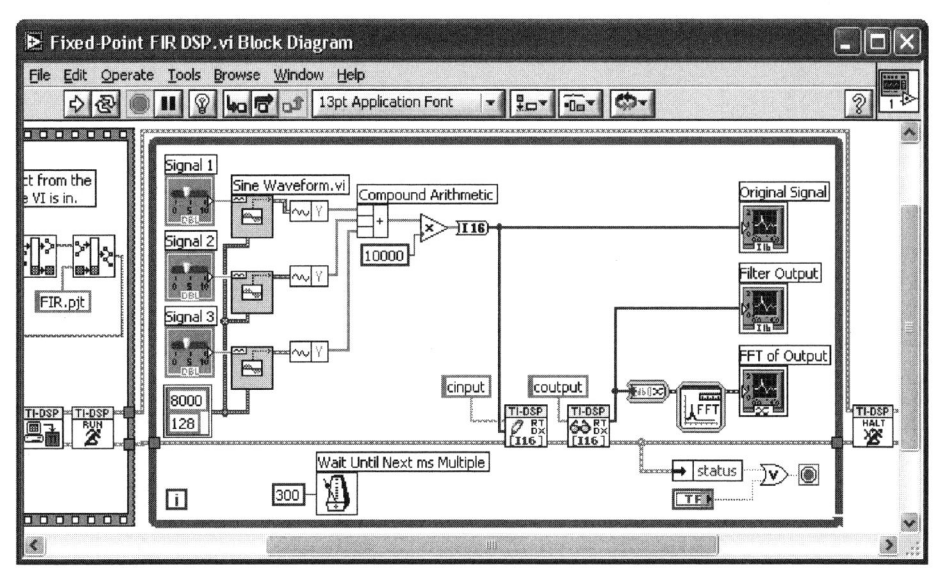

Figure 9-12: BD of fixed-point FIR filtering system with DSP integration.

It is worth mentioning that, if a C6416 DSK is used, the files and libraries to be added to the project are different from those required when using a C6713 DSK. These files include the linker command file *c6416.cmd*, the interrupt service vector *intvecs6416.asm*, the source code *FIR.c*, and the library files *rtdx64xx.lib* (*ti\c6000\rtdx\lib*) and *rts6400.lib* (*ti\c6000\cgtools\lib*). Also, the file path, *ti\examples\dsk6416\rtdx\shared*, should be added in **Include Search Path** of **Build Options**.

The source code of the fixed-point version of the FIR filtering system is shown below. In this code, the filter coefficients originally expressed in floating-point format are first converted into Q15 format. Then, they are scaled by one-half to avoid overflows. The number of scaling is obtained to be one by considering that all the inputs are 1s as discussed in Lab 5.

```
#include <rtdx.h>                  /* RTDX                    */
#include "target.h"                /* TARGET_INITIALIZE()     */

#define kBUFFER_SIZE 128
#define N 15

float b[N] = {-0.008773, 0.0246851, 0.0217041, -0.0396942, -0.0734726,
    0.0560876, 0.305969, 0.437322, 0.305969, 0.0560876, -0.0734726,
    -0.0396942, 0.0217041, 0.0246851, -0.008773};

short samples[N];
short coeff[N];
```

Lab 9

```c
RTDX_CreateInputChannel(cinput);
RTDX_CreateOutputChannel(coutput);

void FIR(short *input, short *output)
{
    int i, j;
    int result;

    for( j = 0; j < kBUFFER_SIZE; j++ )
    {
        for(i = N-1; i > 0; i-- )
            samples[i] = samples[i-1];

        samples[0] = input[j];

        result = 0;
        for( i = 0 ; i < N ; i++ )
            result += ( samples[i] * coeff[i] ) << 1;

        result = result >> 16;

        // Scale the Output to Compensate Scaling of Coefficients.
        output[j] = (short) ( result << 1 );
    }
}

void main()
{
    short input[kBUFFER_SIZE];
    short output[kBUFFER_SIZE];
    int i;

    for( i = 0; i < N ; i++ )
        samples[i] =0;

    // Convert to Q-15
    for( i = 0; i < N ; i++ )
        coeff[i] = b[i] * 0x7fff;

    // Scale by Half
    for( i = 0; i < N ; i++ )
        coeff[i] = coeff[i] >> 1;

    // Target initialization for RTDX
    TARGET_INITIALIZE();

    /*enable RTDX channels*/
    RTDX_enableInput(&cinput);
    RTDX_enableOutput(&coutput);
```

DSP Integration Examples

```
    for (;;)       /* Infinite message loop. */
    {
        while(!RTDX_read(&cinput, input, sizeof(input)));

        FIR(input, output);

        /* Write filtered data back to host. */
        RTDX_write(&coutput, output, sizeof(output));
    }
}
```

The multiplication of two Q15 numbers results in a Q30 format number with an extended sign bit being at the most significant bit. The extended sign bit is removed by shifting left this output number by one bit, which makes it a Q31 format number. To store it in Q15 format, it is right shifted by 16 bits.

The FP corresponding to the fixed-point DSP integration is shown in Figure 9-13. As can be seen from this figure, the displays match those in the floating-point version shown in Figure 9-10.

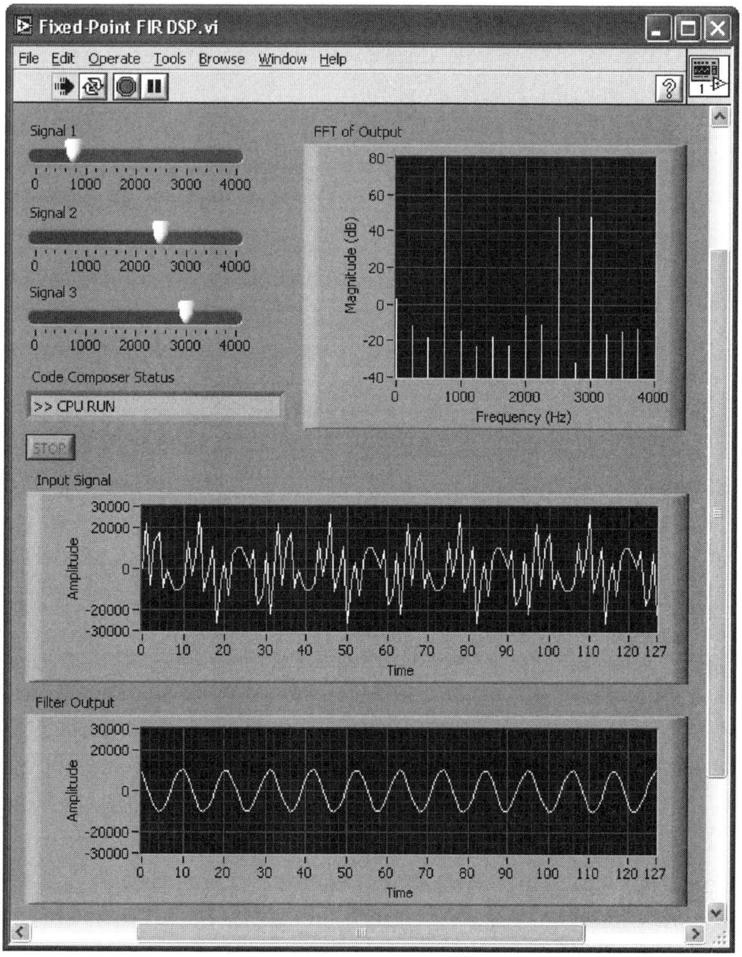

Figure 9-13: FP of fixed-point FIR filtering system with DSP integration.

L9.4 Adaptive Filtering Systems

The DSP integration of the adaptive filtering systems in Lab 6 is presented in this section. Though one can implement adaptive filtering by sending one sample at a time to the DSP, this approach is very inefficient due to the overhead associated with RTDX communication. It is thus more efficient to send an array of input data to the DSP where a point-by-point processing is performed.

L9.4.1 System Identification

An IIR filter is used to act as the unknown system by using the `Butterworth Filter` VI. Note that unlike the `Butterworth Filter PtByPt` VI used in Lab 7, this VI processes an array input. A 64-sample sinusoidal signal is used as the reference input via the RTDX channel `cin1`, and the output of the IIR filter is sent to the DSP via the RTDX channel `cin2`. The output of the LMS FIR filter and the error between the filter output and the desired output are read via the `cout1` and `cout2` channels, respectively.

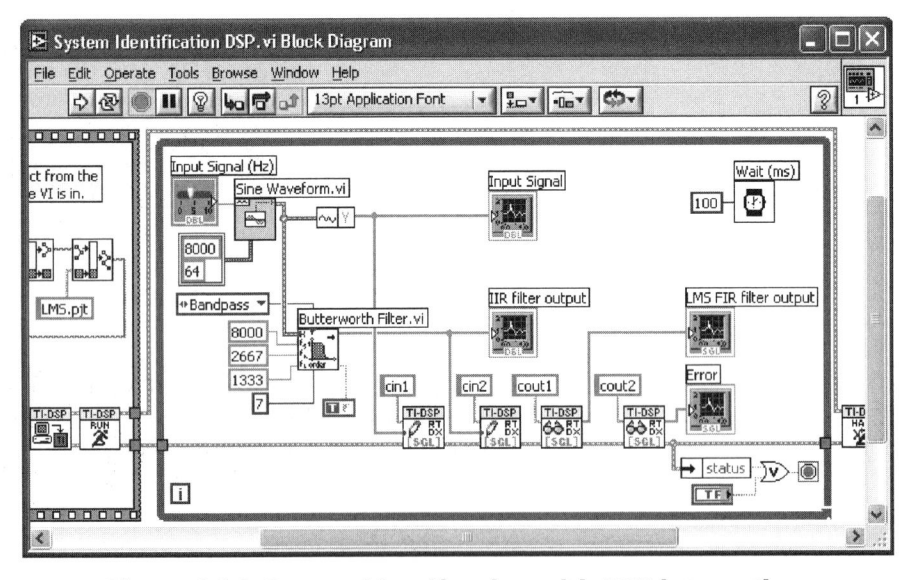

Figure 9-14: System identification with DSP integration.

A `True Constant` is wired to the `init/cont` terminal of the `Butterworth Filter` VI. This disables the initialization of the internal state of the filter, thus avoiding the group delay effect at the beginning of each output array. The BD of the system identification system with DSP integration is shown in Figure 9-14.

DSP Integration Examples

The C code for performing adaptive filtering on the C6x DSP is shown below. This code updates two arrays, consisting of the coefficients and input samples, at each iteration, similar to the LabVIEW implementation.

```c
#include "target.h"
#include <rtdx.h>

#define N 32 //filter length
#define kBUFFER_SIZE 64

float h[N] = {0.0, 0.0, 0.0, 0.0, 0.0, 0.0, 0.0, 0.0, 0.0, 0.0, 0.0, 0.0,
0.0, 0.0, 0.0, 0.0, 0.0, 0.0, 0.0, 0.0, 0.0, 0.0, 0.0, 0.0, 0.0, 0.0, 0.0,
0.0, 0.0, 0.0, 0.0, 0.0};
float samples[N];

RTDX_CreateInputChannel(cin1);
RTDX_CreateInputChannel(cin2);
RTDX_CreateOutputChannel(cout1);
RTDX_CreateOutputChannel(cout2);

void main()
{
    float input1[kBUFFER_SIZE];
    float input2[kBUFFER_SIZE];
    float output[kBUFFER_SIZE];
    float e[kBUFFER_SIZE];

    int i, j;
    float stemp, stemp2;

    for( i = 0 ; i < N ; i++ )
        samples[i] = 0;

    // Target initialization for RTDX
    TARGET_INITIALIZE();

    /*enable RTDX channels*/

    RTDX_enableInput(&cin1);
    RTDX_enableInput(&cin2);
    RTDX_enableOutput(&cout1);
    RTDX_enableOutput(&cout2);

    for (;;)    /* Infinite message loop. */
    {
        /* Wait for input sample */
        while(!RTDX_read(&cin1, input1, sizeof(input1)));
        while(!RTDX_read(&cin2, input2, sizeof(input2)));
```

```c
            for (j = 0; j < kBUFFER_SIZE; j++)
            {
                    // Update array samples
                    for(i = N-1; i > 0; i-- )
                            samples[i] = samples[i-1];

                    samples[0] = input1[j];

                    stemp =0;

                    // FIR Filtering
                    for( i = 0 ; i < N ; i++ )
                            stemp += (samples[i] * h[i]);

                    output[j] = stemp;

                    e[j] =  input2[j] - stemp;

                    stemp = (0.001 * e[j]);

                    // Update Coefficient
                    for(i = 0; i < N; i++)
                    {
                            stemp2 = (stemp * samples[i]);
                            h[i] = h[i] + stemp2;
                    }
            }

            /* Write scaled data back to host. */
            RTDX_write(&cout1, output, sizeof(output));
            RTDX_write(&cout2, e, sizeof(e));

    }
}
```

The output of the IIR filter and the adaptive FIR filter are shown in Figure 9-15. The output of the adaptive FIR filter adapts to the output of the IIR filter (unknown system) when the input is suddenly changed. Notice that the speed of convergence is governed by the step size specified in the C code.

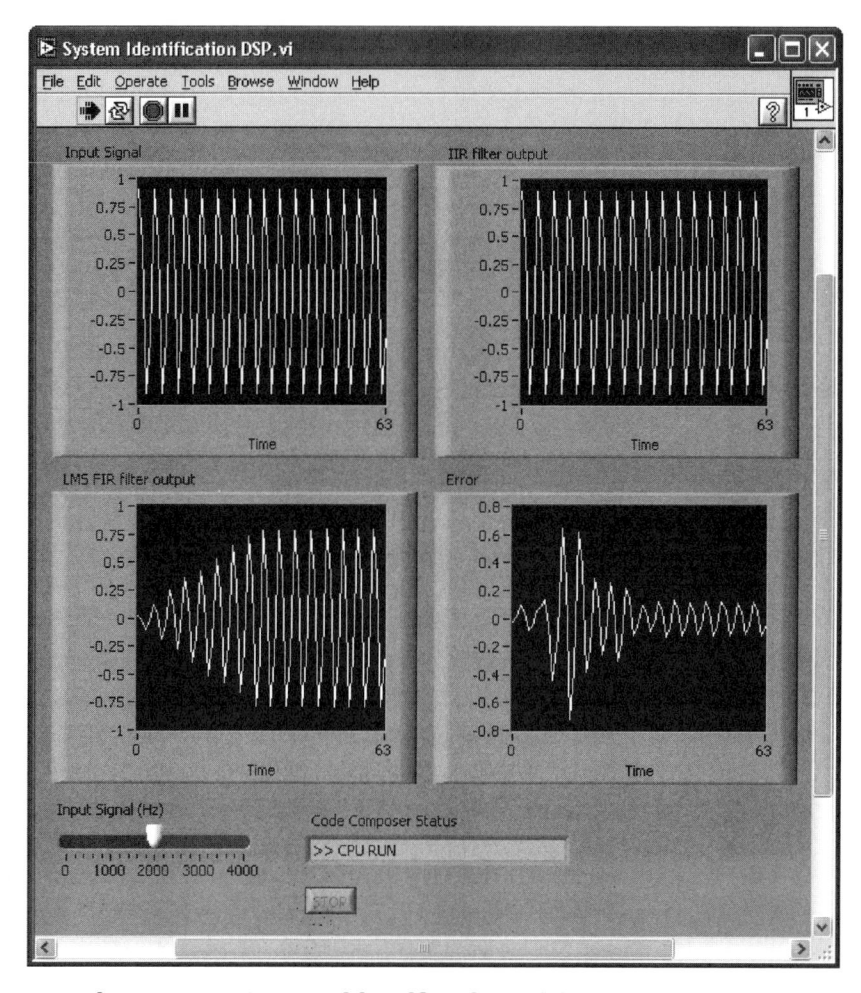

Figure 9-15: System identification with DSP integration.

Lab 9

L9.4.2 Noise Cancellation

For the DSP integration of the noise cancellation system, the same CCS project, *LMS.pjt*, is used here. As shown in Figure 9-16, the noise signal acts as the reference signal and is sent to the DSP via the `cin1` channel. The filtered noise signal, generated by passing the noise signal through a time-varying channel, is sent to the DSP via the `cin2` channel. The LMS filter output then becomes available from the `cout1` channel, and the noise cancelled output signal is read from the `cout2` channel.

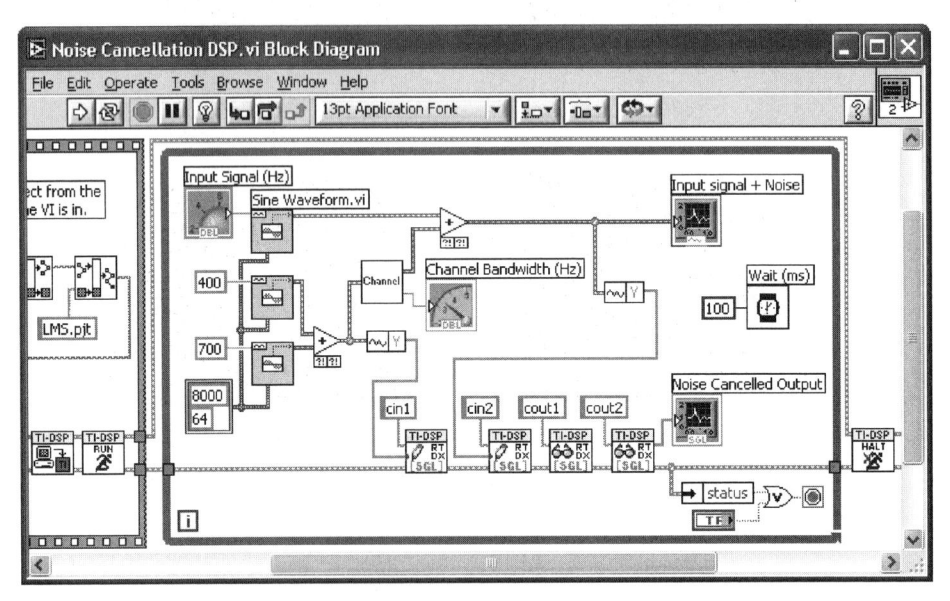

Figure 9-16: BD of noise cancellation with DSP integration.

In the `Channel` VI, introduced in Lab 6, the time duration between the steps is modified. This is done by changing the frequency of the `Basic Function Generator` VI to 25. As shown in Figure 9-17, the LMS filter adapts to the noise signal in such a way that the difference between its output and the noise corrupted signal approaches zero.

DSP Integration Examples

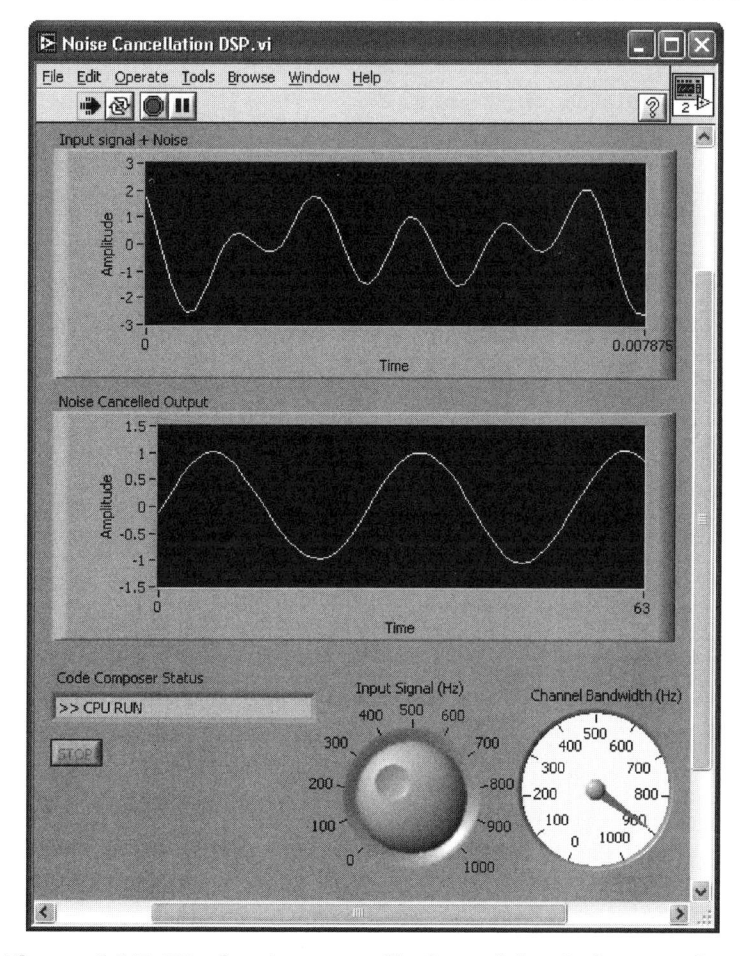

Figure 9-17: FP of noise cancellation with DSP integration.

L9.5 Frequency Processing: FFT

In this section, the DSP integration of the FFT algorithm is presented.

The BD of the combined implementation is shown in Figure 9-18(a). In this BD, a 128-sample sinusoidal signal having a 16-bit integer array format is sent to the DSP. Notice that the samples read from the DSP are in the 32-integer array format since the FFT magnitude values are quite large as indicated in the FP shown in Figure 9-18(b).

Lab 9

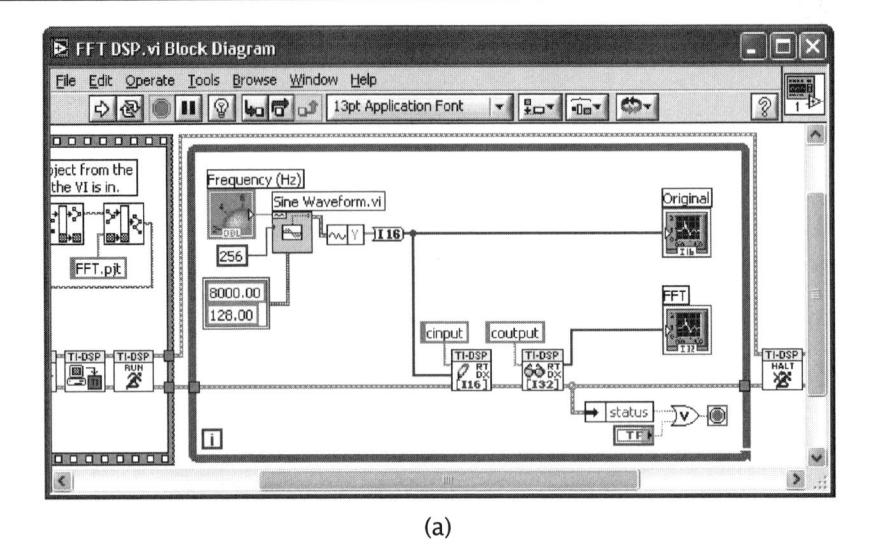

Figure 9-18: FFT DSP integration: (a) BD, and (b) FP.

DSP Integration Examples

For the DSP implementation, it is required to have the C source code of the FFT algorithm presented in Chapter 7. This code appears below and is provided on the accompanying CD [3]:

```c
#include <math.h>
#include "twiddleR.h"
#include "twiddleI.h"

#include <rtdx.h>                          /* RTDX                          */
#include "target.h"                        /* TARGET_INITIALIZE()           */

#define     kBUFFER_SIZE 128
#define     NUMDATA      128    /* number of real data samples */
#define     NUMPOINTS    64     /* number of point in the DFT, NUMDATA/2 */

#define     TRUE   1
#define     FALSE  0
#define     BE     TRUE
#define     LE     FALSE
#define     ENDIAN LE        /* selects proper endianaess. If building
                                code in Big Endian, use BE, else use LE */

#define     PI     3.141592653589793 /* defineition of pi */

typedef struct { /* define the data type for the radix-4 twiddle factors */
short imag;
short real;
} COEFF;

/* BIG Endian */
#if ENDIAN == TRUE

typedef struct {
    short imag;
    short real;
} COMPLEX;

#else

/* LITTLE Endian */
typedef struct {
    short real;
    short imag;
} COMPLEX;

#endif

#pragma DATA_ALIGN(x,NUMPOINTS);
```

Lab 9

```
COMPLEX x[NUMPOINTS+1];          /* array of complex DFT data */
COEFF W4[NUMPOINTS];
short g[NUMDATA];
COMPLEX A[NUMPOINTS];            /* array of complex A coefficients */
COMPLEX B[NUMPOINTS];            /* array of complex B coefficients */
COMPLEX G[2*NUMPOINTS];          /* array of complex DFT result */
unsigned short IIndex[NUMPOINTS], JIndex[NUMPOINTS];
int count;

int magR[NUMDATA];
int magI[NUMDATA];

int output[kBUFFER_SIZE];

void make_q15(short out[], float in[], int N);
void R4DigitRevIndexTableGen(int, int *, unsigned short *, unsigned short *);
void split1(int, COMPLEX *, COMPLEX *, COMPLEX *, COMPLEX *);
void digit_reverse(int *, unsigned short *, unsigned short *, int);
void radix4(int, short[], short[]);
void fft();

RTDX_CreateInputChannel(cinput);
RTDX_CreateOutputChannel(coutput);

void main()
{
   int i,k;
   short tr[NUMPOINTS], ti[NUMPOINTS];

   // Target initialization for RTDX
   TARGET_INITIALIZE();

   /*enable RTDX channels*/
   RTDX_enableInput(&cinput);
   RTDX_enableOutput(&coutput);

   //Read Twiddle factors to COMPLEX array and make Q-15;
   make_q15(tr, TR, NUMPOINTS);   //Data in Header files from Matlab
   make_q15(ti, TI, NUMPOINTS);

   for(i=0;i<NUMPOINTS;i++)
   {
     W4[i].real = tr[i];
     W4[i].imag = ti[i];
   }

   /* Initialize A,B, IA, and IB arrays */
   for(k=0; k<NUMPOINTS; k++)
```

DSP Integration Examples

```c
   {
      A[k].imag = (short)(16383.0 * (-cos(2*PI/(double)(2*NUMPOINTS)*
                 (double)k)));
      A[k].real = (short)(16383.0*(1.0 - sin(2*PI/(double)(2*NUMPOINTS)*
                 (double)k)));
      B[k].imag = (short)(16383.0*(cos(2*PI/
(double)(2*NUMPOINTS)*(double)k)));
      B[k].real = (short)(16383.0*(1.0 + sin(2*PI/(double)(2*NUMPOINTS)*
                 (double)k)));
   }

   /* Initialize tables for FFT digit reversal function */
   R4DigitRevIndexTableGen(NUMPOINTS, &count, IIndex, JIndex);

   for(;;)  /* Infinite message loop. */
   {
      while(!RTDX_read(&cinput, g, sizeof(g)));

      /* Call FFT algorithm */
      fft();

      for (k=0; k<NUMDATA; k++)
      {
         magR[k] = (G[k].real*G[k].real) << 1;
         magI[k] = (G[k].imag*G[k].imag) << 1;

         output[k] = magR[k] + magI[k];
      }

      /* Write scaled data back to host. */
      RTDX_write(&coutput, &output, sizeof(output));
   }
}

void fft()
{
   int n;
   /* Forward DFT */
   /* From the 2N point real sequence, g(n), for the N-point complex
sequence, x(n) */

   for (n=0; n<NUMPOINTS; n++)
   {
      x[n].imag = g[2*n + 1]; /* x2(n) = g(2n + 1) */
      x[n].real = g[2*n]; /* x1(n) = g(2n) */
   }

   /* Compute the DFT of x(n) to get X(k) -> X(k) = DFT{x(n)} */
   radix4(NUMPOINTS, (short *)x, (short *)W4);
```

Lab 9

```c
   digit_reverse((int *)x, IIndex, JIndex, count);
   /* Because of the periodicity property of the DFT, we know that X(N+k) =
X(k) . */
   x[NUMPOINTS].real = x[0].real;
   x[NUMPOINTS].imag = x[0].imag;
   /* The split function performs the additional computations required to
get G(k) from X(k). */

   split1(NUMPOINTS, x, A, B, G);
   /* Use complex conjugate symmetry properties to get the rest of G(k) */
   G[NUMPOINTS].real = x[0].real - x[0].imag;
   G[NUMPOINTS].imag = 0;

   for (n=1; n<NUMPOINTS; n++)
   {
      G[2*NUMPOINTS-n].real = G[n].real;
      G[2*NUMPOINTS-n].imag = -G[n].imag;
   }
}

void radix4(int n, short x[], short w[])
{
   int n1, n2, ie, ia1, ia2, ia3, i0, i1, i2, i3, j, k;
   short t, r1, r2, s1, s2, co1, co2, co3, si1, si2, si3;
   n2 = n;
   ie = 1;
   for (k = n; k > 1; k >>= 2)
   {
      n1 = n2;
      n2 >>= 2;
      ia1 = 0;
      for (j = 0; j < n2; j++)
      {
         ia2 = ia1 + ia1;
         ia3 = ia2 + ia1;
         co1 = w[ia1 * 2 + 1];
         si1 = w[ia1 * 2];
         co2 = w[ia2 * 2 + 1];
         si2 = w[ia2 * 2];
         co3 = w[ia3 * 2 + 1];
         si3 = w[ia3 * 2];
         ia1 = ia1 + ie;
         for (i0 = j; i0 < n; i0 += n1)
         {
            i1 = i0 + n2;
            i2 = i1 + n2;
            i3 = i2 + n2;
            r1 = x[2 * i0] + x[2 * i2];
            r2 = x[2 * i0] - x[2 * i2];
```

DSP Integration Examples

```
            t = x[2 * i1] + x[2 * i3];
            x[2 * i0] = r1 + t;
            r1 = r1 - t;
            s1 = x[2 * i0 + 1] + x[2 * i2 + 1];
            s2 = x[2 * i0 + 1] - x[2 * i2 + 1];
            t = x[2 * i1 + 1] + x[2 * i3 + 1];
            x[2 * i0 + 1] = s1 + t;
            s1 = s1 - t;
            x[2 * i2] = (r1 * co2 + s1 * si2) >> 15;
            x[2 * i2 + 1] = (s1 * co2-r1 * si2)>>15;
            t = x[2 * i1 + 1] - x[2 * i3 + 1];
            r1 = r2 + t;
            r2 = r2 - t;
            t = x[2 * i1] - x[2 * i3];
            s1 = s2 - t;
            s2 = s2 + t;
            x[2 * i1] = (r1 * co1 + s1 * si1) >>15;
            x[2 * i1 + 1] = (s1 * co1-r1 * si1)>>15;
            x[2 * i3] = (r2 * co3 + s2 * si3) >>15;
            x[2 * i3 + 1] = (s2 * co3-r2 * si3)>>15;
        }
      }
   ie <<= 2;
   }
}
void digit_reverse(int *yx, unsigned short *JIndex, unsigned short *IIndex,
int count)
{
   int i;
   unsigned short I, J;
   int YXI, YXJ;
   for (i = 0; i<count; i++)
   {
      I = IIndex[i];
      J = JIndex[i];
      YXI = yx[I];
      YXJ = yx[J];
      yx[J] = YXI;
      yx[I] = YXJ;
   }
}

void R4DigitRevIndexTableGen(int n, int *count, unsigned short *IIndex,
unsigned short *JIndex)
{
   int j, n1, k, i;
   j = 1;
   n1 = n - 1;
```

Lab 9

```c
      *count = 0;
      for(i=1; i<=n1; i++)
      {
         if(i < j)
         {
            IIndex[*count] = (unsigned short)(i-1);
            JIndex[*count] = (unsigned short)(j-1);
            *count = *count + 1;
         }
         k = n >> 2;
         while(k*3 < j)
         {
            j = j - k*3;
            k = k >> 2;
         }
         j = j + k;
      }
}

void split1(int N, COMPLEX *X, COMPLEX *A, COMPLEX *B, COMPLEX *G)
{
   int k;
   int Tr, Ti;

   for (k=0; k<N; k++)
   {
    Tr = (int)X[k].real * (int)A[k].real - (int)X[k].imag * (int)A[k].imag
+
       (int)X[N-k].real * (int)B[k].real + (int)X[N-k].imag * (int)B[k].
imag;
      G[k].real = (short)(Tr>>15);

    Ti = (int)X[k].imag * (int)A[k].real + (int)X[k].real * (int)A[k].imag
+
       (int)X[N-k].real * (int)B[k].imag - (int)X[N-k].imag * (int)B[k].
real;
      G[k].imag = (short)(Ti>>15);
   }
}

void make_q15(short out[], float in[], int N)
{
   int i;

   for(i=0;i<N;i++)
   {
      out[i]=0x7fff*in[i];    //Convert to Q-15, good approximate
   }
}
```

DSP Integration Examples

Care must be taken to avoid overflows if the algorithm is running on a fixed-point DSP. The amplitude of the input signal needs to be scaled properly in order to avoid overflows in the computations of the FFT on the DSP. In the example shown in Figure 9-18, 256 is used as the amplitude of the input signal. Figure 9-19 illustrates an overflowed FFT outcome when the amplitude is set to 4096.

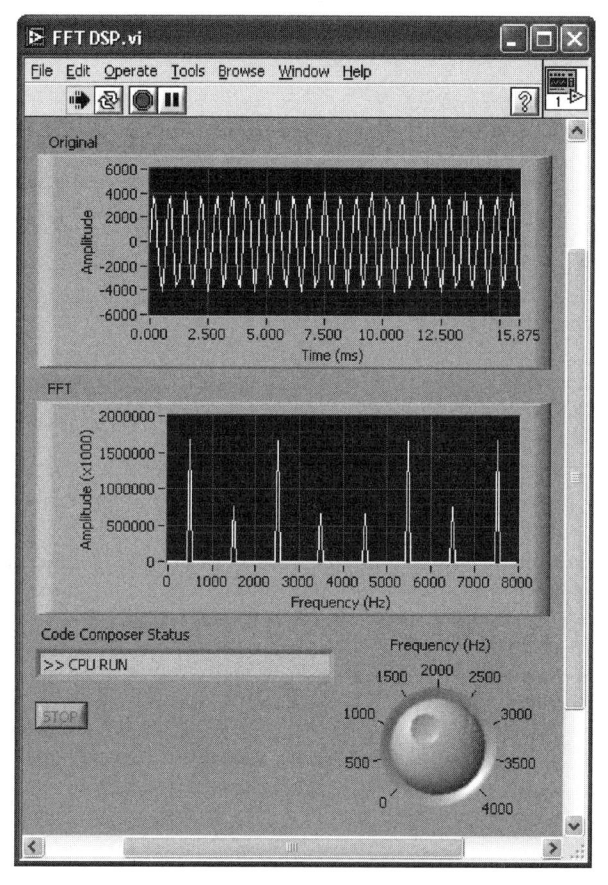

Figure 9-19: Overflow in computing FFT.

It should be noted that the overall timing of a typical DSP integration is often not so efficient due to the overhead associated with RTDX communication. Nevertheless, the discussed DSP integration allows one to examine code execution on the C6x DSP hardware platform.

L9.6 Bibliography

[1] Texas Instruments, *TMS320C6000 Code Composer Studio Tutorial*, Literature ID# SPRU 301C, 2000.

[2] Texas Instruments, *Code Composer Studio User's Guide*, Literature ID# SPRU 328B, 2000.

[3] Texas Instruments, *Application Report SPRA 291*, 1997.

CHAPTER 10

DSP System Design: Dual-Tone Multi-Frequency (DTMF) Signaling

In this and the next two chapters, three DSP system project examples are discussed and designed using LabVIEW. These examples show how a complete DSP system can be built in a relatively short amount of time by using LabVIEW. In increasing order of complexity, they consist of dual-tone multi-frequency signaling, software-defined radio, and MP3 player.

Dual-tone multi-frequency (DTMF) signaling is used extensively in voice communication applications such as voice mail and telephone banking. A DTMF signal is made up of two tones selected from a low and a high tone group. Each pair of tones contains one frequency from the low group (697 Hz, 770 Hz, 852 Hz, 941 Hz) and one frequency from the high group (1209 Hz, 1336 Hz, 1477 Hz). Figure 10-1 shows the frequencies allocated to the telephone pad push buttons.

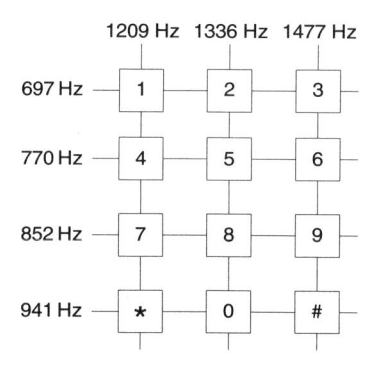

Figure 10-1: Keypad and allocated frequencies.

The implementation of the DTMF receiver system is normally done by using the Goertzel algorithm [1]. This algorithm is more efficient than the FFT algorithm for DTMF detection both in terms of the number of operations and amount of memory

Chapter 10

usage. Furthermore, unlike the FFT, it does not require access to the entire data frame, leading to faster execution. As indicated in Figure 10-2, seven Goertzel filters are used here in parallel to form a DTMF detection system. Each Goertzel filter is designed to detect a DTMF tone. The output from each filter is squared and fed into a threshold detector, where the strongest signals from the low and high frequency groups are selected to identify a pressed digit on the keypad.

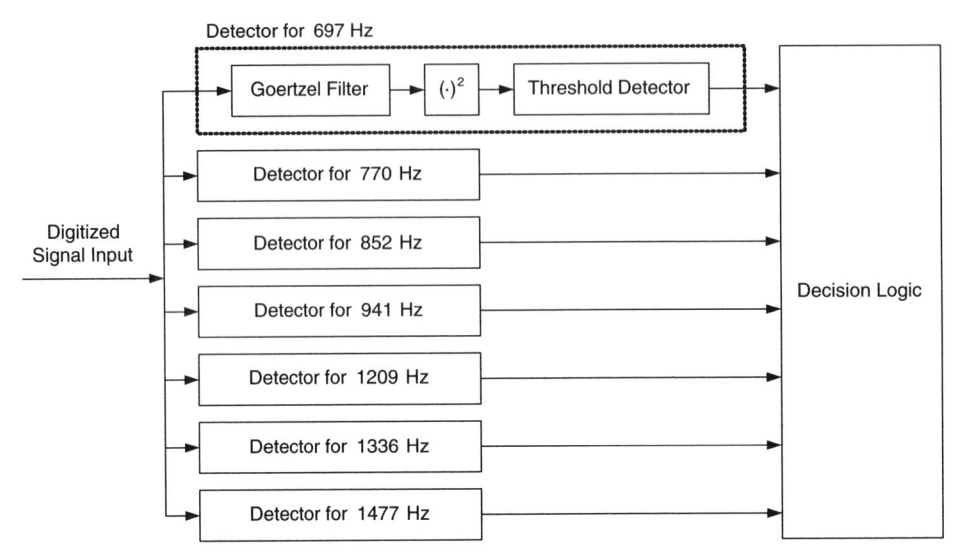

Figure 10-2: DTMF system using Goertzel algorithm.

The difference equations of a second-order Goertzel filter, see Figure 10-3, are given by

$$v_k[n] = 2\cos\left[\frac{2\pi k}{N}\right]v_k[n-1] - v_k[n-2] + x[n], \quad n = 0,1,\ldots,N \tag{10.1}$$

$$y_k[n] = v_k[n] - W_N^k v_k[n-1] \tag{10.2}$$

where $x[n]$ denotes input, $y_k[n]$ output, $v_k[n]$ intermediate output, the subscript k indicates frequency bin, N is the DFT size, and $W_N^k = \exp\left(-j\frac{2\pi k}{N}\right)$.

The initial conditions are assumed to be zero, i.e., $v_k[-1] = v_k[-2] = 0$. Considering that only the magnitude of the signal is required for the DTMF tone detection, the following equation is used to generate magnitude squared outputs:

$$|y_k[N]|^2 = v_k^2[N] + v_k^2[N-1] - 2\cos\left(\frac{2\pi k}{N}\right)v_k[N]v_k[N-1] \tag{10.3}$$

DSP System Design: Dual-Tone Multi-Frequency (DTMF) Signaling

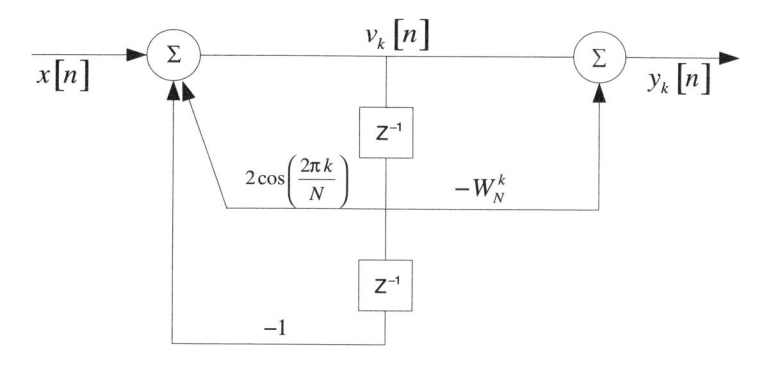

Figure 10-3: Structure of a second-order Goertzel filter.

The coefficients are selected based on the DTMF tones. They are listed in Table 10-1.

Table 10-1: Fundamental frequencies.

Fundamental Frequency (Hz)	Coefficient
697	1.703275
770	1.635585
852	1.562297
941	1.482867
1209	1.163138
1336	1.008835
1477	0.790074

Chapter 10

10.1 Bibliography

[1] E. Ifeachor and B. Jervis, *Digital Signal Processing: A Practical Approach*, Prentice-Hall, 2001.

Lab 10: Dual-Tone Multi-Frequency

In this lab, the DTMF system just explained is built in the LabVIEW graphical programming environment.

L10.1 DTMF Tone Generator System

Before building a BD for the DTMF system, let us begin by creating a keypad as shown in Figure 10-4. The twelve buttons shown are grouped into a cluster so that their outputs are wired via a cluster wire. This is done by placing a `Cluster` shell (**Controls** → **All Controls** → **Array & Cluster** → **Cluster**) on the FP, and then by locating `Text Buttons` (**Controls** → **Buttons & Switches** → **OK Button**) in the cluster area.

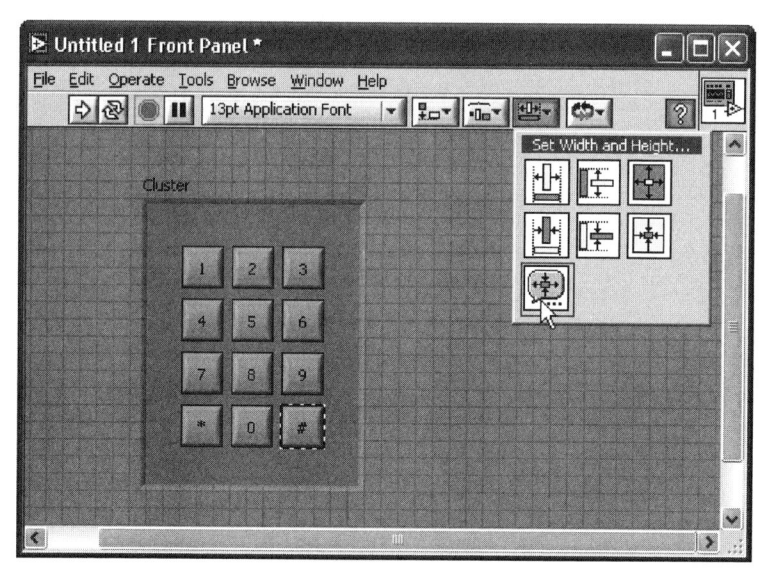

Figure 10-4: Creating a cluster control.

The width and height of a button can be adjusted to have a larger display on the FP. This is achieved by choosing the buttons and then by selecting the option **Set Width and Height** from the **Resize Objects** menu of the FP toolbar as shown in Figure 10-4. In this example, both the width and height of the buttons are set to 30. Also, the mechanical action of the buttons is configured to be **Latch When Released** by right-clicking on the buttons and choosing **Mechanical Action** from the shortcut menu. Once the configuration of a button is complete, the button is copied multiple times

Lab 10

to construct a keypad. Change the Boolean text of the buttons appropriately. Align and distribute the buttons via **Align Objects** and **Distribute Objects** on the FP toolbar.

Next, the output value of the cluster control is specified for each button, in other words the value coming out of the output of the cluster control when one of the buttons is pressed. To accomplish this, right-click on the border of the cluster and choose **Reorder Controls In Cluster** from the shortcut menu. This brings up the window shown in Figure 10-5. The numbers shown in the black background correspond to the modified order and the numbers in the white background to the original order. The number assigned to a key is displayed next to **Click to set to** shown on the toolbar. Click the buttons in sequential order to specify the value shown in the toolbar area. After finishing the assignment of the values to the buttons, click the **OK** button to finish reordering controls, or the **X** button to cancel the changes.

Figure 10-5: Reordering cluster control.

Right-click on the border of the cluster, and choose **Auto Sizing → Size to Fit** to resize the cluster, if desired. Also, rename the label of the cluster as Keypad.

The BD of the built DTMF system is shown in Figure 10-6. Note that the keypad cluster control is shown as an icon on the BD. This VI generates a tone depending on the number pressed on the keypad in the FP. The DTMF decoder based on the Goertzel algorithm can be seen in the lower half of the BD.

To build the BD, wire the output value of the cluster to an array by using the `Cluster to Array` function (**Functions → All Functions → Array → Cluster to Array**). This is done in order to have the value of each button as an element of an array. The array is then wired to a `Search 1D Array` function to search for the `True` values among the array elements. In other words, this is done to check the status of the buttons considering that the index of the array which is greater or equal to zero is returned when a button is pressed, otherwise –1 is returned.

Thus, if the index of the array becomes greater than or equal to zero, that is, any button is pressed, a DTMF signal is generated and the decoding part in the `True` case of the `Case Structure` is executed. In the `False` case of the `Case Structure`, a time delay is included to continue the idle status until a key is pressed.

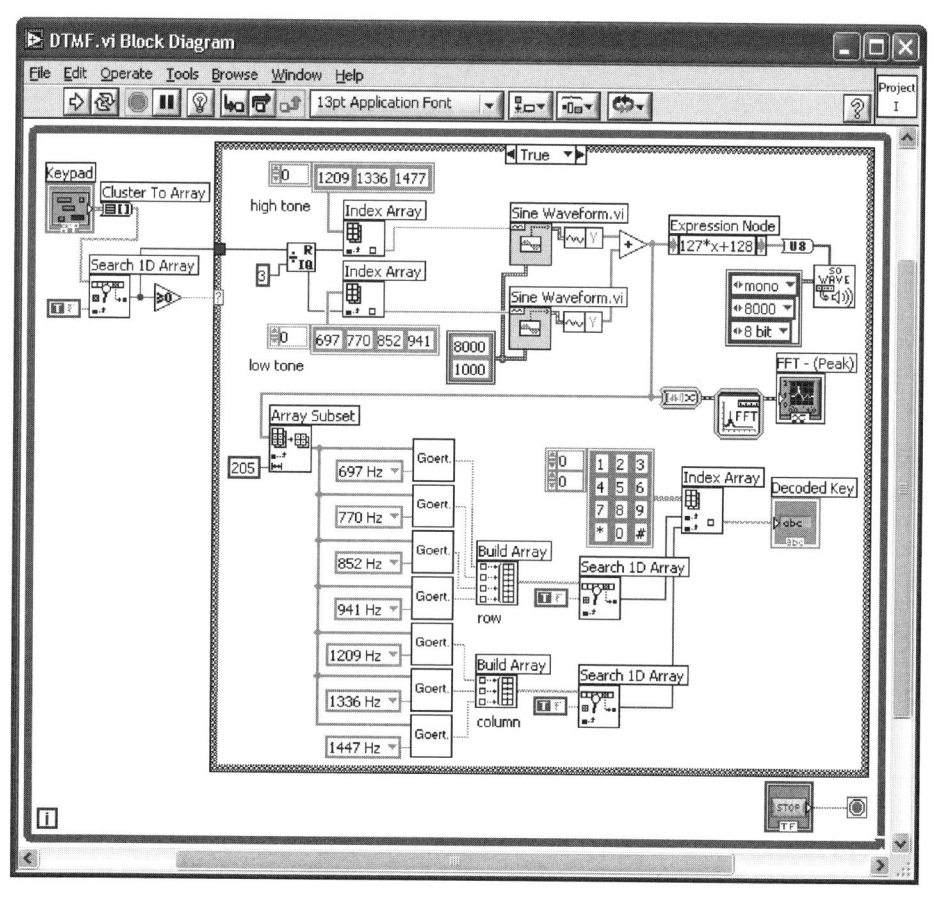

Figure 10-6: BD of DTMF system.

Now, let us go through the DTMF signal generation for the True case of the Case Structure. The value of the array index is wired to the Quotient & Remainder function with 3 as divisor. Since the numbers on the keypad are arranged in three columns and four rows, the remainder of this operation becomes the column index, and the quotient becomes the row index. Based on the column and row indices, a high and a low tone value are chosen using two 1D array constants. The low and high tone values are wired to a Sine Waveform VI to generate a waveform based on the chosen frequencies.

The generated waveform is scaled to 8-bit integer so that it can be played at an audible volume level. An Expression Node (**Functions → All Functions → Numeric → Expression Node**) is used for scaling the waveform. An Expression Node is useful for evaluating a simple equation or expression containing a single variable [1]. A Snd Write Waveform VI (**Functions → All Functions → Graphics & Sound → Sound → Snd Write Waveform**) is located to send out the waveform to the PC sound card. Create a constant on the sound format terminal to specify the sound format including stereo/mono, sampling frequency, and bits per sample. Here, the 8-bit mono format with a sampling frequency of 8000 Hz is used. For spectral analysis of the generated samples, a Spectral Measurement Express VI and a Waveform Graph are used. The Spectral Measurement Express VI is configured as linear amplitude spectrum with no windowing.

At this stage, the DTMF generator is complete. In the next section, the decoding module is covered.

L10.2 DTMF Decoder System

The Goertzel algorithm is used for the decoding of DTMF signals. The BD shown in Figure 10-7 illustrates the Goertzel algorithm described by Equation (10.3). Notice that to access the two previous values of $v_k[n]$, i.e., $v_k[n-1]$ and $v_k[n-2]$, two Shift Registers are stacked by right-clicking on the Shift Register and selecting **Add Element** on the shortcut menu.

The inputs of this subVI consist of a 1D array of 205 samples and the coefficients of the Goertzel algorithm. A Text Ring control (**Controls → All Controls → Ring & Enum → Text Ring**), labeled as Freq/Coeff, is located on the FP. Its data representation and properties are then modified as illustrated in Figure 10-8. This is done by right-clicking on the Text Ring in the FP, and then by choosing **Edit Items...** from the shortcut menu. Note that this Goertzel subVI is designed to incorporate only the first harmonic. It uses 205 input samples and its coefficients are calculated based on 205 frequency bins.

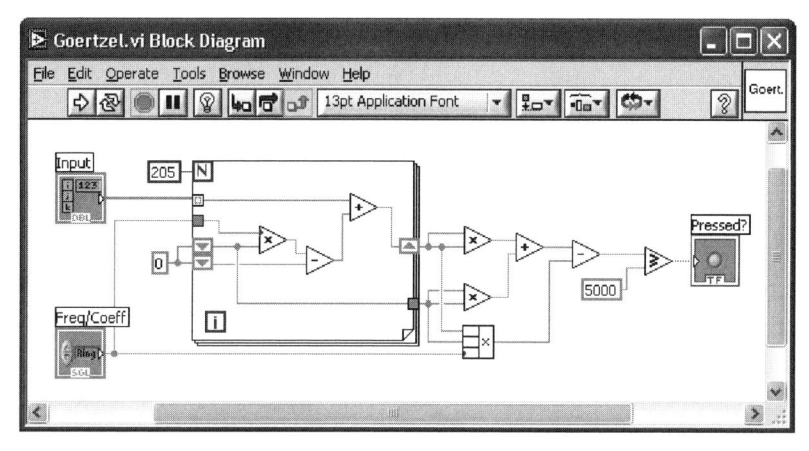

Figure 10-7: BD of Goertzel algorithm.

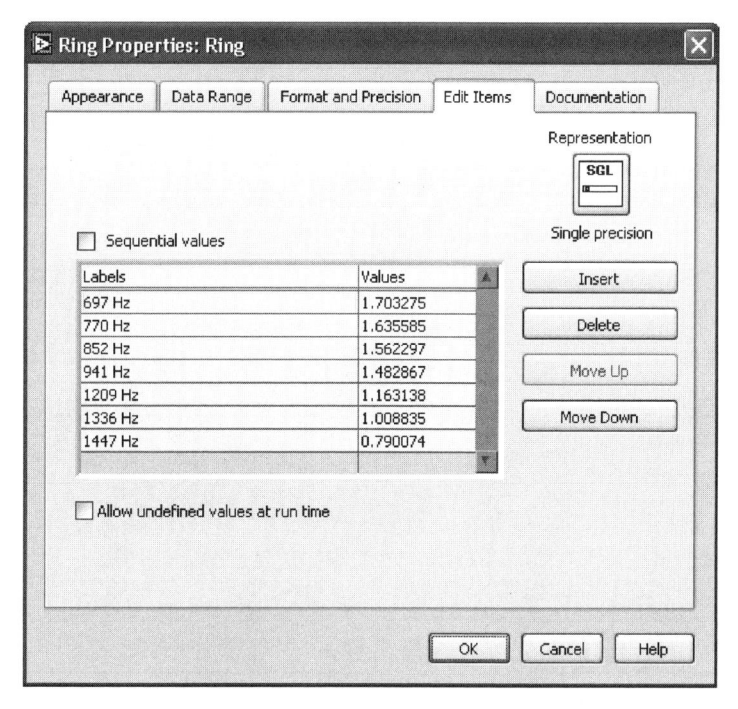

Figure 10-8: Ring Properties.

The output of the `Goertzel` subVI is a Boolean value which indicates whether the specified frequency component is present in the input samples or not. This is decided by comparing the squared output of this subVI with a threshold value. The threshold value here is empirically set to 5000.

Lab 10

In the BD of the DTMF system shown in Figure 10-6, a total of seven Goertzel subVIs are placed to detect each frequency of a DTMF signal. The outputs of the Geortzel subVIs are grouped as two arrays to incorporate the row and column frequencies. From these arrays, indices of the `True` values are searched to determine a pressed key. A string constant is referred to by the indices of the 2D array of string constants.

To create the 2D array of string constants, first place an `Array Constant` shell (**Functions → All Functions → Array → Array Constant**) on the BD. Then, place a `String Constant` (**Functions → All Functions → String → String Constant**) in the `Array Constant`. As a result, a 1D array of string constants is created. In order to increase the dimension of the array, right-click on the `Array Constant` and choose **Add Dimension** from the shortcut menu. Now, enter the corresponding strings in the 2D array.

The output of the DTMF is shown in Figure 10-9. Notice that when the button '#' is pressed, two frequencies are observed at 941 Hz and 1477 Hz in the decoded output. Furthermore, the decoded output matches the expected outcome.

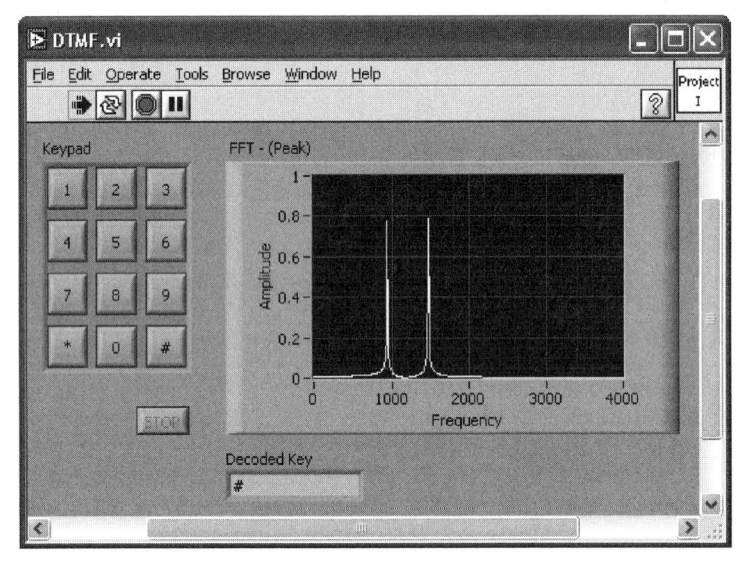

Figure 10-9: FP of DTMF system.

L10.3 Bibliography

[1] National Instruments, *LabVIEW User Manual*, Part Number 320999E-01, 2003.

CHAPTER 11

DSP System Design: Software-Defined Radio

This chapter covers a software-defined radio system built using LabVIEW. A software-defined radio consists of a programmable communication system where functional changes can be made by merely updating software. For a detailed description of software-defined radio, the reader is referred to [1], [2].

4-QAM (Quadrature Amplitude Modulation) is chosen to be the modulation scheme of our software-defined radio system, noting that this modulation is widely used for data transmission applications over bandpass channels such as FAX modem, high speed cable, multi-tone wireless and satellite systems [2]. For simplicity, here the communication channel is considered to be ideal or noise-free.

11.1 QAM Transmitter

For transmission, pseudo noise (PN) sequences are generated to serve as our message signal. A PN sequence is generated with a five-stage linear feedback shift register structure, see Figure 11-1, whose connection polynomial is given by

$$h(D) = 1 + D^2 + D^5 \quad (11.1)$$

where D denotes delay and the summations represent modulo 2 additions.

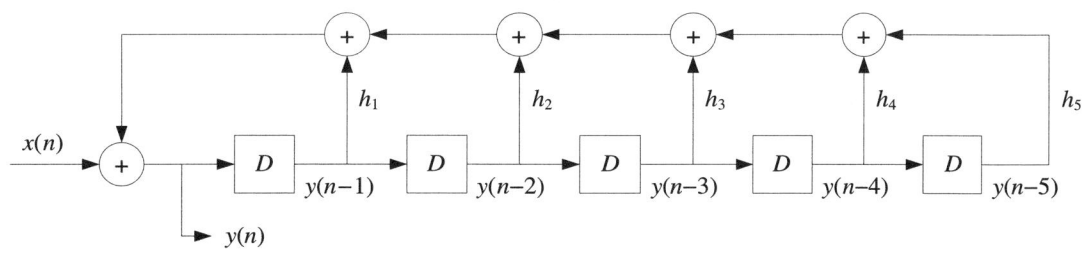

Figure 11-1: PN generation with linear feedback shift register.

Chapter 11

The sequence generated by the above equation has a period of $31(= 2^5 - 1)$. Two PN sequence generators are used in order to create the message sequences for both the in-phase and quadrature phase components. The constellation of 4-QAM is shown in Figure 11-2. For more details of PN sequence generation, refer to [2].

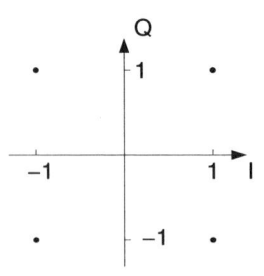

Figure 11-2: Constellation of 4-QAM.

Note that frame marker bits are inserted in front of the generated PN sequences. This is done for frame synchronization purposes, discussed shortly in the receiver section. As illustrated in Figure 11-3, a total of ten frame maker bits are located in front of each period of a PN sequence.

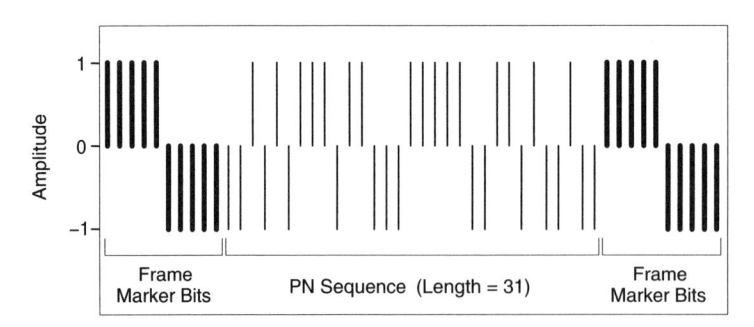

Figure 11-3: PN sequence generator.

The generated message sequences are then passed through a raised-cosine FIR filter to create a band limited baseband signal. The frequency response of the raised cosine filter is given by

$$G(f) = \begin{cases} 1 & \text{for } |f| \leq (1-\alpha)f_c \\ \cos^2\left[\dfrac{\pi}{4\alpha f_c}\left(|f|-(1-\alpha)f_c\right)\right] & \text{for } (1-\alpha)f_c \leq |f| \leq (1+\alpha)f_c \\ 0 & \text{elsewhere} \end{cases} \quad (11.2)$$

where $\alpha \in [0,1]$ denotes a roll-off factor specifying the excess bandwidth beyond the Nyquist frequency f_c.

The output of the raised cosine filter is then used to build a complex envelope, $\tilde{s}(t)$, of a QAM signal expressed by

$$\tilde{s}(t) = \sum_{k=-\infty}^{\infty} c_k g_T(t-kT) \tag{11.3}$$

where c_k indicates a complex message, made up of two real messages a_k and b_k, $c_k = a_k + jb_k$.

By modulating $\tilde{s}(t)$ with $e^{j\omega_c t}$, an analytical signal or pre-envelope, $s_+(t)$, is generated,

$$s_+(t) = \tilde{s}(t)e^{j\omega_c t} = \sum_{k=-\infty}^{\infty} c_k g_T(t-kT) e^{j\omega_c t} \tag{11.4}$$

The transmitted QAM signal, $s(t)$, is thus given by

$$\begin{aligned} s(t) &= \Re\left[s_+(t)\right] \\ &= a(t)\cos(\omega_c t) - b(t)\sin(\omega_c t) \end{aligned} \tag{11.5}$$

where $\Re[\cdot]$ corresponds to the real part of the complex value inside the brackets.

Figure 11-4 illustrates the block diagram of the QAM transmitter just discussed. Notice that the two data paths, indicated by a solid line and a dotted line, represent complex data. Again, the reader is referred to [2] for more theoretical details.

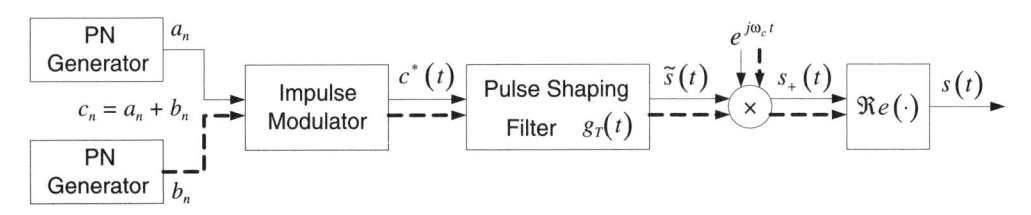

Figure 11-4: QAM transmitter [2].

11.2 QAM Receiver

11.2.1 Ideal QAM Demodulation

Here, it is assumed that the exact phase and frequency information of the carrier is available. The received QAM signal is denoted by $r(t)$. To simplify the system, an ideal channel is assumed between the transmitter and the receiver, i.e., $r(t) = s(t)$.

If $r(nT)$ is considered to be the sampled received signal, the analytic signal $r_+(nT)$ is given by

$$r_+(nT) = r(nT) + j\hat{r}(nT) \qquad (11.6)$$

where $\hat{r}(\cdot)$ indicates the Hilbert transform of $r(\cdot)$. Thus, the complex envelope of the received QAM signal can be expressed as

$$\begin{aligned}\tilde{r}(nT) &= r_+(nT)e^{-j\omega_c nT} \\ &= a(nT) + jb(nT)\end{aligned} \qquad (11.7)$$

Such a QAM demodulation process is illustrated in Figure 11-5.

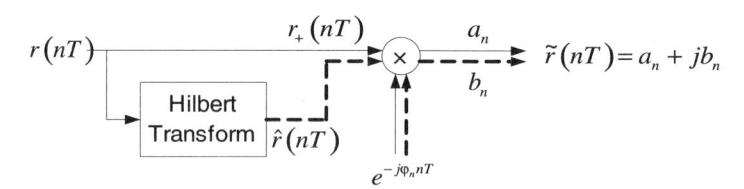

Figure 11-5: Ideal demodulation [2].

11.2.2 Frame Synchonization

Frame synchronization is required for properly grouping transmitted bits into an alphabet. To achieve this synchronization, a similarity measure, consisting of cross-correlation, is computed between the known marker bits and received samples. The cross-correlation of two complex values v and w is given by

$$R_{wv}[j] = \sum_{n=-\infty}^{\infty} \bar{w}[n]v[n+j] \qquad (11.8)$$

where the bar denotes complex conjugate.

An example of the cross-correlation outcome for frame synchronization is shown in Figure 11-6. The maximum value is found to be at the location of index 33. The subsequent message symbols are then framed from this index point.

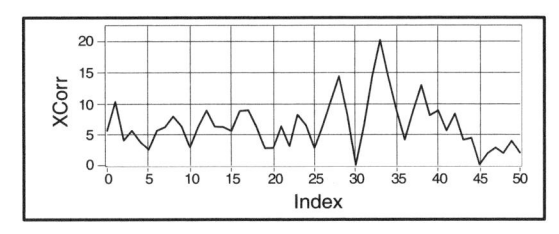

Figure 11-6: Cross-correlation of frame marker bits and received samples.

11.2.3 Decision Based Carrier Tracking

Let us now consider the phase offset, denoted by θ, between the transmitter and the receiver. Based on this offset, the received signal can be written as

$$\tilde{r}(nT) = r_+(nT)e^{-j(\omega_c nT + \theta)}$$
$$= \hat{c}_n e^{-j\theta} \tag{11.9}$$

where $\hat{c}_n$ indicates the output of a slicer mapping a received sample to the nearest ideal reference in the signal constellation. As a result, the baseband error at the receiver is given by

$$\tilde{e}(nT) = \hat{c}_n - \tilde{r}(nT) \tag{11.10}$$

Next, the LMS update method is used to minimize a decision directed cost function, $J_{DD}(\theta)$, consisting of the mean squared baseband error

$$J_{DD}(\theta) = avg\left[|\tilde{e}(nT)|^2\right]$$
$$= avg\left[\tilde{e}(nT)\overline{\tilde{e}(nT)}\right] \tag{11.11}$$

By differentiating $J_{DD}(\theta)$ with respect to θ, we get

$$\frac{dJ_{DD}(\theta)}{d\theta} = avg\left[\frac{d[\tilde{e}(nT)\overline{\tilde{e}(nT)}]}{d\theta}\right]$$
$$= 2avg\left[\Re\left\{\overline{\tilde{e}(nT)}\frac{d\tilde{e}(nT)}{d\theta}\right\}\right] \tag{11.12}$$

Chapter 11

where

$$\frac{d\tilde{e}(nT)}{d\theta} = \frac{d}{d\theta}\left[\hat{c}_n - \tilde{r}(nT)\right] = -\frac{d\tilde{r}(nT)}{d\theta} \tag{11.13}$$

and

$$\frac{d\tilde{r}(nT)}{d\theta} = -j\hat{c}_n e^{-j\theta} = -j\tilde{r}(nT) \tag{11.14}$$

Equation (11.12) can thus be rewritten as

$$\begin{aligned}\frac{dJ_{DD}(\theta)}{d\theta} &= 2\,\text{avg}\left[\Re\left\{\overline{\tilde{e}(nT)j\tilde{r}(nT)}\right\}\right] \\ &= -2\,\text{avg}\left[\Im\left\{\overline{\tilde{e}(nT)\tilde{r}(nT)}\right\}\right] \\ &= -2\,\text{avg}\left[\Im\left\{\overline{\hat{c}_n \tilde{r}(nT)}\right\}\right]\end{aligned} \tag{11.15}$$

where $\Im[\cdot]$ corresponds to the imaginary part of the complex value inside the brackets.

By writing the term $\Im\left\{\overline{\hat{c}_n \tilde{r}(nT)}\right\}$ in polar form, we get

$$\begin{aligned}\Im\left\{\overline{\hat{c}_n \tilde{r}(nT)}\right\} &= \Im\left\{\overline{R_c e^{j\beta_c} R_r e^{j\beta_r}}\right\} \\ &= R_c R_r \sin(\beta_r - \beta_c)\end{aligned} \tag{11.16}$$

Thus,

$$\sin(\beta_r - \beta_c) = \frac{\Im\left\{\overline{\hat{c}_n \tilde{r}(nT)}\right\}}{R_c R_r} \tag{11.17}$$

Note that for small $\beta_r - \beta_c$,

$$\begin{aligned}\sin(\beta_r - \beta_c) &\approx \beta_r - \beta_c \\ R_r &\approx R_c = |c_n|\end{aligned} \tag{11.18}$$

As a result, the phase error $\Delta\theta(n)$ is given by

$$\Delta\theta(n) = \frac{\Im m\{\tilde{e}(nT)\tilde{r}(nT)\}}{|c_n|^2} \tag{11.19}$$

Figure 11-7 shows a block diagram of the above tracking equations.

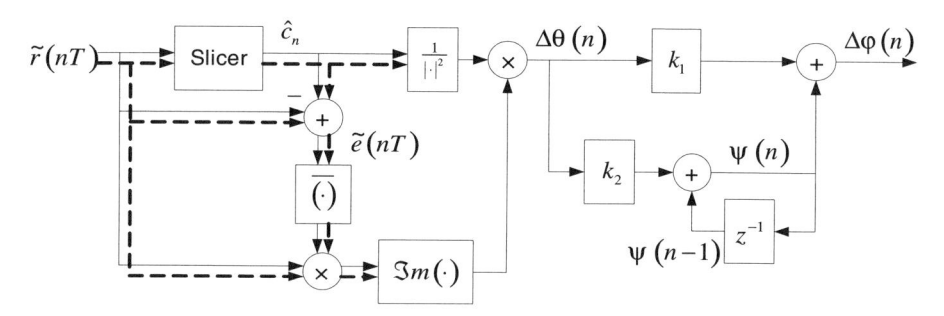

Figure 11-7: Decision directed carrier phase and frequency tracking.

When both phase and frequency tracking are considered, the carrier phase of the receiver becomes

$$\varphi(n+1) = \varphi(n) + \Delta\varphi(n) \tag{11.20}$$

In this case, the phase update $\Delta\varphi(n)$ is given by

$$\Delta\varphi(n) = k_1\Delta\theta(n) + \psi(n) \tag{11.21}$$

where $\psi(n)$ denotes the contribution of frequency tracking, which is expressed as

$$\psi(n) = \psi(n-1) + k_2\Delta\theta(n) \tag{11.22}$$

The scale factors k_1 and k_2 are configured to be small here and usually $k_1/k_2 \geq 100$ is required for phase convergence [2].

11.3 Bibliography

[1] C. Johnson and W. Sethares, *Telecommunication Breakdown: Concepts of Communication Transmitted via Software-Defined Radio*, Prentice-Hall, 2004.

[2] S. Tretter, *Communication System Design Using DSP Algorithms*, Klumer Academic/Plenum Publishers, 2003.

Lab 11: Building a 4-QAM Modem

The design of a 4-QAM modem system is covered in this lab. As shown in Figure 11-8, this system consists of the following functional modules: message source, pulse shape filter, QAM modulator, Hilbert transformer, QAM demodulator, frame synchronizer, and phase & frequency tracker. The system is divided into two parts: transmitter and receiver. The first three modules (message source, pulse shape filter, and QAM modulator) make up the transmitter side and the other modules the receiver side. The building of each functional module is described in the sections that follow.

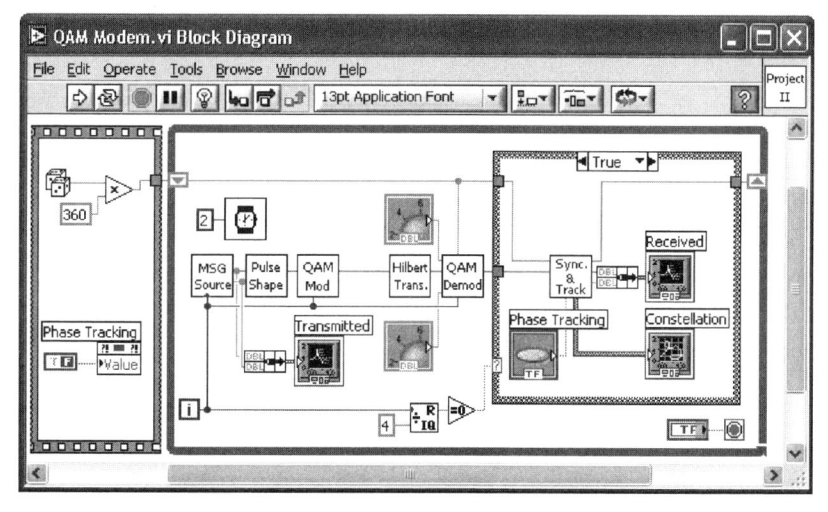

Figure 11-8: System-level VI of 4-QAM modem.

L11.1 QAM Transmitter

The first component of the QAM modem is the message source. Here, PN sequences are used for this purpose. Frame marker bits are inserted in front of these sequences to achieve frame synchronization. The BD of the `Message Source` VI is shown in Figure 11-9.

The generated samples are oversampled 4 times. This is done by comparing with 0 the remainder of the global counter, indicated by n, divided by 4. Thus, out of four executions of this VI, one message sample (frame marker bit or PN sample) is generated. For the remaining three executions of the VI, zero samples are generated. The total length of the message for one period of a PN sequence and frame marker bits

Lab 11

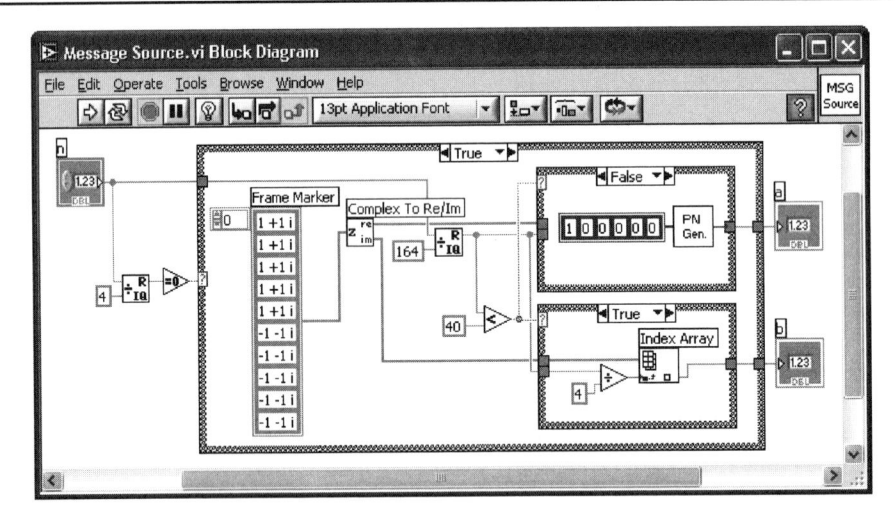

Figure 11-9: Message Source VI.

is 164, which is obtained by 4 (oversampling rate) × [10 (frame marker bits) + 31 (period of PN sequence)]. A constant array of ten complex numbers is used to specify the marker bits. Note that the real parts of the complex values are used as the frame marker bits of the in-phase samples and the imaginary parts as the frame marker bits of the quadrature-phase samples. In order to create complex constants, the representation of a numeric constant is changed by right-clicking on it and choosing **Representation → Complex Double** (or **Complex Single**).

The BD of the PN Generator VI is shown in Figure 11-10. With this subVI, a pseudo noise sequence of length 31 is generated by XORing the values of the second and fifth shift registers.

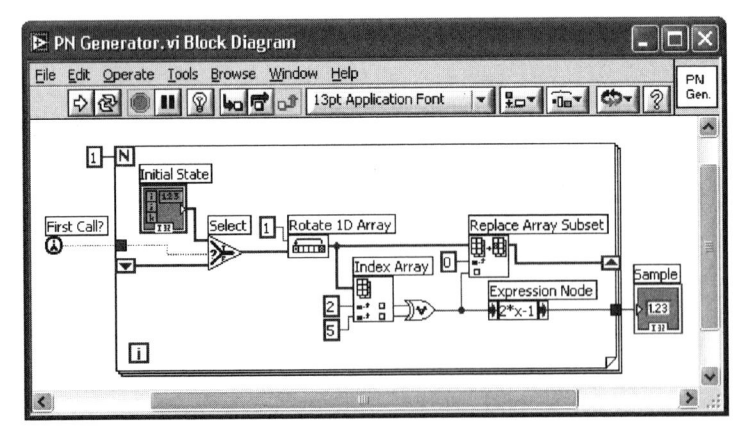

Figure 11-10: PN Generator VI.

Building a 4-QAM Modem

The `Shift Register`, `Rotate 1D Array`, `Index Array`, and `Replace Array Subset` functions are used to compute a new PN sample and to rotate the shift register. A `For Loop` with one iteration and a `First Call?` function (**Functions → All Functions → Advanced → Synchronization → First Call?**) are used in order to pass the shift register value of a current call to a next call of the subVI. The `First Call?` function checks whether a current call is occurring for the first time or not. If that is the case, the shift register values are initialized by their specified initial values. Otherwise, the old values of the shift registers are passed from the previous execution of the subVI. Notice that the `PN Generator` VI shown in Figure 11-10 is built with the consideration of porting the algorithm to a DSP hardware platform. Alternatively, the built-in `Binary MLS` VI, (**Functions → All Functions → Analyze → Signal Processing → Signal Generation → Binary MLS**), can be used for the LabVIEW implementation.

Next, the generated samples are passed to a pulse shape filter shown in Figure 11-11. A raised cosine filter is used to serve as the pulse shape filter. The `FIR Filter PtByPt` VI is utilized for this purpose. The two outputs of the pulse shape filters are combined to construct the pulse shaped message signal by using the `Re/Im to Complex` function (**Functions → All Functions → Numeric → Complex → Re/Im to Complex**).

Figure 11-11:
Pulse shape filter.

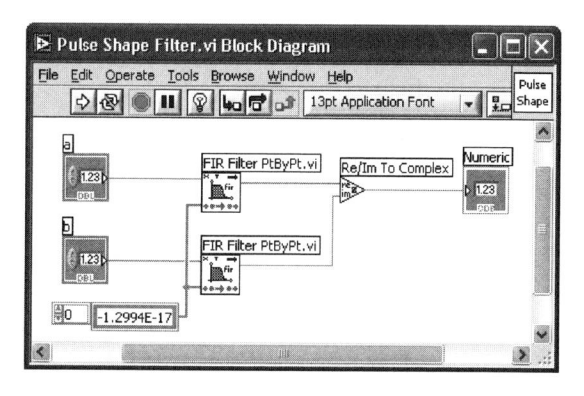

As for the filter coefficients, they can be designed by a filter design tool, such as the one discussed in Lab 4, and stored in an array of constants.

The signal passed through the pulse shape filter is then connected to the QAM modulator shown in Figure 11-12. The QAM modulated signal $s(t)$ is obtained by taking the real part of the pre-envelope signal $s_+(t)$. This is achieved by performing a complex multiplication between the complex input and a complex carrier consisting of a cosine and a sine waveform. This completes the modules of the transmitter. In the next section, the modules of the receiver are built.

Lab 11

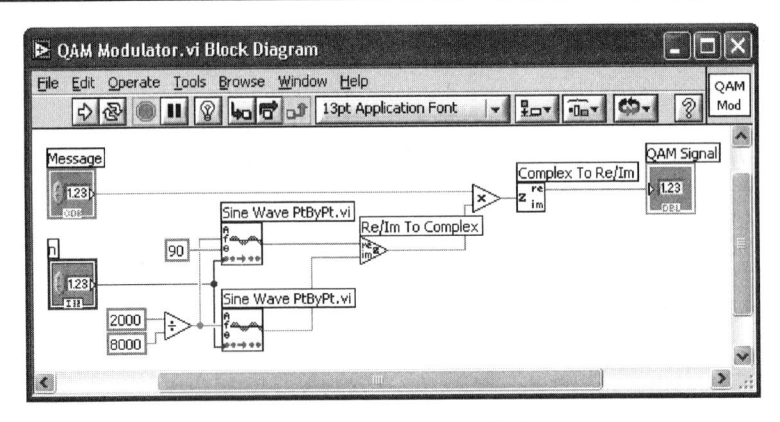

Figure 11-12: QAM modulator.

L11.2 QAM Receiver

The first module on the receiver side is the Hilbert transformer. This module builds the required analytic signal for demodulation based on the transmitted QAM signal.

A Hilbert transformer is built by using the DFD Remez Design VI (**Functions → All Functions → Digital Filter Design → Filter Design → Advance FIR Filter Design → DFD Remez Design**) of the DFD toolkit. To have an integer group delay, an even number, such as 32, is specified as the filter order. The DFD Filter Analysis Express VI is wired to analyze the group delay of the filter as well as its magnitude and phase response, see Figure 11-13.

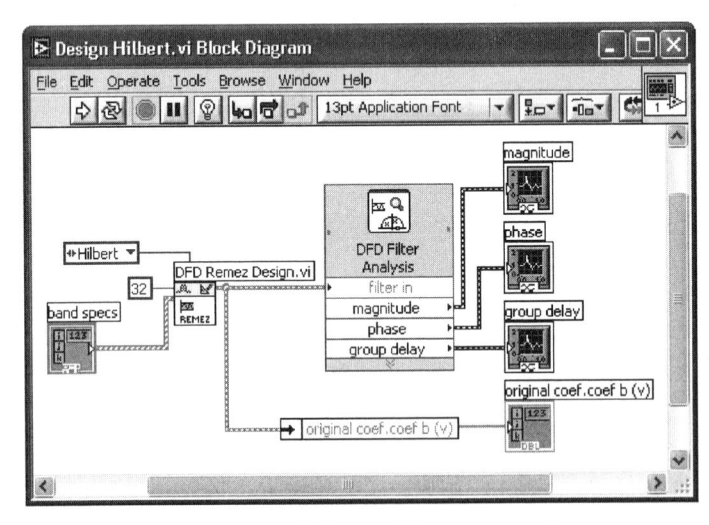

Figure 11-13: Building Hilbert transformer.

Building a 4-QAM Modem

The specifications of the Hilbert transformer are similar to a bandpass filter as indicated in Figure 11-14. Notice that only one element of the cluster array is needed to design the Hilbert transformer. However, when a control is created at the band specs terminal of the DFD Remez Design VI, there are two default cluster values. The second element, indexed at 1, should thus be deleted. To do this, select the element of the cluster array to be deleted, then right-click and choose **Data Operation** → **Delete Element** from the shortcut menu.

By running the VI, the magnitude, phase response and group delay of the Hilbert transformer can be seen as shown in Figure 11-14.

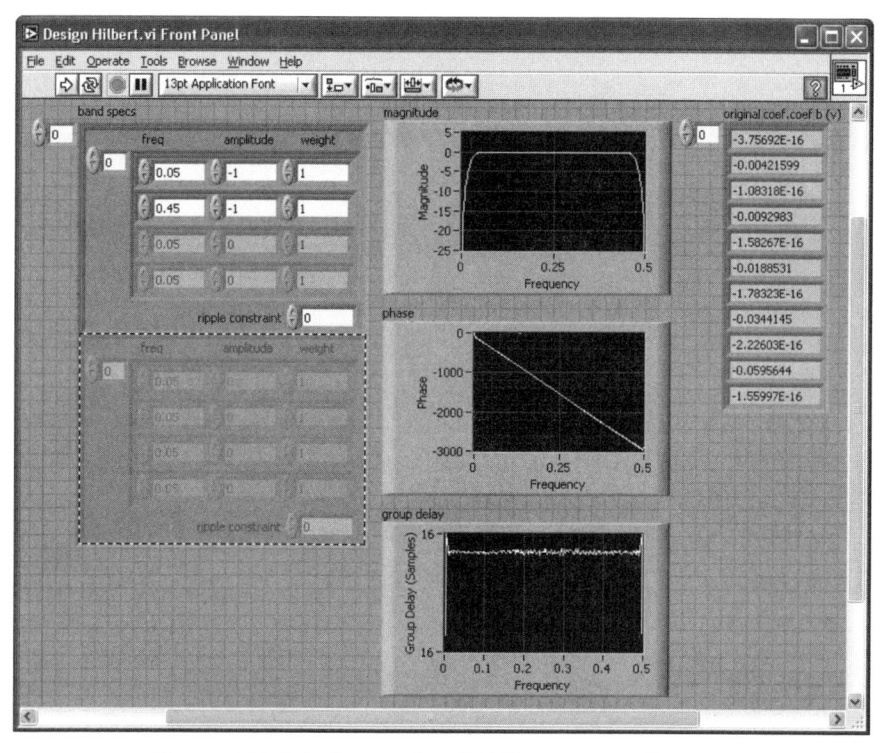

Figure 11-14: Analysis of Hilbert transformer.

The array of indicators corresponding to the Hilbert transform coefficients is converted to an array of constants to be used by the other VIs. Note that the design and analysis of the Hilbert transformer are needed only in the designing phase not in the implementation phase of the modem system.

Lab 11

The BD of the Hilbert transformer using the coefficients obtained from the DFD toolkit is shown in Figure 11-15.

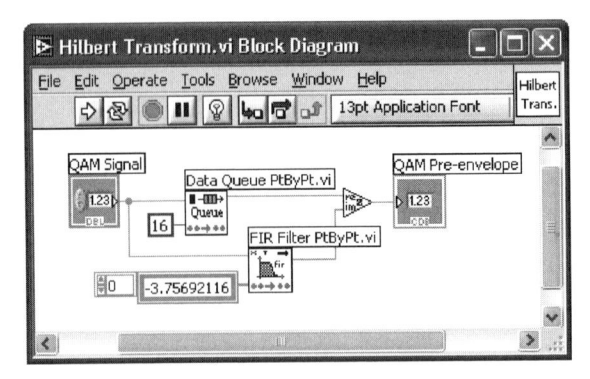

Figure 11-15: Hilbert Transform VI.

A Data Queue PtByPt VI (**Functions → All Functions → Analyze → Signal Processing → Point By Point → Other Functions PtByPt → Data Queue PtByPt**) is employed in order to synchronize the input and output of the Hilbert transformer. In other words, the input samples are delayed until the corresponding output samples become available. This is needed due to the group delay associated with the filtering operation. For an FIR filter of 33 taps, the group delay is 16. An array of numeric constants corresponding to the filter coefficients is set up based on the text file generated by a filter design tool. Here, an FIR filter has been used for the implementation of the Hilbert transformer instead of the built-in VI of LabVIEW. This is done to allow its DSP hardware implementation.

The analytic signal achieved from the Hilbert transformer is demodulated by the QAM demodulator as illustrated in Figure 11-16. The demodulation process is similar to the modulation process illustrated in Figure 11-12 except for the negative frequency part.

Next, the QAM demodulated signal is decimated by 4. To do this, a Case Structure is used so that every fourth sample is selected for processing, as illustrated in Figure 11-8. The decimated signal is sent to the Sync & Tracking VI for frame synchronization and phase/frequency tracking. The Sync & Tracking VI is an intermediate level subVI incorporating several subVIs/functions and operating in two different modes: frame synchronization and phase/frequency tracking. Let us examine the BD of this VI displayed in Figure 11-17. The input samples are passed into the receiver queue, implemented via the Complex Queue PtByPt VI (**Functions → All Functions → Analyze → Signal Processing → Point By Point → Other Functions**

Building a 4-QAM Modem

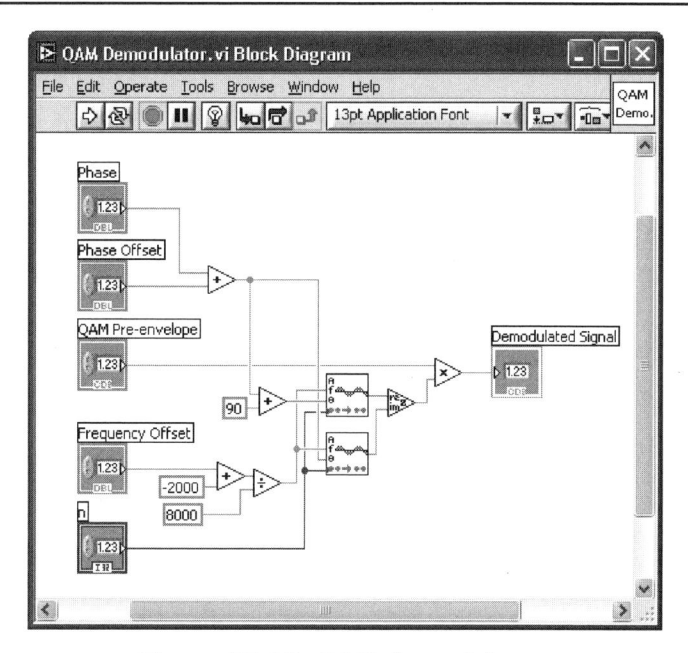

Figure 11-16: QAM demodulator.

PtByPt → Complex Queue PtByPt), in order to obtain the beginning of a frame by cross-correlating the frame marker bits and received samples in the queue. Filling the queue is continued until the queue is completely filled. Extra iterations are done to avoid including any transient samples due to the delays associated with the filtering operations in the transmitter.

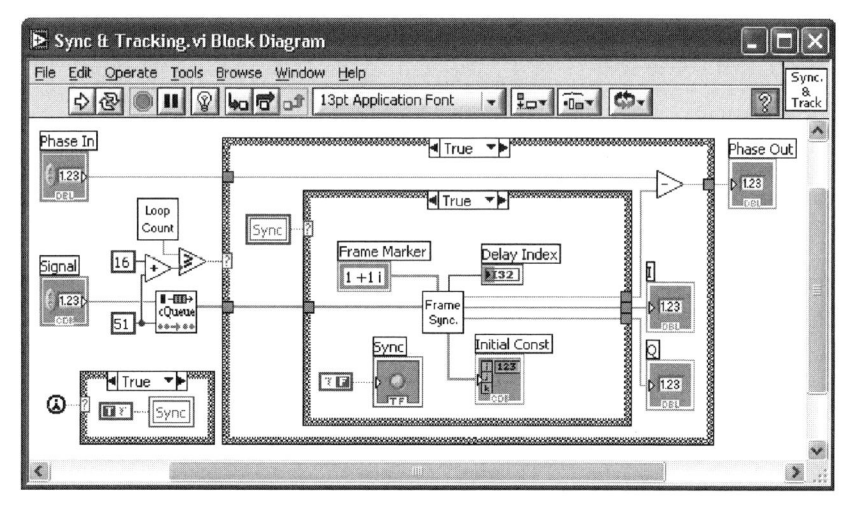

Figure 11-17: Sync & Tracking VI—frame synchronization mode.

Lab 11

The length of the queue is configured to be 51 in order to include the entire marker bits in the queue. This length is decided based on this calculation: 31 (one period of PN sequence) + 2 × 10 (frame marker bits). Also, sixteen extra samples are taken to flush out any possible transient output of the filter as mentioned previously. Bear in mind that the length of the queue or the number of extra reads varies based upon the specification of the transmitted signal such as the length of the frame marker bits and the number of taps of the phase shape filter. A counter, denoted by the Loop Count VI in Figure 11-18, is used to count the number of samples filling the queue. Once the queue is completely filled and extra reads are done, the frame synchronization module is initiated.

Figure 11-18: Loop Counter VI.

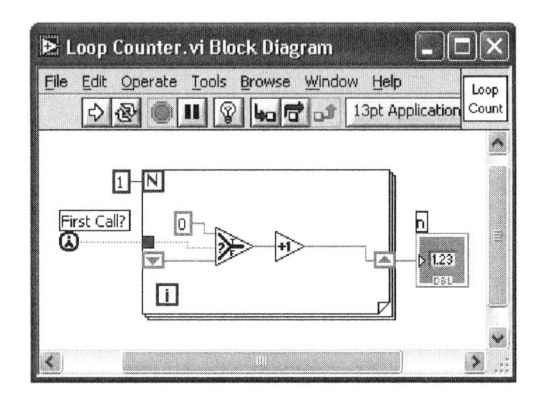

The subVI for frame synchronization is shown in Figure 11-19. In this subVI, the cross-correlation of the frame marker bits and the samples in the receiver queue are computed. The absolute value of the complex output is used to obtain the cross-correlation peak since the location of this peak coincides with the beginning of the

Figure 11-19: Frame Synchronization VI.

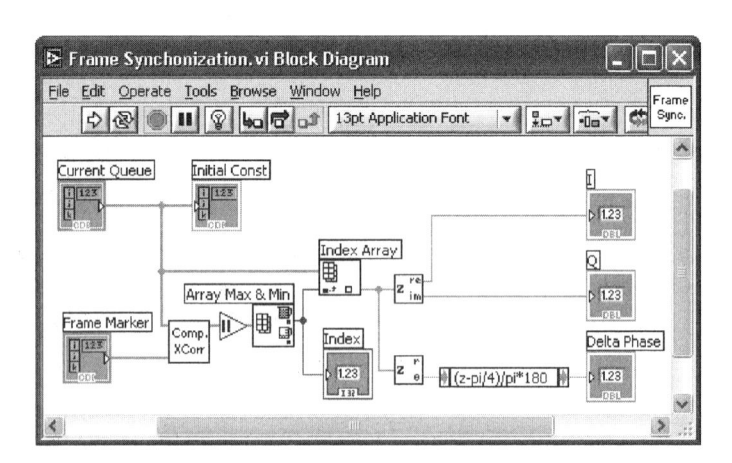

Building a 4-QAM Modem

frame. The `Array Max & Min` function is used to detect the index corresponding to the maximum cross-correlation value.

In Figure 11-20, the `Complex CrossCorrelation` VI is shown. This VI accomplishes the complex cross-correlation operation by evaluating Equation (11.8).

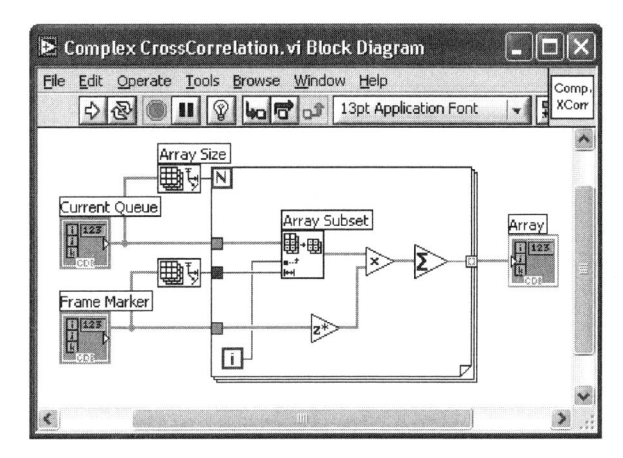

Figure 11-20: Complex CrossCorrelation VI.

Once the index of the maximum cross-correlation value is obtained, all the data samples are taken at this location of the queue. Consequently, the data bits are synchronized.

The initial phase estimation is achieved using the phase of the complex data at the beginning of the marker bits. Considering that the ideal reference is known for the first bit of the frame marker, $1 + i$ in our case, this allows us to obtain the phase difference between the ideal reference and the received frame marker bits. The real and imaginary parts of data at the beginning of the marker bits are also passed to the `Phase and Frequency Tracking` VI to provide the initial constellation.

The subVI of the frame synchronization is now complete. Locate the subVI on the BD of the `Sync & Tracking` VI shown in Figure 11-17. Notice that three local variables are created in order to pass the indicator values to the other parts of the VI which cannot be wired. In the `Sync & Tracking` VI, a `Rounded LED` indicator, labeled as `Sync`, is placed on the FP. A local variable is created by right-clicking either on the terminal icon in the BD or on the `Rounded LED` indicator in the FP and choosing **Create → Local Variable**. Next, a local variable icon is placed on the BD. More details on using local and global variables can be found in [1].

The local variable `Sync` is used to control the flow of data for the frame synchronization. The initial value of the local variable is set to `True` to execute the frame

247

Lab 11

synchronization. Then, it is changed to False within the Case Structure so that it is not invoked again. The other two local variables, Initial Const and Delay Index, are used as the inputs of the phase and frequency tracking module, see Figure 11-21.

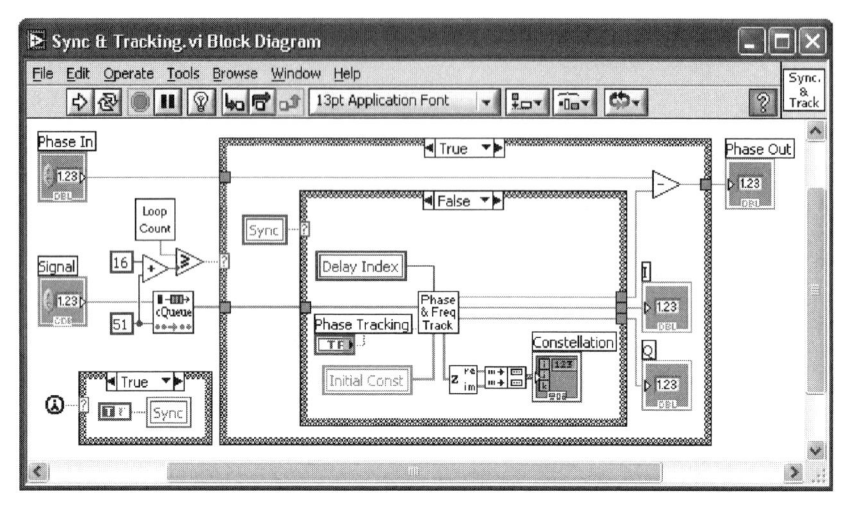

Figure 11-21: Sync & Tracking VI – phase and frequency tracking mode.

Now, let us describe the Phase and Frequency Tracking VI illustrated in Figure 11-22. A Formula Node (**Functions → All Functions → Structures → Formula Node**) is shown in the upper part of the BD, which acts as a slicer to determine the nearest ideal reference based on the quadrant on the I-Q plane. A Formula Node structure is capable of evaluating a script written in text-based C code. There are numerous built-in mathematical functions and variables which can be used in a Formula Node. For example, pi represents π in the formula node script shown in Figure 11-22. Further details on Formula Node can be found in [1].

The phase error, see the BD in Figure 11-22, is computed from Equation (11.19). This error is multiplied by a small scale factor to determine the phase update $\Delta\varphi(n)$ in a second Formula Node implementing Equation (11.20).

Now, all the components of the modem system are completed. As the final step, a Waveform Chart and an XY Graph (**Controls → All Controls → Graph → XY Graph**) are added to the system-level BD shown in Figure 11-8. In Figure 11-23, the FP of the system is shown. After updating the phase, the received signal becomes nearly a perfect reproduction of the transmitted signal except for the time delay. If there exist a phase and a frequency offset with no tracking, the received signal

Building a 4-QAM Modem

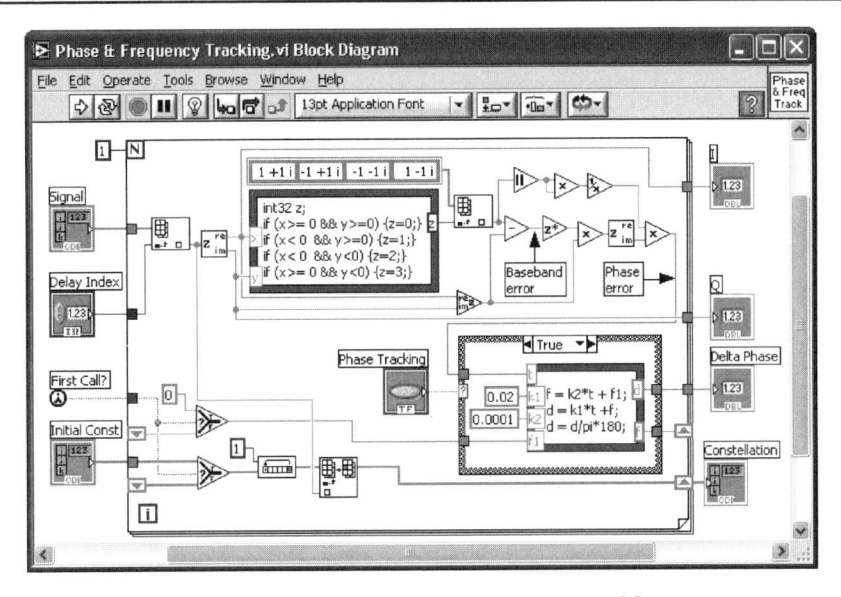

Figure 11-22: Phase & Frequency Tracking VI.

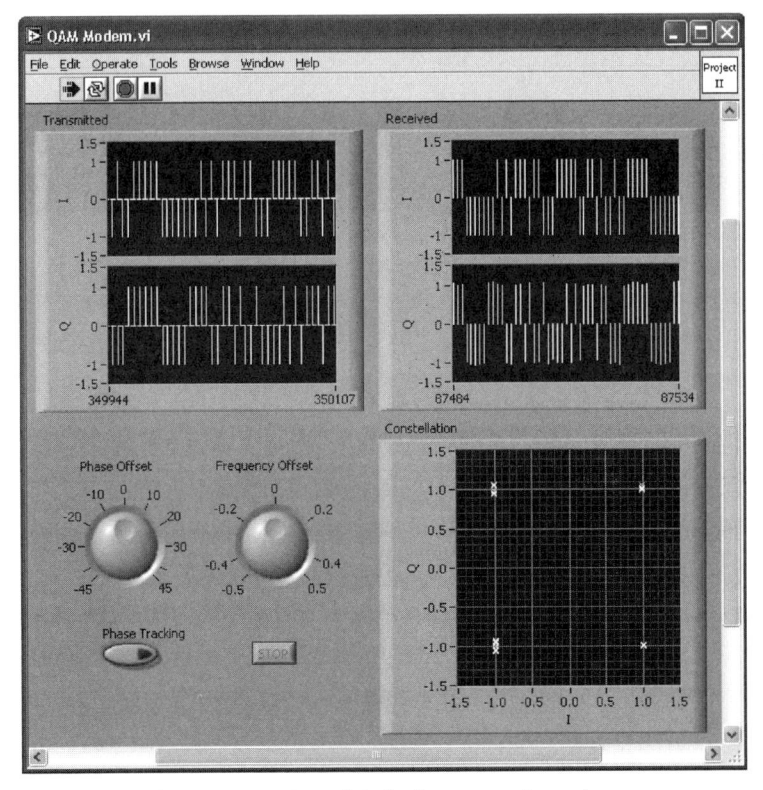

Figure 11-23: Initial phase estimation.

Lab 11

appears as shown in Figure 11-24. As displayed in this figure, the constellation of the received signal is rotated, and the amplitudes of some of the received samples become too small. Obviously, the received signal will change by introducing channel noise.

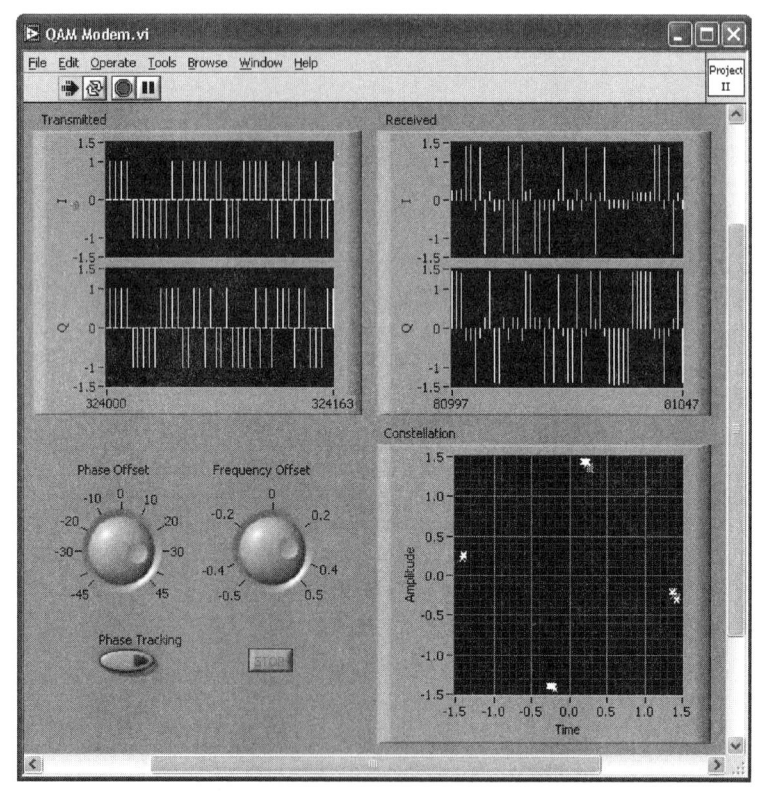

Figure 11-24: Received signal with no phase & frequency tracking.

The change in the constellation via the phase and frequency tracking is illustrated in Figure 11-25. The constellation of the samples in the I-Q plane becomes that of the ideal reference as the tracking operation progresses.

In summary, a 4-QAM transmitter and receiver system is built in LabVIEW by adopting a hierarchical approach. A simplified version of the system hierarchy, displayed by choosing **Browse → Show VI Hierarchy**, is shown in Figure 11-26. Using the phase and frequency tracking module, the phase and/or frequency offset between the transmitter and receiver is successfully compensated.

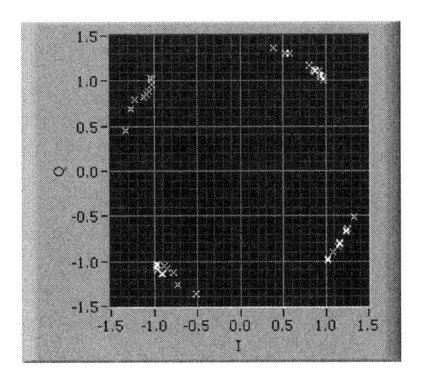

Figure 11-25: Phase and frequency tracking in IQ plane.

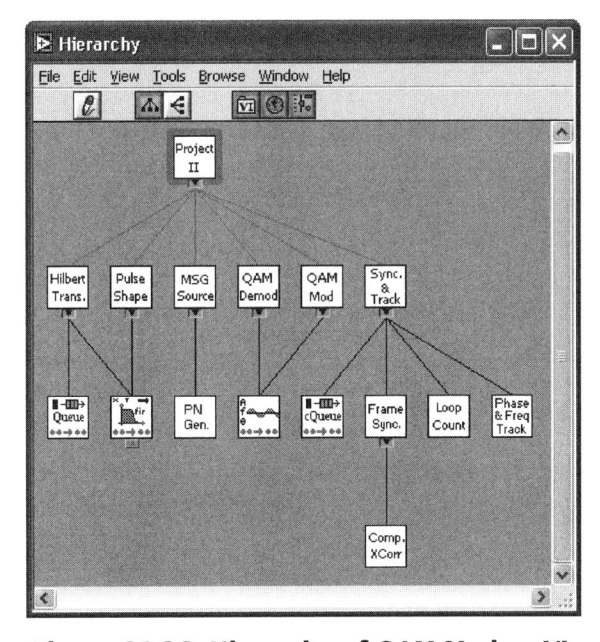

Figure 11-26: Hierarchy of QAM Modem VI.

As the final remark, all the subVIs discussed in this lab can be saved in a LabVIEW Library (LLB) file, such as *Lab 11.llb*. A new LLB file can be created by choosing **New VI Library** and naming it from the Name the VI window, which is brought up during the save operation.

L11.3 Bibliography

[1] National Instruments, *LabVIEW User Manual*, Part Number 320999E-01, 2003.

CHAPTER 12

DSP System Design: MP3 Player

The International Organization of Standardization (ISO) standard MPEG-I Layer-III, known as MP3, is one of the most widely used standards for digital compression and thus storage of audio data. The MP3 standard was developed by the Fraunhofer Institute to provide compression of audio files without any perceptible loss in audio quality [1]. This standard gives a compression ratio of 12:1 and yet preserves CD quality audio. There exist many software tools, in addition to portable MP3 players, that are capable of playing MP3 files.

This chapter presents a LabVIEW software implementation of single channel (mono) MP3 player after providing an overview of its theory. The overview presented below is not meant to be a detailed description of MP3 decoding, rather to provide enough information for one to understand the building components or blocks associated with MP3 decoding. The interested reader can refer to [1-4] for theoretical details on MP3 decoding.

Lab 12 in this chapter covers the LabVIEW implementation of the entire MP3 decoding system. The last part of this lab involves the steps taken for the real-time implementation of the system. The functional blocks associated with an MP3 player are depicted in Figure 12-1. In what follows, the function of each block is briefly mentioned.

Chapter 12

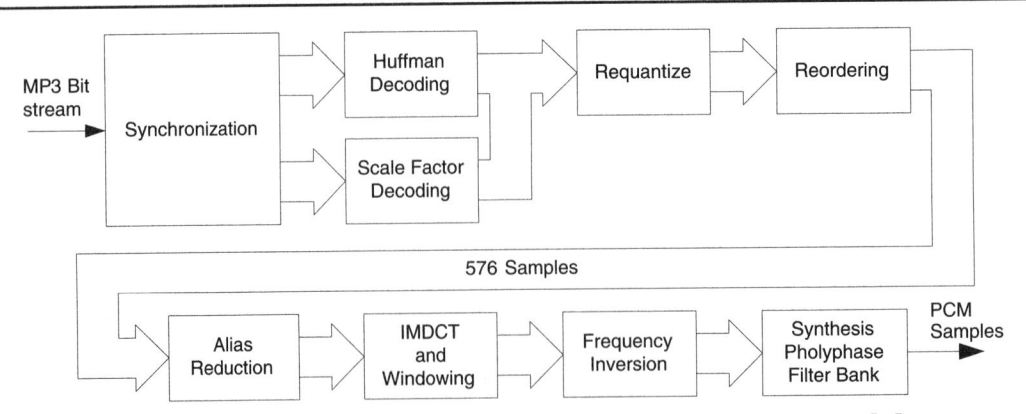

Figure 12-1: Functional blocks associated with MP3 player [2].

12.1 Synchronization Block

The first block is the Synchronization block. This block serves the purpose of receiving the incoming bitstream, extracting certain information from it and passing the extracted information to the succeeding blocks. This information consists of the Header Information, the Cyclic Redundancy Check (CRC) Information, and the Side Information. The Header Information specifies the type of the MP3 file, the bitrate used for transmission, the sampling frequency, and the nature of the audio. During decoding, the Header Information is identified by the occurrence of 12 consecutive '1s' [3]. The CRC Information provides details about the integrity of the data while the Side Information provides the necessary parameters for decoding the data as well as the reconstruction of scale factors.

An MP3 file is divided into smaller units called frames, as shown in Figure 12-2. Each frame is divided into five sections: Header, CRC, Side Information, Main Data, and Ancillary Data. Main Data is the coded audio, while Ancillary Data is optional and

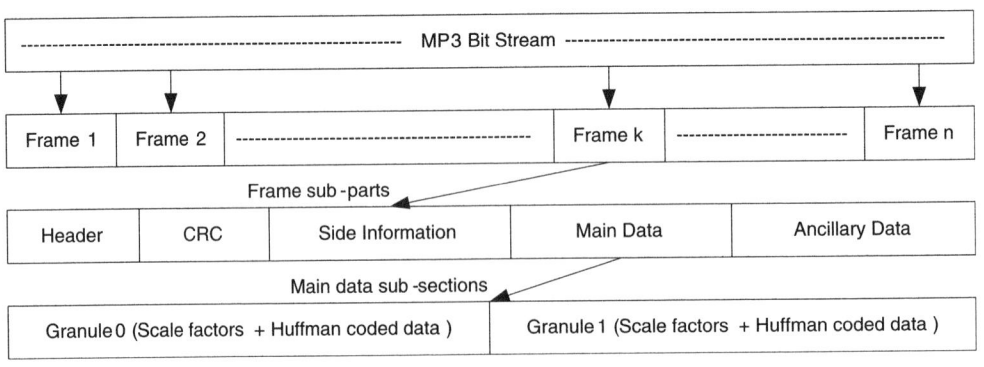

Figure 12-2: Anatomy of an MP3 File.

contains user-defined attributes such as song title, artist name or song genre. Main Data is further divided into two granules: Granule 0 and Granule 1.

Let us first mention some details on the Side Information. The Side Information is divided into four parts: Main Data Begin Pointer, Private bits, SCFSI bits, and Granule 0/Granule 1 information. The length of the Side Information varies depending upon the nature of the audio signal; 17 bytes for mono and 32 bytes for stereo, as indicated in Figure 12-3.

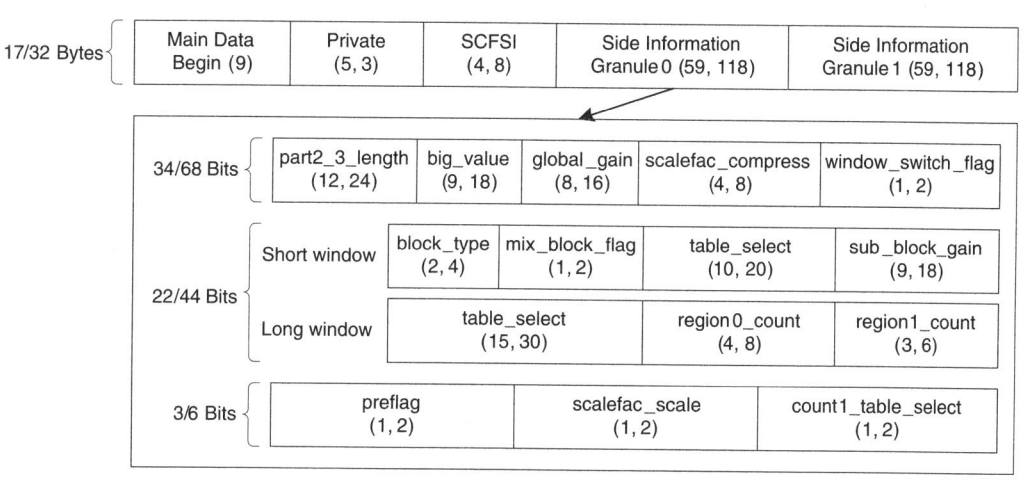

Figure 12-3: Side Information bit mapping (number of bits for mono/stereo) and its fields.

The Main Data Begin Pointer indicates the beginning of Main Data in the bitstream. Noting that the MP3 encoding uses the 'bit reservoir' technique [1], Main Data for a current frame does not necessarily begin after the Side Information for that frame but may reside in another frame, as illustrated in Figure 12-4. This is done to obtain more compression, as data from a current frame may not completely fill that frame. The Main Data Begin Pointer provides a negative offset in bytes from the header of a current frame to the location where Main Data for the current frame begins. This pointer is 9 bits long, so the bit reservoir can be at most $2^9 - 1$ bytes long.

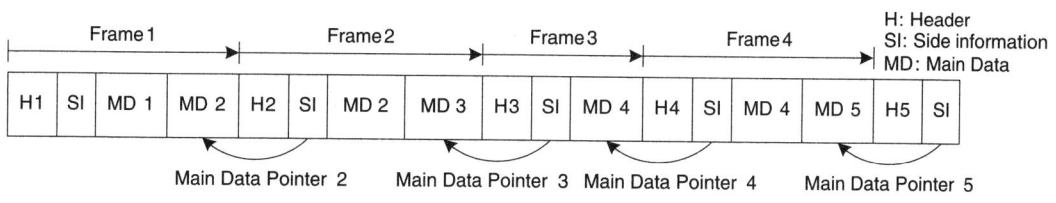

Figure 12-4: Bit reservoir technique [1].

Chapter 12

Private bits are used for private use and not normally used for decoding. The SCale Factor Selection Information (SCFSI) determines whether the same scale factor bands are transmitted for both granules considering that the scale factors for Granule 0 can be sometimes reused by Granule 1 in the Layer III standard. At encoding time, a total of 576 frequency lines of a granule are sorted into four groups of scale factor bands. Each group of scale factor bands corresponds to a bit of the 4-bit long SCFSI as shown in Table 12-1. If the bit corresponding to a group is set, this means the scale factors for that group are common to both granules and are transmitted only once. If it is zero, then they are transmitted separately for the granules.

Table 12-1: Scale factor bands corresponding to SCFSI bits [1].

SCFSI bits	Scale factor band
0	0,1,2,3,4,5
1	6,7,8,9,10
2	11,12,13,14,15
3	16,17,18,19,20

The fields for the Granule 0/Granule 1 side information is illustrated in Figure 12-3. These fields are grouped into three categories based on their functions: Huffman decode (part2_3_length, big_value, table_select, region0_count, region1_count, count1table_select), window selection (window_switch_flag, block_type, mix_block_flag), and requantization (global_gain, scalefac_compress, subblock_gain, scalefac_scale, preflag). These fields are discussed further in the following sections.

12.2 Scale Factor Decoding Block

The Scale Factor Decoding block, see Figure 12-1, decodes the scale factors to allow the reconstruction of the original audio signal. Scale factors are used to mask out the quantization noise during encoding by boosting the sound frequencies that are more perceptible to human ears.

Scale factors are decoded after the coded scale factors portion is separated from Main Data. The number of bits used for coded scale factors is specified by the part2_length field which is obtained from the slen1 and slen2 values. These values are determined from the table of scale factors by using the scalefac_compress field, displayed in Figure 12-3, as an index to the table. Note that the method to calculate part2_length changes depending on the type of window used at encoding. The windowing is done during encoding to reduce the effect of aliasing and the type of window (short or long) is determined by the block type. In the following sections, the scale factor output is referred to as scalefac_l and scalefac_s for long windows and short windows, respectively.

12.3 Huffman Decoder

The Huffman decoding is the most critical block in the decoding process. This is due to the fact that the bitstream consists of contiguous variable length codewords which cannot be identified individually. Once the start of the first codeword is identified, the decoding proceeds sequentially by identifying the start of the next codeword at the end of a previous codeword. Consequently, any error in decoding propagates, in other words the remaining codewords cannot be decoded correctly. To understand better how this block functions, some necessary information regarding the format of Huffman coded bits is mentioned next.

12.3.1 Format of Huffman Code Bits

In the MP3 standard, the frequency lines are partitioned into three regions called rzero, count1 and big_value. As Huffman coding is dependent on the relative occurrence of values, coding in each region is done with the Huffman tables that correspond to the characteristics of that region.

A continuous run of all zero values is counted and grouped as one of the regions called rzero. This region is not coded, because its size can be calculated from the size of the other two regions. The second region, count1, comprises a continuous run of −1, 0, or 1. The two Huffman tables for this region encode four values at a time, so the number of values in this region is a multiple of 4. Finally, the third region, big_value, covers all the remaining values. The 32 Huffman tables for this region encode the values in pairs. This region is further subdivided into three sub-regions: region0, region1, and region2. The region boundaries are determined during encoding. Figure 12-5 depicts the output of the Huffman Decoder splitted into the regions.

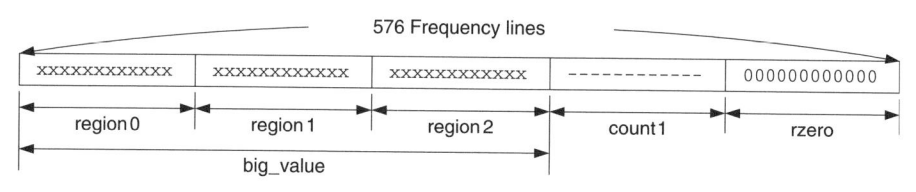

Figure 12-5: Regions of Huffman Decoder output [4].

In the big_value region, a parameter called Escape is used in order to improve the coding efficiency. In this region, values exceeding 15 are represented by 15, and the difference is represented by the Escape value. Notice that the number of bits required to represent the Escape value is called Linbits, which is associated with the Huffman table used for encoding. A sign bit follows Linbits for nonzero values.

12.3.2 Huffman Decoding

The Huffman Decoding block consists of two components: Huffman Information Decoding and Huffman Decoding. Huffman Information Decoding uses the Side Information to set up the fields for Huffman Decoding. It acts as a controller and controls the decoding process by providing information on Huffman table selection, codeword region, and how many frequency lines are decoded. This decoding is illustrated in Figure 12-6.

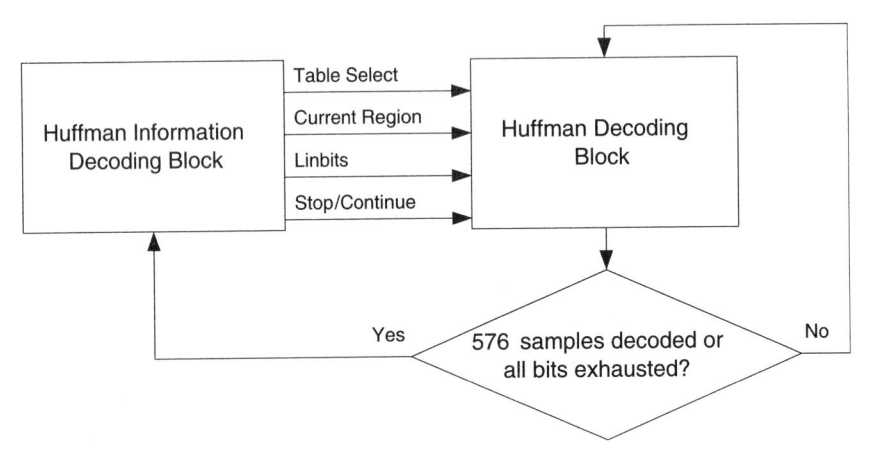

Figure 12-6: Huffman Information Decoding block as a controller.

Huffman Information Decoding starts by determining the part3_length value, which indicates the number of Huffman coded bits present in the current granule. This value is obtained by subtracting part2_length of the Scale Factor Decoder from part2_3_length of the Side Information.

The next step involves determining the codeword region considering that the selection of the Huffman tables is region specific. The decoding always starts with region0 of the big_value region. The start of region1 and region2 is determined using the region0_count and region1_count fields of the Side Information. The start of the count1 region is not explicitly defined and begins after all the codewords in the big_value region have been decoded. The count1 region ends when the number of bits exceeds part3_length.

The third step consists of obtaining the table numbers for each region. The table_select field of the Side Information gives the table numbers for all the three regions of the big_value region. For the count1 region, 32 is added to the count1_table_select field of the Side Information to get the table number for this region. This block terminates the decoding process if 576 lines are decoded or part3_length bits are used. If the decoding stops before 576 lines are decoded, zeros are padded at the end so that 576 lines are generated.

Huffman Decoding requires 34 Huffman tables for decoding Main Data. Since two of the 34 tables (table number 4 and 14) are not used, only 32 tables are required for decoding. Two tables out of these 32 tables are used for the count1 region while the rest are used for the big_value region. The process of Huffman Decoding can be understood better with the help of Figure 12-7, which shows the pattern of the Huffman coded bitstream.

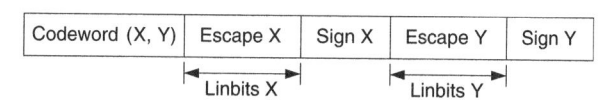

Figure 12-7: Bit format of Huffman coded bits.

The decoder fetches one bit at a time and compares it with all the codewords in the selected table. The fetched bits represent the codeword(x, y) in the figure. If there is a match, then the corresponding value is returned. If there is no match, then another bit is fetched and the process is repeated. If the returned value for the big_value region equals 15, the next bits are fetched and the number represented by them is added to the decoded value. As discussed earlier, the Linbits field is determined by the Huffman table in which a codeword match exists.

As the last step, the sign of the decoded codeword is determined. The same procedure is carried out for the codewords in the count1 region with the difference that Linbits are not used and one codeword from this region gives four decoded values.

12.4 Requantizer

The MP3 encoder incorporates a quantizer block that quantizes the frequency lines so that they can be Huffman coded. The output of the quantizer is multiplied by the scale factors to suppress the quantization noise. The function of the requantizer block is to combine the outputs of the Huffman Decoder and Scale Factor Decoder blocks, generating the original frequency lines. Figure 12-8 illustrates the function of the Requantization block.

Chapter 12

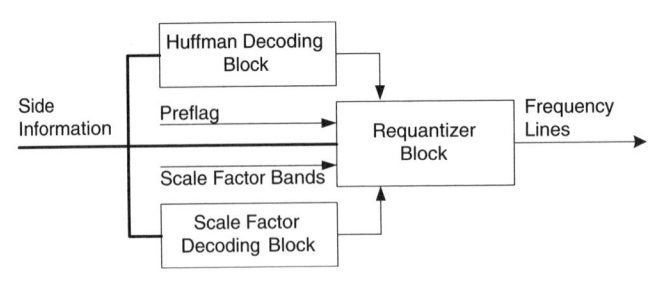

Figure 12-8: Block diagram of Requantization block.

The Requantizer block is one of the most computationally intensive blocks of MP3 decoding, since every output sample from the Huffman Decoder block needs to be raised to the power 4/3. This power is the inverse of the one used in the encoder quantization process, i.e., 0.75. The result is then multiplied by the sign of the Huffman decoded value and logarithmically quantized.

Requantization can be described by two equations, one for long and one for short windows, which are stated below [1]

$$\text{For long window,} \quad xr_i = sign(is_i) \cdot abs(is_i)^{\frac{4}{3}} \cdot \frac{2^{\frac{1}{4}(global\_gain[gr]-210)}}{2^{(scalefac\_mul(scalefac\_l[gr][sfb]+preflag[gr]+pretab[sfb]))}} \quad (12.1)$$

$$\text{For short window,} \quad xr_i = sign(is_i) \cdot abs(is_i)^{\frac{4}{3}} \cdot \frac{2^{\frac{1}{4}(global\_gain[gr]-210-8 \cdot subblock\_gain[gr][sfb][window])}}{2^{(scalefac\_mul \cdot scalefac\_s[gr][sfb][window])}} \quad (12.2)$$

where xr_i's, is_i's, and gr denote requantized arrays, Huffman decoded arrays, and granule, respectively. The remaining terms are explained below. The preflag and global_gain parameters are obtained from the Side Information. The value 210 is a system constant and is defined in the ISO/IEC 11172-3 document [1]. The scalefac_mul parameter depends on the scalefac_scale field of the Side Information. If scalefac_scale is 0, then scalefac_mul is 0.5. If scalefac_scale is 1, then scalefac_mul is 1. Thus, in the above equations, the Huffman values are scaled by 2 or $\sqrt{2}$. The parameter pretab is used for further amplification of higher frequencies and is specified as follows:

$$pretab[21] = \{0, 0, 0, 0, 0, 0, 0, 0, 0, 0, 1, 1, 1, 1, 2, 2, 3, 3, 3, 2\} \quad (12.3)$$

The 21 elements in this array correspond to the 21 scale factor bands. The parameters scalefac_1 and scalefac_2 represent the decoded scale factors for long and short windows, respectively. The parameter sfb indicates the current scale factor band. The selection of the scale factor bands depends on the sampling frequency. One thing that must be ensured before requantization is that the scale factor bands should cover all the 576 frequency lines.

12.5 Reordering

Huffman coding gives better results if its inputs are ordered in an increasing order or have similar values. This is the reason the frequency lines are ordered in increasing order of frequency during encoding as values closer in frequency have similar values. Normally, the output of the Modified Discrete Cosine Transform (MDCT) in the encoder is ordered into subbands with increasing frequency values. However, for short blocks, the output samples are ordered into subbands first by increasing windows and then by frequency. In order to remove this dependency of the window type on the output samples, the output of the MDCT is ordered first by subband then by frequency and lastly by window. Figure 12-9 illustrates the effect of the reordering block on the frequency lines.

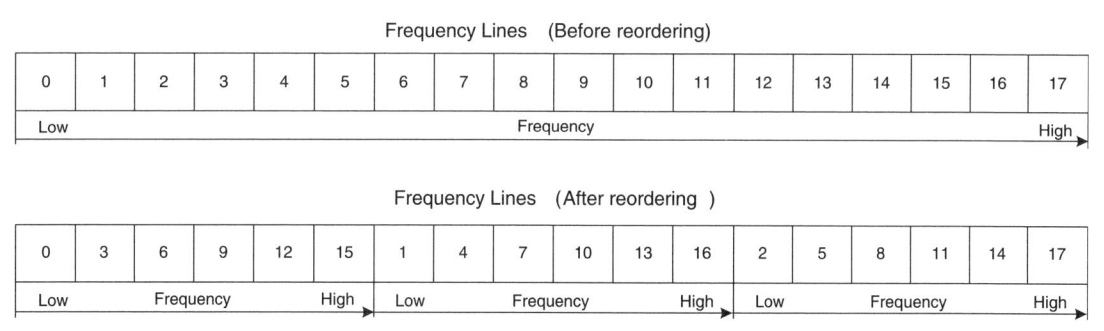

Figure 12-9: Reordering of frequency lines for short blocks [4].

Note that reordering is done only for subbands with short windows. Hence, the main task of the reordering block is to search for short windows to reorder the frequency lines. The output of the requantizer, for short windows, gives 18 samples in a subband. These samples are not dependent on the window used. The reordering block simply picks up the samples and reorders them in groups of six for each window, thereby generating them as they were before reordering.

12.6 Alias Reduction

During the encoding process, the pulse code modulated (PCM) samples are filtered into subbands using bandpass filters. However, due to the non-ideal nature of the bandpass filters, aliasing effects occur. To minimize aliasing artifacts, windowing is done after the MDCT block. That is why during the decoding process, an alias

Chapter 12

reduction block is used to generate the frequency lines similar to those generated by the MDCT in the encoder. This block adds the alias components to each frequency line to produce the original frequency lines. The alias components are specified in the ISO/IEC 11172-3 document [1]. Alias reduction is applied to all windows other than short windows.

Basically, the Alias Reduction block consists of eight butterfly calculations, described in [2], per subband, similar to that in the FFT calculation. The alias components correspond to the scale factors with which the frequency lines are scaled. The formula to calculate the two scale factors ca_i and cs_i is stated below,

$$cs_i = \frac{1}{\sqrt{1+c_i^2}}, \quad ca_i = cs_i \cdot c_i, \quad 0 \leq i \leq 7 \tag{12.4}$$

where ca_i and cs_i denote the aliased components and c_i's are defined in [1].

12.7 IMDCT and Windowing

The Inverse MDCT (IMDCT) block is responsible for generating samples which serve as the input to the Polyphase filter. The IMDCT takes in 18 input values and generates 36 output values per subband in each granule. The reason for generating twice as many output values is that the IMDCT contains a 50% overlap. This means that only 18 out of 36 values are unique, and the remaining 18 values are generated by a data copying operation. For a fast implementation of the IMDCT block, its symmetry property is used, which is described by the following equation:

$$x[i] = \begin{cases} -x\left[\dfrac{n}{2} - i - 1\right], & i = 0, \ldots, \dfrac{n}{4} - 1 \\ -x\left[\dfrac{3n}{2} - i - 1\right], & i = 0, \ldots, \dfrac{3n}{4} - 1 \end{cases} \tag{12.5}$$

After performing IMDCT, windowing is done so as to generate time samples that are similar to those obtained after the filter bank in the encoder. This is because windowing is carried out on the output of the filter bank which provides the input to the MDCT block during encoding. The type of window used depends on the block_type field in the Side Information being long or short. The window functions that are used for different blocks are stated as follows [1], for block type = 0,

$$z[i] = x[i] \cdot \sin\left(\frac{\pi}{36}\left(i+\frac{1}{2}\right)\right), \quad 0 \le i \le 35 \quad (12.6)$$

for block type = 1,

$$z[i] = \begin{cases} x[i] \cdot \sin\left(\frac{\pi}{36}\left(i+\frac{1}{2}\right)\right), & 0 \le i \le 17 \\ x[i], & 18 \le i \le 23 \\ x[i] \cdot \sin\left(\frac{\pi}{12}\left(i-18+\frac{1}{2}\right)\right), & 24 \le i \le 29 \\ 0, & 30 \le i \le 35 \end{cases} \quad (12.7)$$

for block type = 2,

$$z[win][i] = x[win][i] \cdot \sin\left(\frac{\pi}{12}\left(i-18+\frac{1}{2}\right)\right), \quad 0 \le i \le 11, 0 \le win \le 2 \quad (12.8)$$

for block type = 3,

$$z[i] = \begin{cases} 0, & 0 \le i \le 5 \\ x[i] \cdot \sin\left(\frac{\pi}{12}\left(i-6+\frac{1}{2}\right)\right), & 6 \le i \le 11 \\ x[i], & 12 \le i \le 17 \\ x[i] \cdot \sin\left(\frac{\pi}{36}\left(i+\frac{1}{2}\right)\right), & 18 \le i \le 35 \end{cases} \quad (12.9)$$

After the output of the IMDCT is multiplied with a window function, the 36 output values per subband are overlapped and added to produce 18 output values for every subband of a granule. The upper 18 values of the previous subband are stored and added to the lower 18 values of the current subband. Figure 12-10 illustrates the flow diagram for this overlap and add operation.

Chapter 12

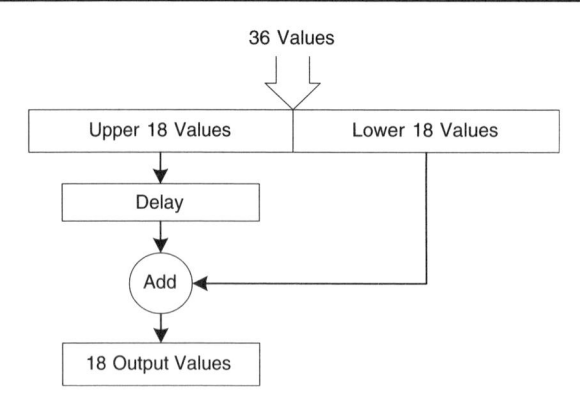

Figure 12-10: Overlap and Add operation.

12.8 Polyphase Filter Bank

Finally, a Polyphase filter bank is used to transform the 32 subbands, each with 18 time samples from every granule, to 18 bands of 32 PCM samples. PCM is a standard format of storing digital data in uncompressed format, CD audio being the prime example. PCM samples are defined depending on the sampling frequency and bitrate. A higher sampling frequency implies that higher frequencies are present and a higher bitrate produces a better resolution. Generally, CD audio uses 16 bits at 44.1 kHz. Here, after a brief overview of the PCM format, the generation of PCM samples is briefly explained with the help of the flow diagram shown in Figure 12-11.

12.8.1 MDCT

The MDCT block transforms the 32 input samples, one from each subband, to a 64-point V vector given by the following equation:

$$V[i] = \sum_{k=0}^{31} \cos\left[\pi(16+i)(2k+1)/2n\right] S[k], \qquad 0 \leq i \leq 63 \qquad (12.10)$$

where $S[k]$ denotes 32 input samples and $n = 64$.

For a fast implementation of this block, as mentioned earlier, the symmetry property of MDCT is used which requires only the computation of 50% of the values. An alternative way for a fast computation of MDCT is via the Lee's method [5], which uses a FFT type approach. In this method, an n-point MDCT is calculated using two $n/2$-point MDCTs. Such a reduction is repeated till a single point is reached.

DSP System Design: MP3 Player

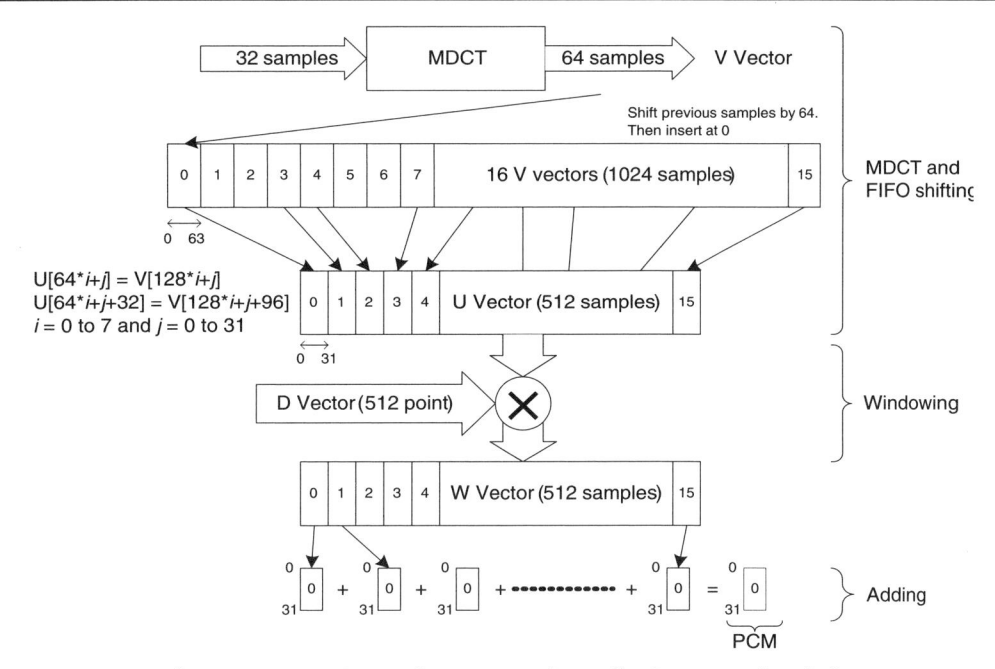

Figure 12-11: Steps in generation of PCM samples [4].

12.8.2 FIFO Shifting

The 64 output values (V vector) from the MDCT block are then fed to a 1024 sample first-in-first-out (FIFO) shift register. The MDCT operation is repeated 18 times per granule. However, due to the size of the shift register, only the last 16 V vectors are saved. Each time a V vector is generated, the shift register is shifted by 64 places to accommodate for a new V vector. The shift register is reset only at the beginning of the decoding process and not during the decoding of each frame. The last part of the IMDCT block involves the generation of a 512-sample U vector. This vector is generated by selecting alternate subbands from the shift register.

12.8.3 Windowing and Adding

As the last part towards generating PCM samples, the windowing shown in Figure 12-11 is done. This consists of multiplying the U vector by a 512-point window function. The resulting 512 point vector is transformed into 16 vectors consisting of 32 values each. The 16 vectors are then added sample-wise to generate one subband of PCM samples. The sum is then represented in 16-bit format. This operation is repeated 18 times to generate the 18 subbands of the 32 PCM samples.

12.9 Bibliography

[1] ISO/IEC 11172-3, Information Technology- Coding of Moving Pictures and Associated Audio for Digital Storage media at 1.5 Mbits/s- Part 3: Audio, first edition, August 1993.

[2] K. Lagerström, *Design and Implementation of an MPEG-1 Layer III Audio Decoder*, Master's Thesis, Chalmers Institute of Technology, Sweden, 2001.

[3] http://www.id3.org/mp3frame.html.

[4] S. Gadd and T. Lenart, *A Hardware Accelerated MP3 Decoder with Bluetooth Streaming Capabilities*, Master's Thesis, Lund Institute of Technology, Sweden, 2001.

[5] B. Lee, A New Algorithm to Compute Discrete Cosine Transform, *IEEE Trans. Accoust., Speech, Signal Processing*, vol. 32, pp. 1243-1245, Dec 1984.

Lab 12: Implementation of MP3 Player in LabVIEW

In this lab, the entire MP3 decoder system for mono MP3 files is implemented in LabVIEW. The last section of this lab covers the modifications made in order to obtain a real-time throughput from the decoder.

L12.1 System-Level VI

Figure 12-12 illustrates the system-level BD of the implemented MP3 decoder. The subVIs located in the `For Loop` are repeated as many as number of frames.

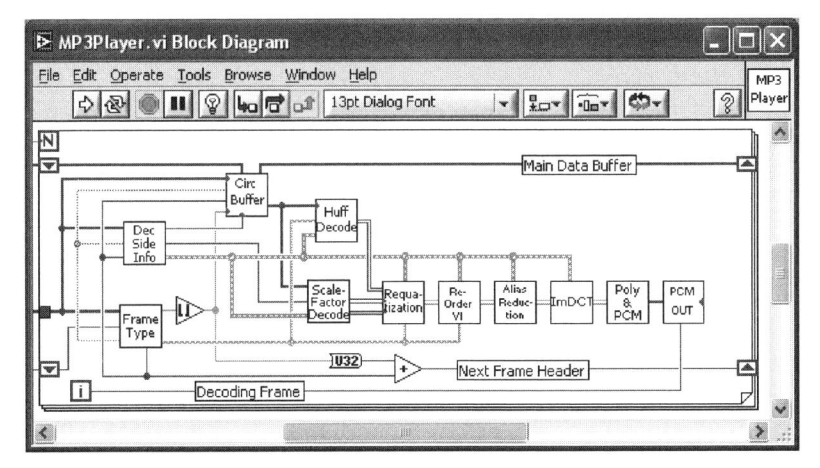

Figure 12-12: System-level Block Diagram of MP3 decoder.

The `Frame Type` VI finds the frame header from the bit stream and extracts the decoding bitrate and sampling frequency. The location of the frame header is passed to the `Dec Side Info` VI to identify its beginning. The `Dec Side Info` VI decodes the Side Information and bundles the decoded parameters into a cluster for easy access by the other VIs.

A `Shift Register` is placed to serve as a buffer for Main Data as part of the bit reservoir technique. The `Circ Buffer` VI is used to fill the buffer with Main Data, to calculate a pointer marking the start of Main Data of a current frame, and to provide Main Data and its starting index as outputs to the other VIs.

Lab 12

The `Scale Factor Decode` VI decodes the scale factors that are used to suppress quantization and any other noise. The output of this VI consists of two arrays: `scalefac_l` and `scalefac_s`. The `Huffman Decode` VI incorporates a number of subVIs to determine the required information about Huffman decoding, to calculate the length of Huffman coded bits, and to decode different regions of Huffman coded data. The `Requantization` VI combines the output of the `Scale Factor Decode` and `Huffman Decode` VIs. This VI implements the requantization equations.

The `Reorder` VI arranges the frequency lines of short blocks in the same order as in the MDCT block of the encoder. The `Alias Reduction` VI computes the antialias coefficients and weighs the frequency lines accordingly. The `IMDCT` VI computes IMDCT, does windowing on the IMDCT output, and performs overlap/add on the windowed output to generate the polyphase filter input.

The `Poly & PCM` VI carries out three operations. It multiplies every odd sample of each odd subband by '–1' (referred to as Frequency Inversion), implements the polyphase filter, and generates PCM samples.

L12.2 LabVIEW Implementation

L12.2.1 MP3 Read

As the first step of the MP3 decoding process, an MP3 file is opened and read by the `MP3 Read` VI, see Figure 12-13. This VI reads an MP3 file as specified by the `File Path` control via the `Read Characters From File` VI (**Functions → All Functions → File I/O → Read Characters From File**). The data stream is converted to unsigned integer byte using the `String to Byte Array` function (**Functions → All Functions → String → String/Array/Path Conversion → String to Byte Array**).

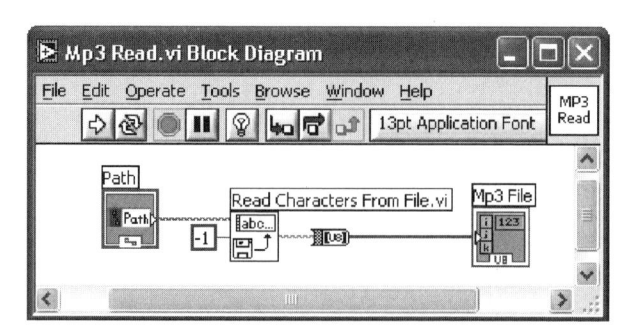

Figure 12-13: BD of MP3 Read VI.

Implementation of MP3 Player in LabVIEW

L12.2.2 MP3 File Info

The `MP3 File Info` VI extracts and displays the information about an MP3 file that is present in the file header, see Figure 12-14. It consists of two VIs: `Frame Type` and `MP3 Info Display`. The former finds the header and extracts the decoding information, while the latter displays this information in an ordered fashion.

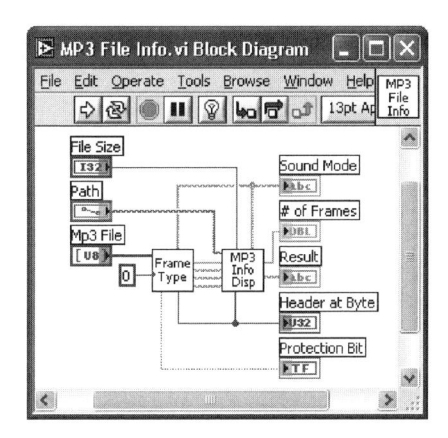

Figure 12-14: BD of MP3 File Info VI.

The `Frame Type` VI, see Figure 12-15, uses the `Find Header` subVI to find the new location of the header by searching for twelve consecutive ones. Once the header is found, three bytes of the header which contain the frame information are passed to the `Find Type` VI. Next, the `Find Type` VI performs the actual decoding of the header information using table look-ups.

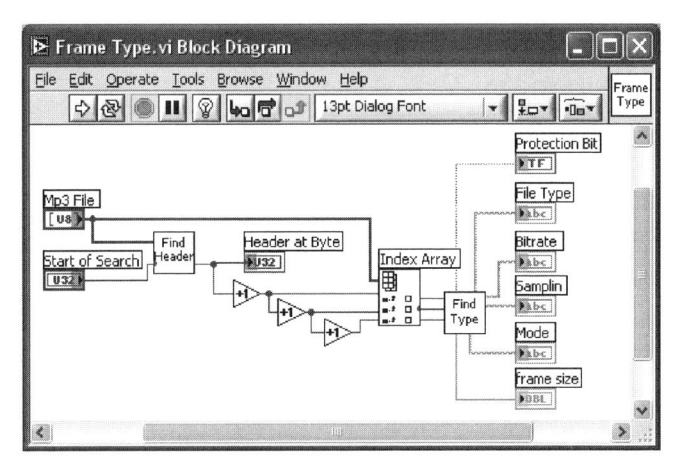

Figure 12-15: BD of Frame Type VI.

Lab 12

The `MP3 Info Display` VI generates a string array based on the output of the `Frame Type` VI. The output string array of this VI is displayed on the FP.

L12.2.3 Dec Side Info

The `Dec Side Info` VI extracts the 17 bytes of the Side Information from the bit stream and decodes it in a sequential order. Figure 12-16 illustrates the extraction of the 17 bytes of the Side Information using the `Array Subset` function. In case an optional CRC is present, indicated by the Protection bit, the starting location of the Side Information is moved by two bytes after the header. Next, each byte of the 17-byte long Side Information is wired to the `Get Bits` subVIs in order to extract the individual parameters.

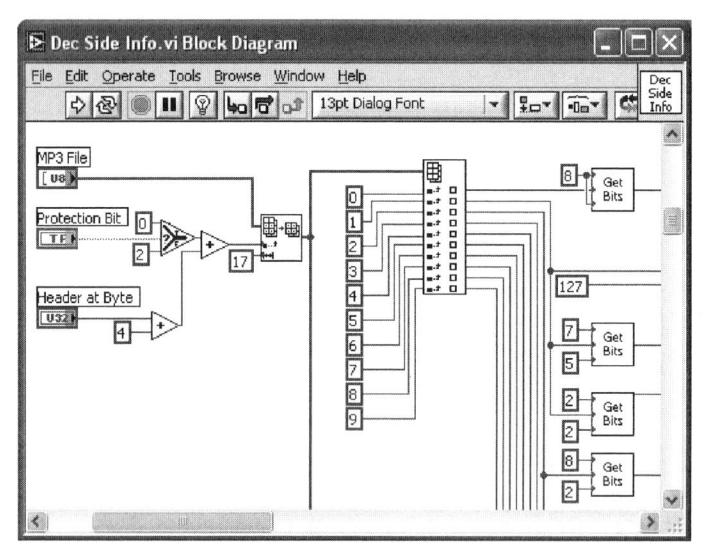

Figure 12-16: A BD section of Dec Side Info VI.

The `Get Bits` subVI performs 'logical AND' to extract bits from the input byte. This VI provides a mask that contains ones at the positions to be extracted from the input byte. The length and location of the ones in the mask are specified by the input parameters to the VI. Notice that the MSB and LSB of the byte are indicated by the indices 8 and 1, respectively. If a field is split into two bytes, the bits extracted from the higher byte is logically shifted and added to the bits from the lower byte to form one value. All the parameters of the Side Information extracted from the bit stream are bundled to form a cluster, indicated by `Side Info`.

Implementation of MP3 Player in LabVIEW

L12.2.4 Circ Buffer

Main Data of a frame can be obtained from previous frames as well as a current frame depending on the value of the Main Data Begin Pointer. The `Circ Buffer` VI, see Figure 12-17, extracts the Main Data section of the current frame and fills a 1024 point buffer. Thus, a total of 1024 Main Data samples from the previous frames and current frame are stored in the buffer. This buffer is rotated in such a way that a new Main Data section is inserted to the end of the buffer. In order to fetch Main Data for decoding of a current frame, the index of the start of Main Data is provided by subtracting the sum consisting of the size of the new Main Data section and `Main Data Ptr` from the length of the buffer.

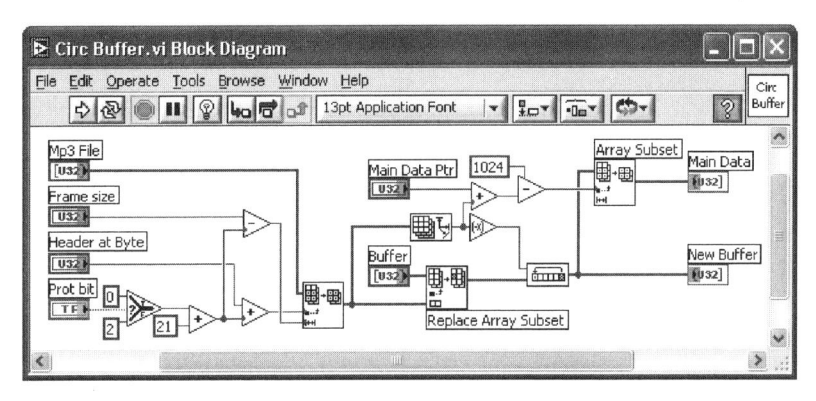

Figure 12-17: BD of Circular Buffer VI.

L12.2.5 Scale Factor Decode

The `Scale Factor Decode` VI consists of two VIs: `Scale Factor Decode0` and `Scale Factor Decode1`, which are used for decoding Granule0 and Granule1, respectively. Considering that the basic components of these VIs are the same, here only a general description for the `Scale Factor Decode0` VI is mentioned. A BD section of the VI is illustrated in Figure 12-18.

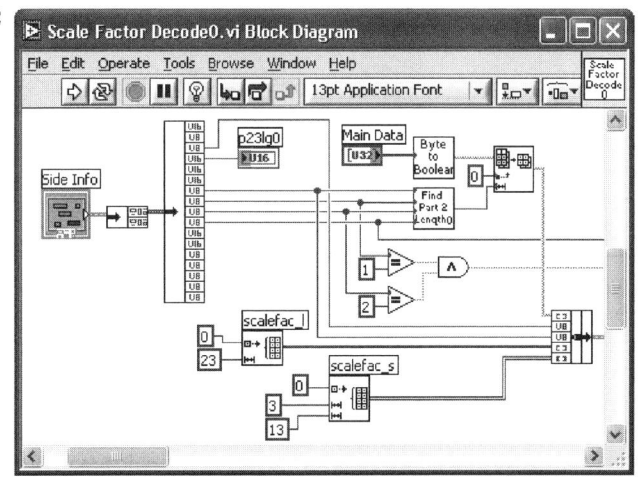

Figure 12-18: A BD section of Scale Factor Decode0 VI.

Lab 12

The Side Information input of the VI is unbundled to obtain the individual fields. Find Part2 Length0 VI calculates part2_length, which forms the length of the coded scale factors in Main Data. Once part2_length is calculated, the scale factors are extracted. Note that the equivalent C code for extracting the scale factors is provided as part of the BD on the accompanying CD. The Byte to Boolean Array VI converts each element of the integer array to Boolean digits and concatenates them to form a Boolean array.

Once the scale factor decoding for Granule0 is done, the scale factor decoding for Granule1 is done creating two scale factor outputs, scalefac_l and scalefac_s.

L12.2.6 Huffman Decode

The Huffman Decode VI decodes coded data by performing a table look-up using thirty-four standard Huffman tables. This VI consists of the following subVIs: Part2 Length, Byte to Boolean Array, Huffman Info, Big Value Search, Big Value Sign, Count1 Search, and Count1 Sign. In what follows, this VI and its subVIs are described. Considering that the Part2 Length and Byte to Boolean Array subVIs are already explained, we begin by the Huffman Info VI.

The Huffman Info VI, see Figure 12-19, extracts the individual table_select for each of the three sub-regions of the big_value region from the table_select field of the Side Information.

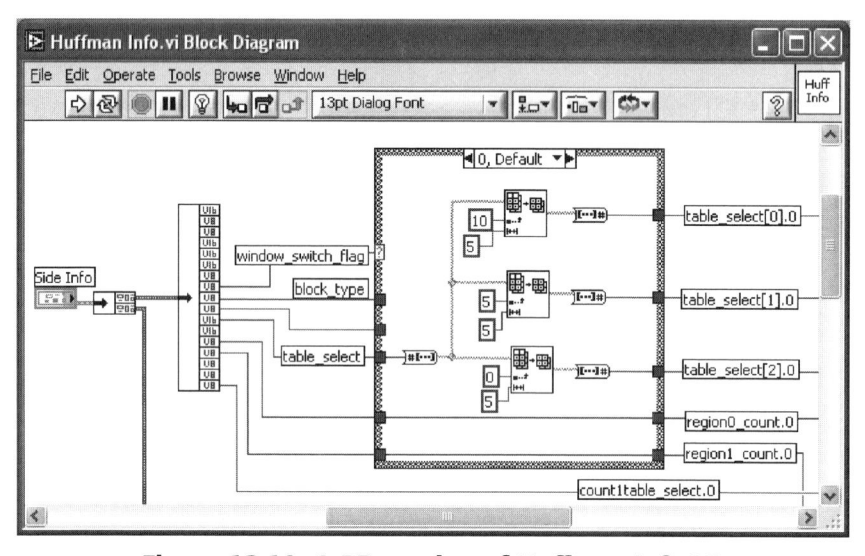

Figure 12-19: A BD section of Huffman Info VI.

Implementation of MP3 Player in LabVIEW

Note that the two parameters region0_count and region1_count are used to determine the boundaries of the big_value region. Once table_select and the boundaries are determined in the `Huffman Info` VI, the decoding of the Big Value and Count1 regions begins.

The big_value stage of the `Huffman Decode` VI consists of the `search loop` section which is located in the `While Loop` shown in Figure 12-20. This section searches for a codeword match in the big_value region. Before this search takes place in the `search loop`, the current region to be decoded must be determined. This is achieved by using the `Formula Node` exhibited in Figure 12-20. This is important, as the table_select value is region dependent. The `search loop` extracts one bit at a time, appends it to the previously extracted bit, and passes it to the Big Value Search VI. The `search loop` terminates when a match is found.

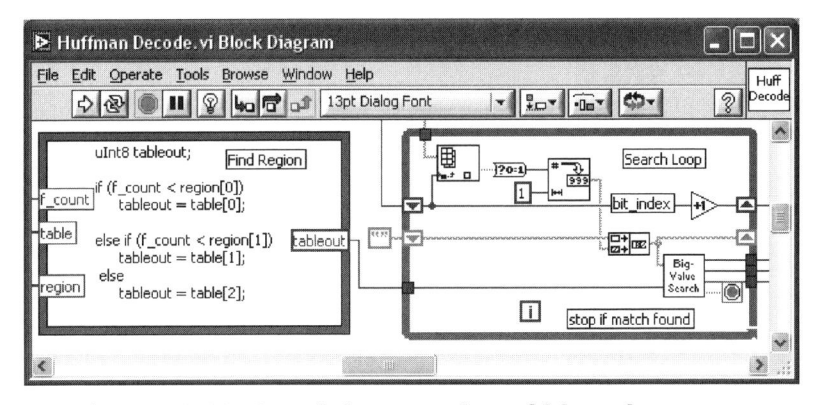

Figure 12-20: Search loop section of big_value stage.

The `Big Value Search` VI performs the search for the bit string generated in the `search loop` using the tables and generates the decoded values as its output. Figure 12-21 illustrates the BD of the `Big Value Search` VI. It uses the standard Huffman tables with `table_select` being used as an index.

The `Search 1D Array` function (**Functions → All Functions → Array → Search 1D Array**) is used to search for the codewords in the selected tables. Once a match is found, the VI extracts the decoded values, X and Y, from the selected table at the index produced by the `Search 1D Array` function. The Boolean indicator, denoted by Match, is used to terminate the `search loop` of the `Huffman Decode` VI.

Next, the `Big Value Sign` VI, see Figure 12-22, is used to determine the Escape and the sign value for X and Y. As mentioned earlier, the Escape value is only evaluated when the decoded value is equal to fifteen. This value is determined by fetching

Lab 12

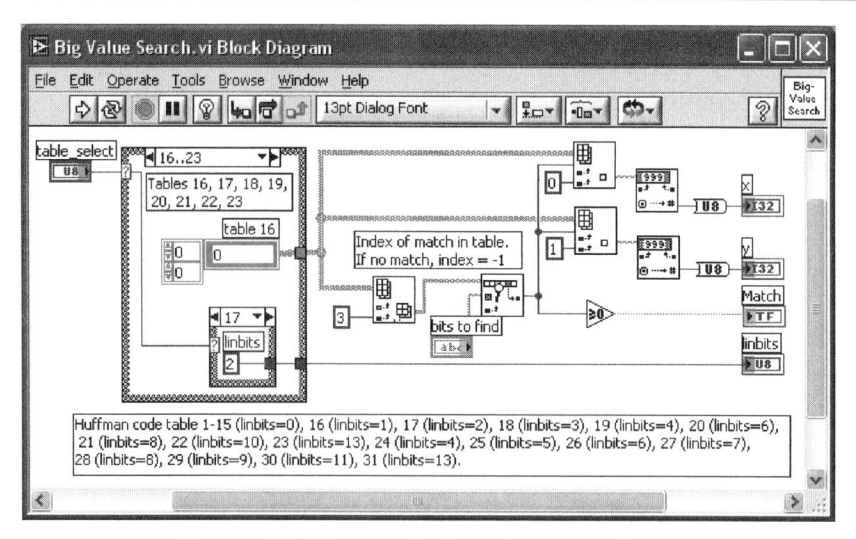

Figure 12-21: BD of Big Value Search VI.

Linbits number of bits. Then, the value represented by the fetched bits is added to the decoded outputs as shown in the `Escape Value` case structure in the BD. The decoded outputs are updated with the Escape and sign value to form the final output.

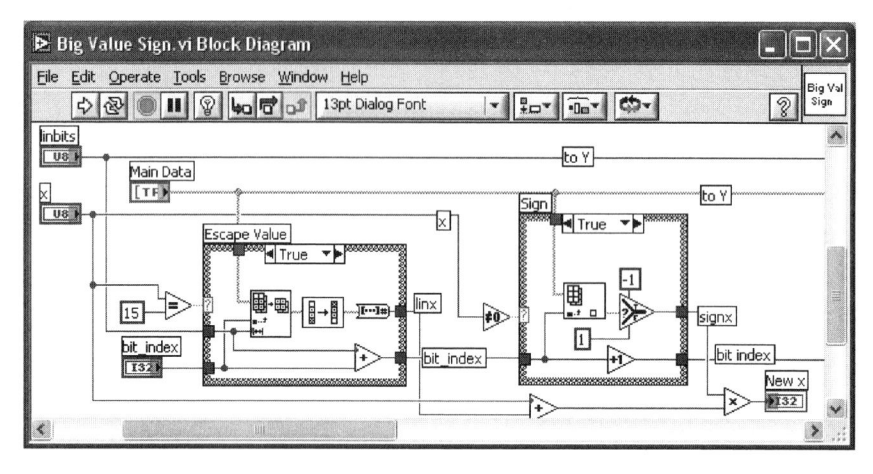

Figure 12-22: Escape and Sign bit decoding.

After the big_value stage, the decoding of the count1 stage follows. This stage is similar to the big_value decoding, except that this time four decoded outputs, V, W, X, and Y are obtained. The process of determining the sign bit is similar to the big_value sign process. Figure 12-23 illustrates the count1 stage of the Huffman decoding.

Implementation of MP3 Player in LabVIEW

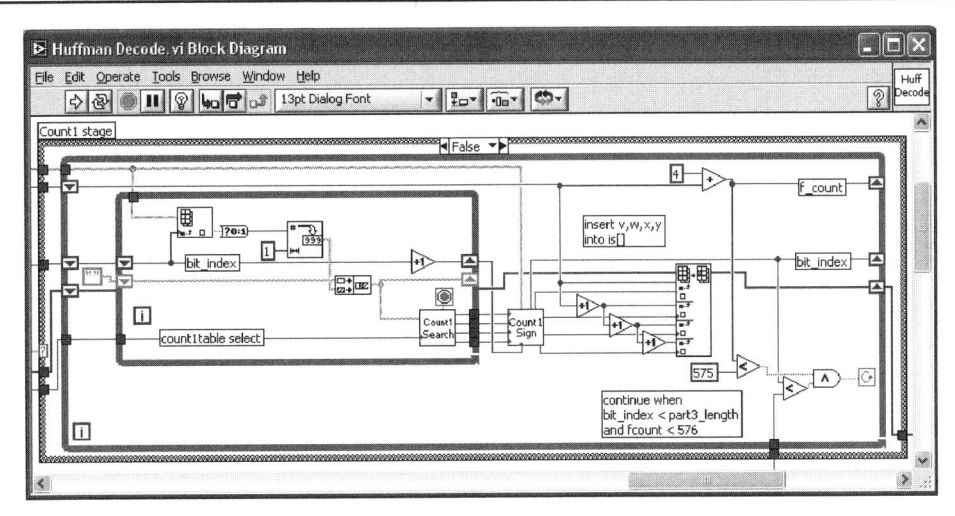

Figure 12-23: Count1 sage of Huffman Decoding.

L12.2.7 Requantization

The Requantization VI combines the outputs of the Scale Factor Decode and Huffman Decode VIs to implement the requantization equations given by Equations (12.2) and (12.3). Figure 12-24 shows one section of the Requantization VI, which illustrates the implementation of the requantization equation for long blocks.

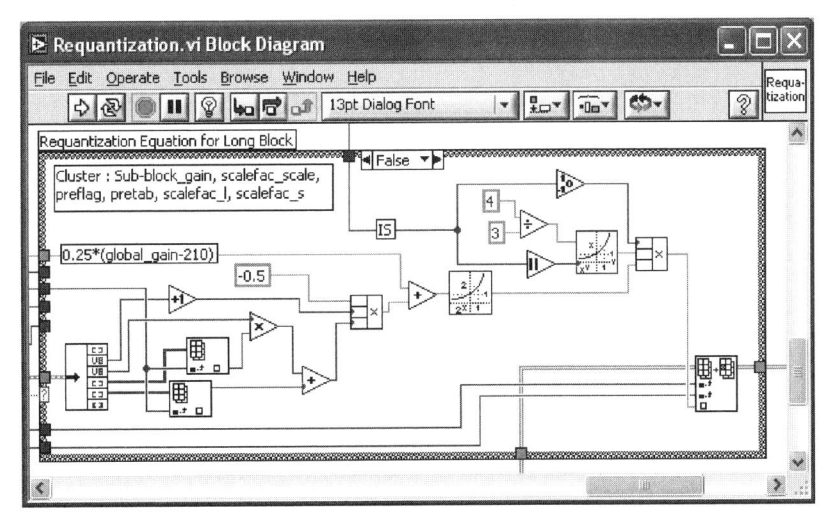

Figure 12-24: A BD section of Requantization VI.

Lab 12

This VI calculates the overall scaling factor and raises the Huffman decoded outputs to the power 4/3. Then, it multiplies them with the overall scaling factor and the sign of the Huffman decoded outputs. An array consisting of all the outputs is built to serve as the output of the requantization block.

In the other section of the `Requantization` VI shown in Figure 12-24, the `MainSI Params` VI is used. This VI is created to extract the frequently used fields of the Side Information including `global_gain`, `window_switch_flag`, `block_type`, `mix_block_flag`, `scalefac_scale`, and `preflag`. The `MainSI Params` subVI is used in the other decoding stages, e.g. `IMDCT` VI, as well as in the requantization block.

L12.2.8 Reordering

The `Reorder` VI changes the order of frequency lines or the output of the requantization block. Since reordering is required for short blocks and mix blocks only, the `Reorder` VI first identifies the block type before performing reordering. The BD of the `Reorder` VI for short blocks is illustrated in Figure 12-25. Note that this VI simply copies the frequency lines to a new array having a reordered index.

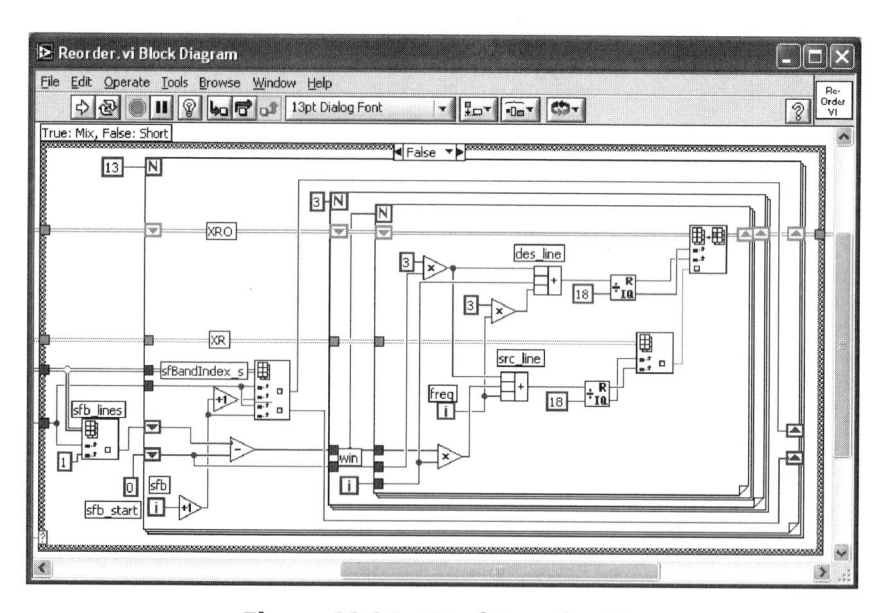

Figure 12-25: BD of Reorder VI.

For mix blocks, the reordering is not carried out for the first two subbands, as they consist of long blocks.

Implementation of MP3 Player in LabVIEW

L12.2.9 Alias Reduction

The `Alias Reduction` VI scales the reordered frequency lines with the alias coefficients. Since alias reduction is performed only for long blocks, this VI identifies such blocks and performs alias reduction on them. The BD shown in Figure 12-26 carries out the butterfly calculation. Note that the alias coefficients are defined in the arrays `ca` and `cs`, see Figure 12-26.

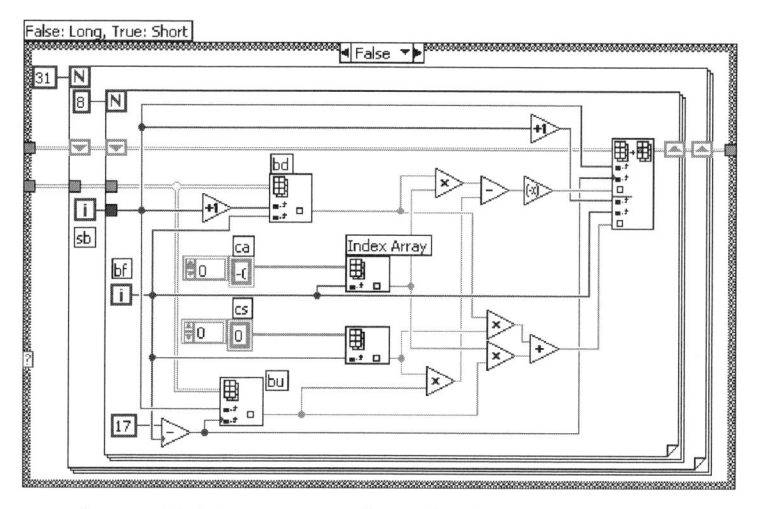

Figure 12-26: A BD Section of Alias Reduction VI.

L12.2.10 IMDCT

The `IMDCT` VI converts the frequency domain samples to time domain samples, thereby providing the input samples to the Polyphase filter. This VI also performs windowing and an overlap/add operation on the output samples as shown in Figure 12-27. The global variable `PrevBlck` is used to pass the output values used for the overlap/add operation from one frame to another. Note that this value is initialized only once

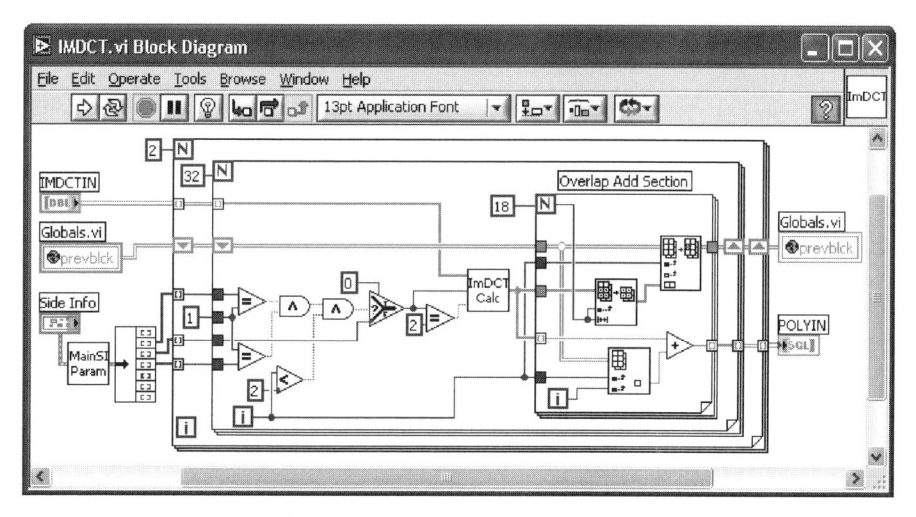

Figure 12-27: BD of IMDCT VI.

Lab 12

at the beginning of decoding. The `MainSI Params` VI, mentioned earlier, is located to obtain the main Side Information fields used for the IMDCT operation.

The `IMDCT Calc` subVI performs the actual IMDCT computations. Since the formula used for the calculation of IMDCT is different for long and short blocks, the implementation of each case is done separately.

The BD for calculating IMDCT for long blocks is illustrated in Figure 12-28. Input values are convolved with the IMDCT coefficients denoted by `Cos_l`. The windowing is performed on the convolution output. This is done by multiplying the IMDCT output with the predefined window coefficients, denoted by `win`.

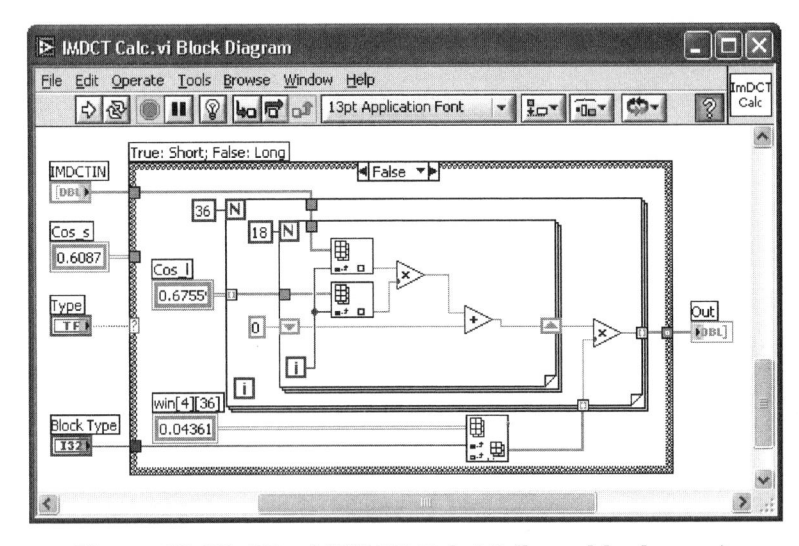

Figure 12-28: BD of IMDCT Calc VI (long block case).

For short blocks, the three outputs of IMDCT are overlapped and added for the final IMDCT array.

L12.2.11 Poly & PCM

The `Poly & PCM` VI generates the final output of the MP3 decoding process, i.e., PCM samples. This VI transforms the thirty-two subbands of eighteen samples each to eighteen subbands of thirty-two PCM samples. It consists of three subVIs: `FreqInv`, `MDCT & Wvec`, and `Add & PCM`. Figure 12-29 displays the BD of the `Poly & PCM VI` and its associated subVIs. Two global variables, a 1024 point `Vvector` and `bufferOffset`, are created for transferring the data of the current frame to the next frame.

Implementation of MP3 Player in LabVIEW

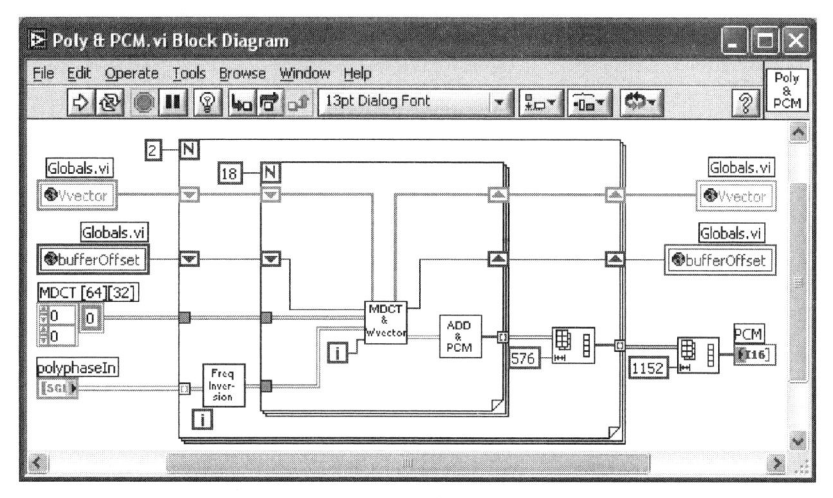

Figure 12-29: BD of Poly & PCM VI.

The `FreqInv` VI performs frequency inversion on the input samples by negating every odd sample of every odd subband. This inversion is done to compensate for the negation of values during the MDCT stage.

The `MDCT & Wvec` VI, see Figure 12-30, computes the MDCT array and generates the windowed vector, indicated by `Wvector`. The global variable `Vvector` plays a circular buffer role where the elements are shifted circularly and a new element is inserted at the index specified by `bOf`. Sixteen 32-point arrays of the replaced `Vvector` are extracted to form the pre-windowed vector. This vector is then multiplied with the window function to form the `Wvector` as illustrated in Figure 12-30.

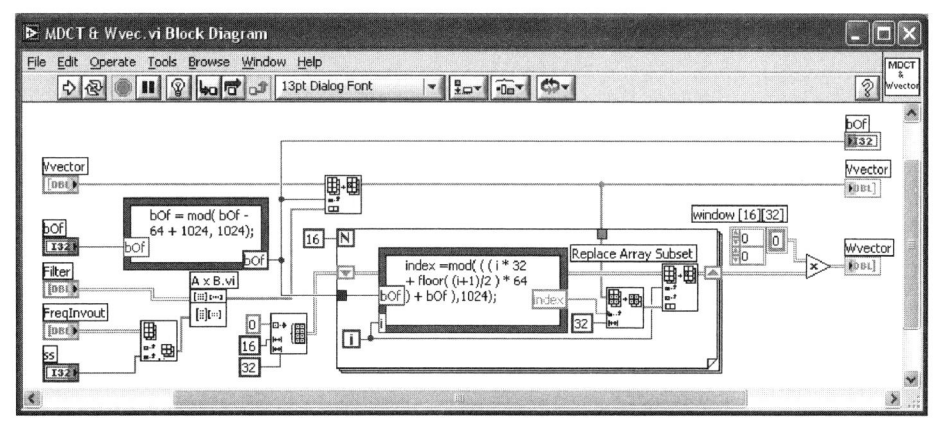

Figure 12-30: BD of MDCT & Wvec VI.

279

Lab 12

The `Add and PCM` VI adds the elements of the `Wvector`. The addition of the 16×32 `Wvector` takes place column wise so that the sixteen values in each column are added to generate one value for the final output array as previously illustrated in Figure 12-11. This operation is repeated eighteen times to form the final 18×32 output array. This array is then converted to the I16 format for the final PCM samples.

L12.2.12 PCM Out

This VI writes the output of the `Poly & PCM` VI to a file. It uses the `Build Path` function (**Functions → All Functions → File I/O → Build Path**) and the `Write to I16 File` VI (**Functions → All Functions → File I/O → Binary File VIs → Write to I16 File**). The controls are connected to the appropriate functions as illustrated in Figure 12-31.

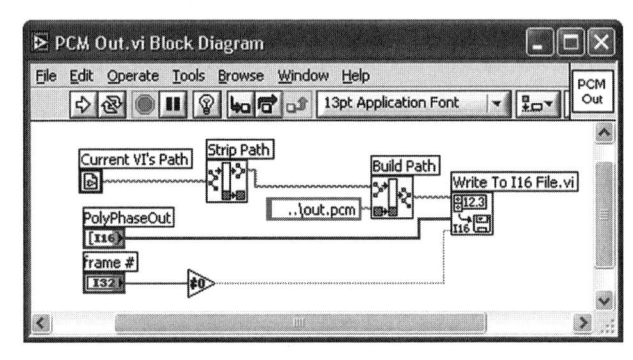

Figure 12-31: BD of PCM Out VI.

L12.2.13 MP3 Player

The `MP3 Player` VI is the top-level VI of the MP3 decoder system. It integrates all the discussed VIs. A new VI is opened and the `MP3 Read` VI is placed in its BD. The file path of the MP3 file is obtained from the path terminal of the `File Dialog` function (**Functions → All Functions → File I/O → Advanced File Functions → File Dialog**). The output of the `MP3 Read` VI is wired to an `Array Size` function and the `MP3 Frame Info` VI. The output of the `Array Size` function is wired to the `MP3 frame Info` VI. Indicators for all the outputs of the `MP3 Frame Info` VI are created to display the MP3 file information on the FP. A 1024 point array, `Buffer`, and the global variable used in the `IMDCT` and `Poly & PCM` VIs are initialized with zeros. The `bufferOffset` global variable in the `Poly & PCM` VI is initialized to 64.

Finally, a `For loop` is created and the `Frame #` output of the `MP3 Frame Info` VI is wired to the loop count N. All the VIs are wired together as shown in

Figure 12-12. Two `Shift Registers` are located in the `For Loop`. The initialized `Buffer` array is wired to one of the shift registers and a constant '0' to the other.

Running the VI brings up a file dialog to choose an MP3 file to play. The decoded PCM file can be played by any audio application software, such as Adobe® Audition™. Figure 12-32 displays the FP of the MP3 Player VI.

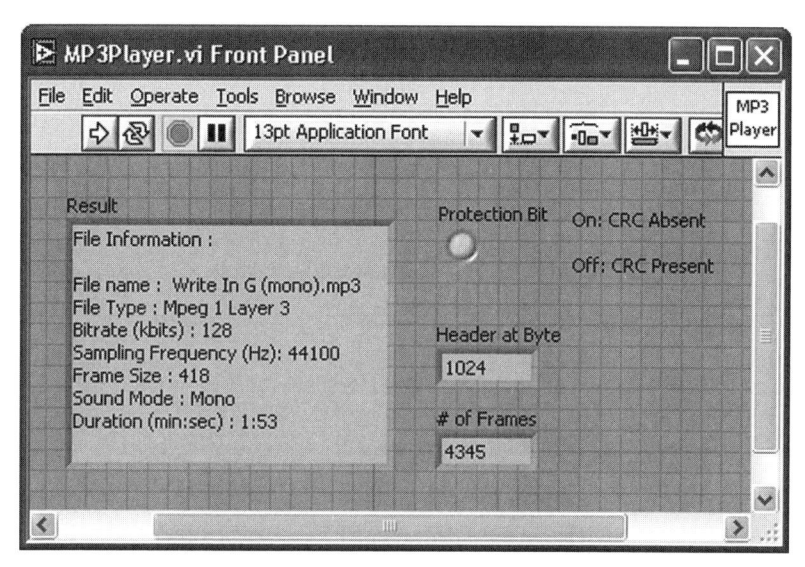

Figure 12-32: FP of MP3 Player VI.

L12.3 Modifications to Achieve Real-Time Decoding

By profiling the VI built in the previous section, one can find that four of the subVIs account for over 93% of the execution time. The time consuming subVIs along with their time usage (in percentage) are listed in Table 12-2. The overall decoding time for an examined file of 21 seconds takes 58 seconds, which means not having a real-time throughput. The following VI modifications are thus carried out in order to decode MP3 files within their play times.

Table 12-2: Processing time percentages for the time consuming VIs.

SubVI	Relative time consumption (%)
Huffman Decode VI	47.34
Requantization VI	4.50
IMDCT VI	8.90
Polyphase VI	32.28
Other VIs	6.8

L12.3.1 Huffman Decode

In the above version of this VI, the section for the codeword table look-up in the Huffman tables is quite slow. Also, the search algorithm employed is not efficient, which results in a very slow execution. In this section, a faster way of implementing the Huffman Decoding block is presented. The outer structure of the VI remains the same. The changes made are in the `search loop`, the `Big Value Search` VI, and the `Count1 Search` VI.

In the `search loop`, the search is done at bit level. However, a faster version can be obtained by performing the search on multiple bits of Huffman coded data. The length of the bits to be extracted is determined by the predefined codeword lengths in the selected table. As a result, all the redundant searches for codewords whose lengths are not specified in the selected table are avoided. A `Case Structure` is created where its contents are arrays corresponding to the codeword lengths of the tables. The array for a table is selected using the `tableout` parameter. Figure 12-33 illustrates the implementation of this new `search loop`.

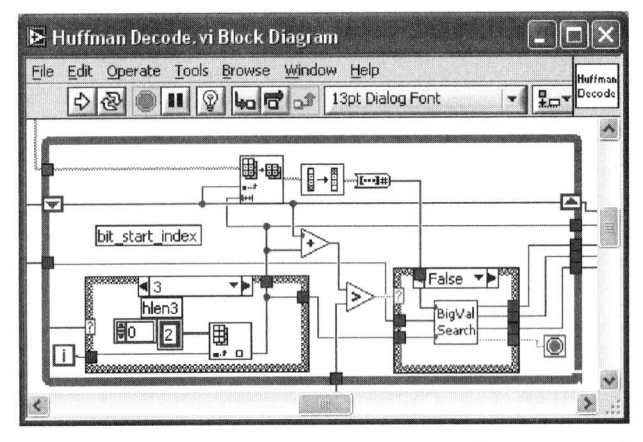

Figure 12-33: Faster search loop for Huffman Decode VI.

The `Big Value Search` VI creates the 2D Huffman tables consisting of integers instead of strings, thereby improving the speed of this VI by a factor of ten. This VI uses a new style of look-up tables (LUTs), whose description follows next.

Divide the Huffman tables such that codewords with the same length occur together. Construct a 2D array with two columns and rows given by the number of codewords of the selected length. The elements are specified by the decoded values, X and Y. Construct another 1D array with the integer values of the codewords. For example, the codeword '111001' is represented in the array as 57. Combine the two arrays into a cluster and form a 2D array of all such clusters. The clusters in the 2D array are indexed by `table_select` and `length`. As a result, the search is only done on the table specified by `table_select` and `length`. This makes the VI run quite fast. Also, create a 1D array whose elements are `Linbits` for each table.

After indexing the correct table, codeword searching is carried out using the `Search 1D Array` VI. This VI returns '–1' for no match. Figure 12-34 shows the BD of the `Big Value Search` VI.

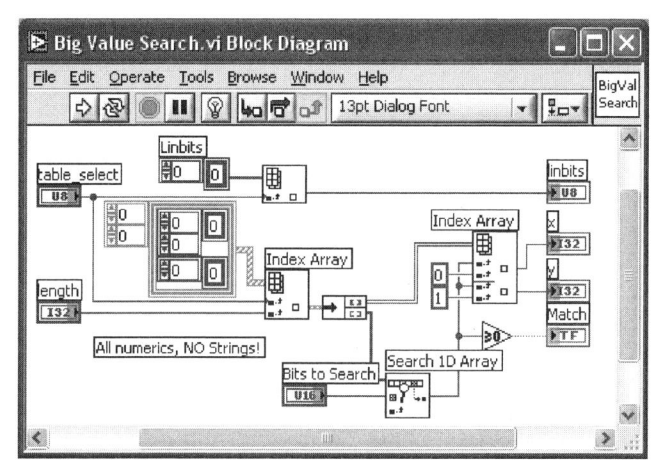

Figure 12-34: BD of faster Big Value Search VI.

Now, create three controls: `table_select`, `length`, and `Bits to search`. Next, place three functions: `Index Array`, `Unbundle`, and `Search 1-D Array`. Implement the `Big Value Search` VI using these functions as illustrated in Figure 12-34. Link the `Big Value Search` VI to the `Big Value Sign` VI. Make the same changes in the `search loop` of the count1 region as done in the big_value region earlier. Similarly, place the `Count1 Search` VI and repeat the above steps to construct a cluster of LUTs. The difference is that the 2D array in the cluster has four columns. The four columns correspond to the decoded values `V`, `W`, `X`, and `Y`. Search for the bits as before and index out elements from the 2D array at the row index given by the output of the `Search 1-D Array` function and columns 0, 1, 2, and 3. Place all the indicators and complete the updated version of the `Huffman Decode` VI by linking this VI to the `Count1 Sign` VI.

L12.3.2 IMDCTDLL

The `IMDCT` VI consumes an excessive amount of execution time because of the number of loops in the code. Also, reading and writing to a global variable adds to the overall time of the VI. Therefore, to speed up the `IMDCT` VI, the `Call Library Function Node` feature of LabVIEW is utilized here. This function improves the speed by a factor of four.

Open a new VI and save it as *IMDCTDLL.vi*. Place all the controls and connect them to the functions in the VI as shown in Figure 12-35. Place the `Call Library Function Node` function (**Functions → All Functions → Advanced → Call Library Function**). Before using this function, a Dynamic Link Library (DLL) must be created to carry out the desired operation. To create a DLL, use Visual C++. Open a new **Win32 Dynamic Link Library** project (**File → New → Projects → Win32 Dynamic Link Library**). Give the project a name and click **OK**. Select **Empty DLL Project** in the next window.

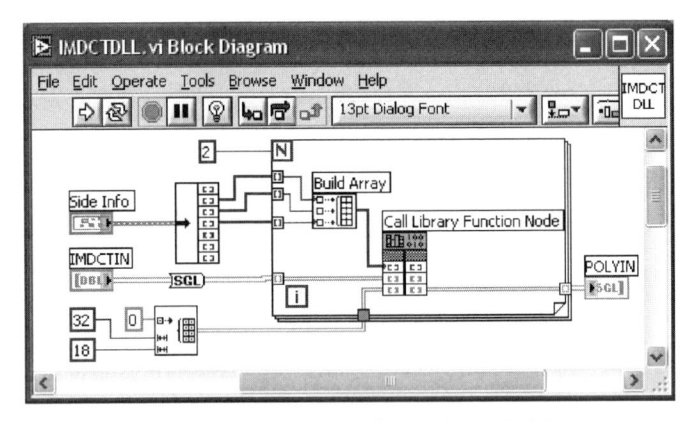

Figure 12-35: BD of IMDCTDLL VI.

The source code of the `IMDCTDLL` project is provided on the accompanying CD. Add the files from the `IMDCT` folder in the project (**Project → Add to Project → Files...**). Build the project (**Build → Rebuild All**). Then, double-click on the `Call Library Function Node` function. Select the DLL just created using the `Browse` button. The function in the DLL will be displayed in the **Function Name** field. Select C in the **Calling Convention** field. Click on **Add Parameters After** in that window and provide a name to be entered in the **Parameter** field. Select the parameter type (array, number or string etc.) and the data type. Also, select the method of passing arguments to the function. For numbers, select `Pass by Value` or `Pass by Reference`. However, for arrays, select `Array Data Pointer`. Add all the controls as defined in the C function. Click **OK** when finished. Attention should

Implementation of MP3 Player in LabVIEW

be paid to the data type matching in LabVIEW and C. For example, `Float` in C corresponds to `Single Precision` in LabVIEW. Also, when outputs are arrays, an array initialized with zeros must be passed to the `Call Library Function Node` function as input. For more details on the use of DLL in LabVIEW, refer to [1].

L12.3.3 Poly & PCM

The `Poly & PCM` VI is also modified using the `Call Library Function Node` function to improve its speed. First, open a new VI. Then, implement this VI as illustrated in Figure 12-36 with the use of the `Call Library Function Node` function. Now, create a project that builds a DLL to perform the function of the `Poly & PCM` VI as just described in the `IMDCTDLL` VI. The original code used to build the DLL is provided on the accompanying CD. Add the files from the `Polyphase` folder and build the project to create the DLL. After the completion of the DLL project, add appropriate parameters to the `Call Library Function Node` function as done in the `IMDCTDLL` VI. Use the `Reshape Array` function to convert the 2D output of the library function to a 1D array of length 1152 as shown in Figure 12-36. Place an indicator, `POLYOUT`, at the output of the first `Shift Register`.

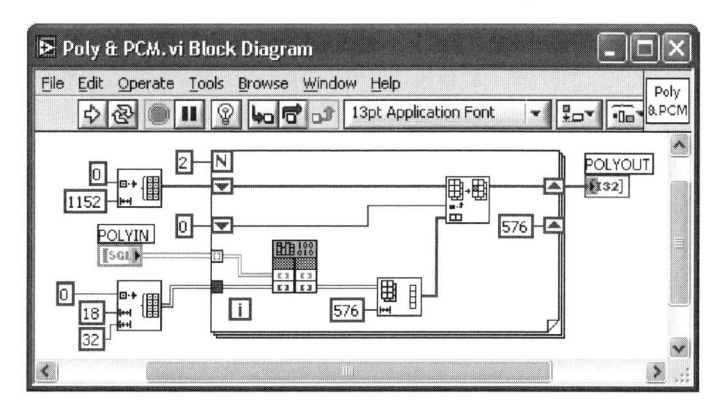

Figure 12-36: Polyphase VI using Call Library Function Node.

The above modifications allow the 21-second MP3 file to be decoded in 3.2 seconds. Table 12-3 provides five other MP3 files and their decoded times. Note that all these files are decoded in shorter times than their durations.

Table 12-3: Decoding times for different MP3 files.

MP3 File Specs	Play time (sec)	Decoding time (sec)
32 kHz, 48 kbps, 330 Frames	11	3.0
44.1 kHz, 56 kbps, 1075 Frames	27	9.2
44.1 kHz, 128 kbps, 454 Frames	11	4.0
44.1 kHz, 128 kbps, 2573 Frames	67	47.4
48 kHz, 128 kbps, 1099 Frames	26	15.9

L12.4 Bibliography

[1] National Instruments, *LabVIEW User Manual*, Part Number 320999E-01, 2003.

Index

Symbols
.D unit 157
.L unit 157
.M unit 157
.S unit 157
2's-complement number 95

A
A/D converter 1
adaptive filtering 123, 191
adaptive filtering systems 206
   DSP integration of 206
Align object 26
analog-to-digital (A/D) converter 46
analog-to-digital signal conversion 43
   sampling 43
Analysis VI 78
analytical signal 233
application program interfaces (APIs) 188
   RTDX_CreateInputChannel() 188
   RTDX_CreateOutputChannel() 188
   RTDX_enableInput() 188
   RTDX_enableOutput() 188
   RTDX_readNB() 188
   TARGET_INITIALIZE() 188
approximation coefficients 143
array 12
   tunnel 27
array-based processing 127
Array Constant shell 230
array functions
   Replace Array Subset 130
   Rotate 1D Array 130
Array shell 109
Array Size function 67
assembly 163
assembly optimizer 164
assembly source file 167
auto-indexing 27
automatic tool selection 7
Autoscale 36, 129
auto indexing 67

B
Basic Function Generator VI 37
basic types of wires 11

Binary MLS VI 241
block diagram (BD) 5
   align object 26
   conditional terminal 25
   distribute object 26
   expandable nodes 10
   loop iteration terminal 25
   node 6
   remove broken wires 18
   structure 6, 12
      case 12
      conditional terminal 12
      count terminal 12
      For Loop 12
      iteration terminal 12
   terminal icon 6, 11
      data type terminal icon 11
   view as icon 19
   wires 6
Boolean to (0,1) function 68
breakpoints 13, 164, 180
Build Array function 38
Build Path function 191
Build Waveform function 57
Butterworth 74
Butterworth Filter PtByPt VI 128

C
C6416 DSK 164
C64x DSP 160
C6713 DSK 164
C6xxx Simulator 164
C6x architecture 157
case selector 12
Case Structure 57
case structure 12
   case selector 12
   selector terminal 12
CCS Automation 184
CCS Automation VIs 192
   CCS Build VI 192
   CCS Download Code VI 192
   CCS Open Project VI 192
   CCS Reset VI 192
   CCS Run VI 192
   Simple Error Handler VI 192

CCS Build VI 186, 192
CCS Close Project VI 186, 193
CCS Communication 184
CCS Communication VIs 192
   CCS RTDX Read VI 192
   CCS RTDX Write VI 192
CCS Download Code VI 186, 192
CCS Halt VI 186, 193
CCS Open Project VI 186, 192
CCS Reset VI 186, 192
CCS RTDX Read Array SGL 194
   timeout value 195
CCS RTDX Read VI 192
CCS RTDX read VI 184
CCS RTDX Write Array SGL 194
CCS RTDX Write VI 192
CCS RTDX write VI 184
CCS Run VI 186, 192
Chart History Length 131
Chebyshev 74
Chirp Pattern VI 149
cluster 13
Cluster shell 225
Cluster to Array function 227
Code Composer Studio (CCS or
   CCStudio) 164
code efficiency 163
coding effort 163
coefficient quantization error 76
coercion dot 105
compiler 165
Complex Queue PtByPt VI 244
Concatenate Inputs 154
conditional terminal 12, 25
connector pane 6, 20
controls 5, 8
controls palette 7
Convolution VI 71
count terminal 12
Create Histogram Express VI 65
cross-correlation 247
Current VI's Path function 191
Cursor Movement Tool 62
custom probe 29
C compiler optimizer 163

Index

D
Data Queue PtByPt VI 244
data type terminal icon 11
decision directed cost function 235
decomposition filter bank 143
detail coefficients 143
DFD Build Filter from Cascaded Coef VI 90
DFD Build Filter from TF VI 90, 116
DFD Classical Filter Design Express VI 77, 79, 83, 114
DFD Convert Structure VI 119
DFD Filtering VI 83
DFD Filter Analysis Express VI 79
DFD FXP Coef Report VI 115, 120
DFD FXP Quantize Coef VI 114, 119
DFD FXP Simulation Report VI 115
DFD FXP Simulation VI 115
DFD Get Cascaded Coef VI 86
DFD Get TF VI 81
DFD Pole-Zero Placement Express VI 77
DFD Pole-Zero Plot control 79
digital filtering 191
Digital Filter Design (DFD) toolkit 73, 77
Digital FIR Filter VI 91
digital frequency 43
digital signal processing (DSP) 1
digital signal processors 1
direct-form II 76
direct-form implementation 76
Disable indexing 27
discrete Fourier transform (DFT) 46, 139
discrete wavelet transform (DWT) 139
 approximation coefficients 143
 decomposition filter bank 143
 detail coefficients 143
 filter bank 143
 frequency resolution 142
 inverse DWT 143
 multi-level decomposition 144
 reconstruction filter bank 144
 scaling function 143
 time-frequency localization 142
 time resolution 142
 wavelet function 143
 wavelet transform 142
Discrete Wavelet Transform Ex VI 152, 155
Distribute object 26
Dolph-Chebyshev window 73
double precision (DP) 100
DSK (DSP Starter Kit) 3
DSP integration examples 191
 adaptive filtering 191
 digital filtering 191
 frequency processing 191
 integer arithmetic 191
DSP processor 157
DSP starter kit (DSK) 161
DSP Test Integration for TI DSP 183
DTMF tone detection 222
dual-tone multi-frequency (DTMF) signaling 221
Cluster shell 225
 Cluster to Array function 227
 Mechanical Action 225
 Reorder Controls In Cluster 226
 Goertzel algorithm 221
dynamic data type 33
dynamic range 96

E
elliptic method 74, 85
enumerated constant 128
enumerated type control 37
enumerate control 148
Enum Constant 148
equi-ripple method 73, 79
executable file 164
execution highlighting 13
expandable node 10
Expression Node 57, 228
Express VI 10
 create histogram 65
 DFD classical filter design 77, 79, 83
 DFD filter analysis 79
 DFD pole-zero placement 77
 scaling and mapping 32
 simulate signal 32
 spectral measurements 60
 time delay 27
extended sign bit 98, 205
external memory 163

F
fast Fourier transform (FFT) 48, 139, 140
Feedback Node 131
FFT algorithm 211
 DSP integration of 211
 overflows 219
FFT VI 151
filter bank 143
finite impulse response (FIR) filter 73
finite word length 99
finite word length quantization effect 99
First Call? function 241
FIR filtering system 203
 fixed-point version 203
FIR Filter PtByPt VI 130, 241
fixed-point 95
fixed-point arithmetic 202
fixed-point digital filtering 114
fixed-point processor 75, 95
Fixed-Point Tools VI 78
floating-point arithmetic 95
floating-point processor 95
Formula Node 248
Formula Waveform VI 65
For Loop 12, 27
 count terminal 12
 iteration terminal 12
Fourier series 46
Fourier transform pair for analog signals 45
Fourier transform pair for discrete signals 45
frame marker bit 232, 239
frame synchronization 232, 234, 239

cross-correlation 234, 246
frequency processing 191
frequency resolution 142
front panel (FP) 5
 numeric controls 16
 numeric indicator 16
 terminal icon 11
functions 10
functions palette 7
fundamental frequency 47

G
Get Waveform Components function 40
Goertzel algorithm 221, 228
Graph Palette 62
 Cursor Movement Tool 62
 Panning Tool 62
 Zoom 62
Greater or Equal? function 24
group delay 244

H
hand optimized assembly 163
Hanning window 151
Hanning window VI 151
header file 167
Hilbert transform 234
Hilbert transformer 243
Horizontal Pointer Slide control 82

I
implied or imaginary binary point 96
Impulse Pattern VI 149
Index Array function 67
indicators 5, 8
infinite impulse response (IIR) filter 73, 74
 Butterworth method 74
 Chebyshev method 74
 coefficient quantization error 76
 direct-form II 76
 direct-form implementation 76
 elliptic method 74
 inverse Chebyshev method 74
 second-order cascade form 76
 transfer function 76
initializer terminal 131
Initialize Array function 132
instruction 159
 decoding 159
 execution 159
 fetching 159
integer arithmetic 191
Integer number of cycles 62
Intensity Graph 151
internal memory 163
International Organization of Standardization (ISO) 253
interrupt service vector 203
Inverse Chebyshev 74
Inverse Discrete Wavelet Transform Ex VI 152, 155
inverse DWT (IDWT) 143, 144
In Range and Coerce function 109

Index

iteration terminal 12

J
Joint Time-Frequency Analysis (JTFA) 144

K
Kaiser window 73

L
labeling tool 16
LabVIEW 2, 5
  numeric data types 102
    128-bit time stamp 103
    array 103
    Boolean 103
    byte signed integer 103
    byte unsigned integer 103
    cluster 103
    complex double-precision, floating-point 103
    complex extended-precision, floating-point 103
    complex single-precision, floating-point 103
    digital 104
    digital waveform 103
    double-precision, floating-point 103
    dynamic 103
    enumerated type 103
    extended-precision, floating-point 103
    I/O name 104
    long signed integer 103
    long unsigned integer 103
    path 103
    picture 104
    reference number 104
    single-precision, floating-point 103
    string 103
    variant 104
    waveform 103
    word signed integer 103
    word unsigned integer 103
LabVIEW 7.1 3, 15
LabVIEW Library (LLB) file 251
least mean square (LMS) algorithm 124
least significant bit (LSB) 50
levels of optimization 172
library file 167
linear assembly 163
linker command file 164, 167
linking 165
loop iteration terminal 25
Loop Tunnel 67
Loose Fit 34

M
Main Data Begin Pointer 255
Mechanical Action 225
memory map 162
Menu Ring control 148
Merge Signals function 33
Merge Signal function 119
million instructions per second (MIPS) 157

modulo 2 additions 231
MP3 decoding 253
  Add and PCM VI 280
  Alias Reduction VI 277
  Big Value Search VI 273
  Big Value Sign VI 273
  Build Path function 280
  Call Library Function Node function 285
  Circ Buffer VI 271
  Dec Side Info VI 270
  File Path control 268
  Find Type VI 269
  Frame Type VI 269, 270
  FreqInv VI 279
  Huffman Decode VI 272
  Huffman Decoding block 258
    Big Value Search VI 282
    Huffman decoding 257
    Huffman Information Decoding 258
    Huffman tables 257
  IMDCT VI 277, 284
  MainSI Params VI 278
  MDCT & Wvector VI 279
  MP3 File Info VI 269
  MP3 Frame Info VI 280
  MP3 Info Display VI 270
  MP3 Player VI 280
  MP3 Read VI 268, 280
  Poly & PCM VI 278, 280
  Read Characters From File VI 268
  Reordering VI 276
  Scale Factor Decode VI 271
  Search 1D Array function 273
  String to Byte Array function 268
  Write to I16 File VI 280
MP3 encoding
  'bit reservoir' technique 255
  Requantizer 259
MP3 file
  Main Data 254
    granules 255
      SCale Factor Selection Information (SCFSI) 256
    Main Data Begin Pointer 255
MP3 player 253
  Alias Reduction 262
  Alias Reduction VI 268
  Circ Buffer VI 267
  Dec Side Info VI 267
  Frame Type VI 267
  Huffman Decode VI 268
  IMDCT VI 268
  Inverse MDCT (IMDCT) 262
  Polyphase filter bank 264
  Poly & PCM VI 268
  Reordering 261
    Modified Discrete Cosine Transform (MDCT) 261
  Reordering VI 268
  Requantization VI 268
  Scale Factor Decode VI 268
  Synchronization block 254

Cyclic Redundancy Check (CRC) Information 254
  Header Information 254
  Side Information 254
MPEG-I Layer-III 253
multi-level decomposition 144
Multiply function 35
Multirate Filter Design VI 78

N
NI Example Finder 184
node 6
noise cancellation 123
noise cancellation system 210
  DSP integration of 210
nonpipelined CPU 160
normalization 96
number of frequency bins 152
number of scaling 112
Number to Boolean Array function 67
Numeric Controls 16
numeric data types in LabVIEW 102
Numeric Indicators 16
Nyquist rate 46

O
object file 164
off-chip or external memory 163
on-chip or internal memory 163
overflow 102, 105, 107, 219
oversampling 46

P
packed data processing 160
Panning Tool 62
Path Constant 192
phase and frequency tracking 237
pipelined CPU 159
Plot Interpolation 66
point-by-point processing 127, 206
polymorphic VI 110
Polymorphic VI Selector 193
probe 28
  custom probe 28, 29
probe point 164, 174
probe tool 13, 28
profiler 164
Profile Memory Usage 41
profile tool 13, 41
profile window 41, 179
project file 167
property node 152
pseudo noise (PN) 231
pulse shape filter 241

Q
Q-format or fractional representation 96, 109
QAM modem
  system hierarchy 250
    Show VI Hierarchy 250
QAM receiver 234
  complex envelope 234

Index

Hilbert transform 234
   phase and frequency tracking 237
   slicer 235
QAM transmitter 233
   frame marker bit 239
   frame synchronization 239
      cross-correlation 246, 247
      global variables 247
      local variables 247
   Hilbert transformer 243
      group delay 244
   pulse shape filter 241
   raised cosine filter 241
Quadrature Amplitude Modulation (4-QAM) 231
quantization 49
quantization interval 49
quantization noise 49
quantized filter coefficients 115

R

raised-cosine FIR filter 232
raised cosine filter 241
Random Number (0-1) function 27
Re/Im to Complex function 241
real-time data exchange (RTDX™) 183
   continuous mode 183
   noncontinuous mode 183
Real FFT PtByPt VI 128
reconstruction filter bank 144
Remez algorithm 73
Remez Design VI 242
Remove Broken Wires 18
Reorder Controls In Cluster 226
Ring Constant 148
roll-off factor 233
RTDX communication 219
run-time support library 171

S

sampling 43
   digital frequency 43
sampling frequency 43
sampling time interval 43
scaling 100, 102, 111
   number of 112
Scaling and Mapping Express VI 32
scaling function 143
scope chart 131
second-order cascade form 76
selector terminal 12
Shift Register 132
short-time Fourier transform (STFT) 139, 140
Signal Processing Toolset (SPT) 144
signal reconstruction 51
Simple Error Handler VI 192
Simulate Signal Express VI 32
sinc function 51
Sine Waveform VI 82, 149
Sine Wave PtByPt VI 128
Sine Wave VI 56
single precision (SP) 100

Snd Write Waveform VI 228
software-defined radio 231
   pseudo noise (PN) 231
      frame maker bits 232
      frame synchronization 232
      linear feedback shift register 231
      modulo 2 additions 231
   Quadrature Amplitude Modulation (4-QAM) 231
Special Filter Design VI 77
Spectral Measurements Express VI 60
spectrogram 141
Stacked Sequence structure 186
Step Out 180
STFT VI 151
strip chart 131
Strip Path function 191
structure 6, 12
   case 12
      case selector 12
      selector terminal 12
   For Loop 12
      count terminal 12
      iteration terminal 12
   While Loop 12, 25
      conditional terminal 12
subVI 5, 23
   Greater or Equal? function 24
subVI icon 22
Super-resolution Spectral Analysis (SRSA) 144
sweep chart 131
system identification 123

T

terminal icon 5, 6, 11
   data type terminal icon 11
Text Ring control 228
time-frequency localization 142
time-frequency resolution 141
time-frequency tiling 143
time-frequency window 149
timeout value 195
Time Delay Express VI 27
time resolution 142
Timing Statistics 41
TMS320C6000 2
tools palette 7, 16
   labeling tool 16
To Double Precision Float function 107
To Unsigned Byte Integer function 65
To Word Integer function 107
transfer function of an IIR filter 76
truncation error 99
tunnel 27

U

universal serial bus (USB) 183

V

Vertical Pointer Slide control 35
very long instruction word (VLIW) architecture 159

View As Icon 19
virtual instrument (VI) 5
   Analysis VI 78
   block diagram (BD) 5
   breakpoints 13
   build waveform function 57
   connector pane 6, 20
   controls 5, 8
   execution highlighting 13
   Express VI 10
   Fixed-point Tools VI 78
   front panel (FP) 5
      numeric controls 16
   functions 10
   Greater or Equal? function 24
   indicators 5, 8
   labeling tool 16
   Multirate Filter Design VI 78
   probe tool 13, 28
   profile memory usage 41
   profile tool 13, 41
   profile window 41
   sine wave 56
   special filter design 77
   subVI 23
   timing statistics 41
   VI icon 6
VI icon 6

W

Wait (ms) function 39
waveform 38
   data values 38
   time interval 38
   trigger time 38
Waveform Chart 128, 131
   chart history length 131
   Scope chart 131
   Strip chart 131
   Sweep chart 131
waveform data type 38
waveform decomposition tree 154
Waveform Graph 33
   autoscale 36
Waveform Graph Properties 58
   Fill to 58
   Plot Interpolation 58
   Point Style 58
Wavelet Analysis 144
Wavelet Filter VI 152
wavelet function 143
wavelet transform 142
While Loop 12, 25
   conditional terminal 12
wires 6
   basic types of 11

X

XY Graph 248

Z

Zoom 62

ELSEVIER SCIENCE CD-ROM LICENSE AGREEMENT

PLEASE READ THE FOLLOWING AGREEMENT CAREFULLY BEFORE USING THIS CD-ROM PRODUCT. THIS CD-ROM PRODUCT IS LICENSED UNDER THE TERMS CONTAINED IN THIS CD-ROM LICENSE AGREEMENT ("Agreement"). BY USING THIS CD-ROM PRODUCT, YOU, AN INDIVIDUAL OR ENTITY INCLUDING EMPLOYEES, AGENTS AND REPRESENTATIVES ("You" or "Your"), ACKNOWLEDGE THAT YOU HAVE READ THIS AGREEMENT, THAT YOU UNDERSTAND IT, AND THAT YOU AGREE TO BE BOUND BY THE TERMS AND CONDITIONS OF THIS AGREEMENT. ELSEVIER SCIENCE INC. ("Elsevier Science") EXPRESSLY DOES NOT AGREE TO LICENSE THIS CD-ROM PRODUCT TO YOU UNLESS YOU ASSENT TO THIS AGREEMENT. IF YOU DO NOT AGREE WITH ANY OF THE FOLLOWING TERMS, YOU MAY, WITHIN THIRTY (30) DAYS AFTER YOUR RECEIPT OF THIS CD-ROM PRODUCT RETURN THE UNUSED CD-ROM PRODUCT AND ALL ACCOMPANYING DOCUMENTATION TO ELSEVIER SCIENCE FOR A FULL REFUND.

DEFINITIONS

As used in this Agreement, these terms shall have the following meanings:

"Proprietary Material" means the valuable and proprietary information content of this CD-ROM Product including all indexes and graphic materials and software used to access, index, search and retrieve the information content from this CD-ROM Product developed or licensed by Elsevier Science and/or its affiliates, suppliers and licensors.

"CD-ROM Product" means the copy of the Proprietary Material and any other material delivered on CD-ROM and any other human-readable or machine-readable materials enclosed with this Agreement, including without limitation documentation relating to the same.

OWNERSHIP

This CD-ROM Product has been supplied by and is proprietary to Elsevier Science and/or its affiliates, suppliers and licensors. The copyright in the CD-ROM Product belongs to Elsevier Science and/or its affiliates, suppliers and licensors and is protected by the national and state copyright, trademark, trade secret and other intellectual property laws of the United States and international treaty provisions, including without limitation the Universal Copyright Convention and the Berne Copyright Convention. You have no ownership rights in this CD-ROM Product. Except as expressly set forth herein, no part of this CD-ROM Product, including without limitation the Proprietary Material, may be modified, copied or distributed in hardcopy or machine-readable form without prior written consent from Elsevier Science. All rights not expressly granted to You herein are expressly reserved. Any other use of this CD-ROM Product by any person or entity is strictly prohibited and a violation of this Agreement.

SCOPE OF RIGHTS LICENSED (PERMITTED USES)

Elsevier Science is granting to You a limited, non-exclusive, non-transferable license to use this CD-ROM Product in accordance with the terms of this Agreement. You may use or provide access to this CD-ROM Product on a single computer or terminal physically located at Your premises and in a secure network or move this CD-ROM Product to and use it on another single computer or terminal at the same location for personal use only, but under no circumstances may You use or provide access to any part or parts of this CD-ROM Product on more than one computer or terminal simultaneously.

You shall not (a) copy, download, or otherwise reproduce the CD-ROM Product in any medium, including, without limitation, online transmissions, local area networks, wide area networks, intranets, extranets and the Internet, or in any way, in whole or in part, except that You may print or download limited portions of the Proprietary Material that are the results of discrete searches; (b) alter, modify, or adapt the CD-ROM Product, including but not limited to decompiling, disassembling, reverse engineering, or creating derivative works, without the prior written approval of Elsevier Science; (c) sell, license or otherwise distribute to third parties the CD-ROM Product or any part or parts thereof; or (d) alter, remove, obscure or obstruct the display of any copyright, trademark or other proprietary notice on or in the CD-ROM Product or on any printout or download of portions of the Proprietary Materials.

RESTRICTIONS ON TRANSFER

This License is personal to You, and neither Your rights hereunder nor the tangible embodiments of this CD-ROM Product, including without limitation the Proprietary Material, may be sold, assigned, transferred or sub-licensed to any other person, including without limitation by operation of law, without the prior written consent of Elsevier Science. Any purported sale, assignment, transfer or sublicense without the prior written consent of Elsevier Science will be void and will automatically terminate the License granted hereunder.

TERM

This Agreement will remain in effect until terminated pursuant to the terms of this Agreement. You may terminate this Agreement at any time by removing from Your system and destroying the CD-ROM Product. Unauthorized copying of the CD-ROM Product, including without limitation, the Proprietary Material and documentation, or otherwise failing to comply with the terms and conditions of this Agreement shall result in automatic termination of this license and will make available to Elsevier Science legal remedies. Upon termination of this Agreement, the license granted herein will terminate and You must immediately destroy the CD-ROM Product and accompanying documentation. All provisions relating to proprietary rights shall survive termination of this Agreement.

LIMITED WARRANTY AND LIMITATION OF LIABILITY

NEITHER ELSEVIER SCIENCE NOR ITS LICENSORS REPRESENT OR WARRANT THAT THE INFORMATION CONTAINED IN THE PROPRIETARY MATERIALS IS COMPLETE OR FREE FROM ERROR, AND NEITHER ASSUMES, AND BOTH EXPRESSLY DISCLAIM, ANY LIABILITY TO ANY PERSON FOR ANY LOSS OR DAMAGE CAUSED BY ERRORS OR OMISSIONS IN THE PROPRIETARY MATERIAL, WHETHER SUCH ERRORS OR OMISSIONS RESULT FROM NEGLIGENCE, ACCIDENT, OR ANY OTHER CAUSE. IN ADDITION, NEITHER ELSEVIER SCIENCE NOR ITS LICENSORS MAKE ANY REPRESENTATIONS OR WARRANTIES, EITHER EXPRESS OR IMPLIED, REGARDING THE PERFORMANCE OF YOUR NETWORK OR COMPUTER SYSTEM WHEN USED IN CONJUNCTION WITH THE CD-ROM PRODUCT.

If this CD-ROM Product is defective, Elsevier Science will replace it at no charge if the defective CD-ROM Product is returned to Elsevier Science within sixty (60) days (or the greatest period allowable by applicable law) from the date of shipment.

Elsevier Science warrants that the software embodied in this CD-ROM Product will perform in substantial compliance with the documentation supplied in this CD-ROM Product. If You report significant defect in performance in writing to Elsevier Science, and Elsevier Science is not able to correct same within sixty (60) days after its receipt of Your notification, You may return this CD-ROM Product, including all copies and documentation, to Elsevier Science and Elsevier Science will refund Your money.

YOU UNDERSTAND THAT, EXCEPT FOR THE 60-DAY LIMITED WARRANTY RECITED ABOVE, ELSEVIER SCIENCE, ITS AFFILIATES, LICENSORS, SUPPLIERS AND AGENTS, MAKE NO WARRANTIES, EXPRESSED OR IMPLIED, WITH RESPECT TO THE CD-ROM PRODUCT, INCLUDING, WITHOUT LIMITATION THE PROPRIETARY MATERIAL, AN SPECIFICALLY DISCLAIM ANY WARRANTY OF MERCHANTABILITY OR FITNESS FOR A PARTICULAR PURPOSE.

If the information provided on this CD-ROM contains medical or health sciences information, it is intended for professional use within the medical field. Information about medical treatment or drug dosages is intended strictly for professional use, and because of rapid advances in the medical sciences, independent verification of diagnosis and drug dosages should be made.

IN NO EVENT WILL ELSEVIER SCIENCE, ITS AFFILIATES, LICENSORS, SUPPLIERS OR AGENTS, BE LIABLE TO YOU FOR ANY DAMAGES, INCLUDING, WITHOUT LIMITATION, ANY LOST PROFITS, LOST SAVINGS OR OTHER INCIDENTAL OR CONSEQUENTIAL DAMAGES, ARISING OUT OF YOUR USE OR INABILITY TO USE THE CD-ROM PRODUCT REGARDLESS OF WHETHER SUCH DAMAGES ARE FORESEEABLE OR WHETHER SUCH DAMAGES ARE DEEMED TO RESULT FROM THE FAILURE OR INADEQUACY OF ANY EXCLUSIVE OR OTHER REMEDY.

U.S. GOVERNMENT RESTRICTED RIGHTS

The CD-ROM Product and documentation are provided with restricted rights. Use, duplication or disclosure by the U.S. Government is subject to restrictions as set forth in subparagraphs (a) through (d) of the Commercial Computer Restricted Rights clause at FAR 52.22719 or in subparagraph (c)(1)(ii) of the Rights in Technical Data and Computer Software clause at DFARS 252.2277013, or at 252.2117015, as applicable. Contractor/Manufacturer is Elsevier Science Inc., 655 Avenue of the Americas, New York, NY 10010-5107 USA.

GOVERNING LAW

This Agreement shall be governed by the laws of the State of New York, USA. In any dispute arising out of this Agreement, you and Elsevier Science each consent to the exclusive personal jurisdiction and venue in the state and federal courts within New York County, New York, USA.